AF769601

# Contents

## 5. BSIM-BULK Charge and Capacitance Model



## 6. Noise and RF Modeling



## 7. Junction Diode and Layout Dependent Parasitic Model



## 8. Compact Modeling of High Voltage Transistors



## 9. Self-Heating and Thermal Effects

CHAPTER

# 1

# Compact Models and the Road Leading to BSIM-BULK

## 1.1 Circuit simulation and compact models

An integrated circuit contains millions to billions of transistors. Its functionality and performance must be verified by computer simulation before it is committed to expensive fabrication. Circuits are simulated by a method known as SPICE (Simulation Program with Integrated Circuits Emphasis), which was developed at the University of California, Berkeley, in the early 1970s. SPICE solves the nodal and branch differential equations of a circuit numerically.

SPICE uses compact models to represent the detailed behaviors of transistors. BSIM (Berkeley Short-channel IGFET Model) became the first industry-standard compact model for simulating integrated circuits in 1997. BSIM model codes are copyrighted by the University of California and provided royalty-free to all users. BSIM models have been used in the design of most IC products worldwide since 1998 through a dozen generations of CMOS technologies and for every application.

Circuits may be simulated and designed with SPICE and compact models directly. For instance, an entire memory chip may be simulated with SPICE. Other circuits may be designed with cell library methodology. A circuit can also be assembled from a library of building blocks known as standard cells. A standard cell library may have thousands of standard cells that are carefully optimized and characterized with SPICE simulation and transistor compact models for every transistor iteration of every technology generation. Basically, a compact model translates the physical transistors into computer models that the computer-aided IC design tools used to design all ICs, from the simplest to the most complex.

A MOSFET compact model represents a transistor's behaviors in the form of f($V_g$, $V_d$, $V_s$, $V_b$, L, W, $T_{ox}$, etc.), where f stands for the currents at each transistor terminal ($I_d$, $I_s$, $I_g$, $I_b$) and the charges associated with each terminal ($Q_d$, $Q_s$, $Q_g$, $Q_b$). The time derivative of a terminal charge

*BSIM-Bulk MOSFET Model for IC Design - Digital, Analog, RF and High-Voltage*
https://doi.org/10.1016/B978-0-32-385677-5.00005-6

is the capacitive current at that terminal. The compact model must capture the effects of physical dimensions such as channel length (L), channel width (W), gate oxide thickness ($T_{OX}$), other geometry parameters, and channel doping concentration so that it can model manufacturing variations. A model may have many thousands of lines of code, usually in the Verilog-A language. The complexity of the model has a significant impact on model accuracy (close match with measured transistor behaviors) and the circuit simulation time. It is a technically daunting task to create a compact model that computes the currents and charges for any bias voltage within microseconds, with $\geq 99\%$ accuracy from pA to mA (per transistor) and accurate from DC to hundreds of GHz. Besides accuracy and speed, a good compact model must meet the quality criteria that it is robust in any circuit configuration and condition so that the SPICE simulation does not fail to complete.

Compact model equations contain adjustable model parameters. The model parameters are adjusted during model parameter extraction by modeling engineers to customize a compact model, e.g., BSIM, to the measured transistor characteristics of each flavor (n-type, p-type, low-power, high-performance, etc.) of each technology generation of each IC manufacturer. In this sense, each transistor technology is represented by a unique set of BSIM model parameters. Industry-standard compact models facilitate collaborations in the semiconductor ecosystem among IC foundries, IC EDA companies, and IC product design companies. An integrated design and manufacturing company may also adopt industry standard models.

## 1.2 The beginnings of BSIM and industry-standard compact model

BSIM's genesis can be traced back to 1984 [1]. This work produced BSIM1 and then a successor version BSIM2 [2]. From the beginning, BSIM emphasized the creation of original model equations based on an accurate quantitative understanding of device physics beyond the textbook understanding.

In the 1990s semiconductor technology entered the sub-micron regime, and the transistor started to deviate from the textbook behaviors mode more profoundly. The BSIM style of compact modeling matched the semiconductor industry's new needs well. The BSIM researchers fabricated in the UC Berkeley Micro Lab transistors with more aggressively scaled channel length and gate oxide thickness, and much lower yield, than commercially produced MOSFETs at that time. This experiment-based research allowed them to gradually build a series of quantitative models of MOSFET's short-channel effects important to the IC designers, such as drain-induced barrier lowering, channel electric field, substrate current, velocity saturation, and overshoot, drain-induced barrier lowering, output conductance, mobility degradation, threshold voltage roll-off, reverse short-channel effect, short-channel intrinsic capacitances, and fringing-field capacitance. Besides the more accurate intrinsic IV and QV (or CV) models, BSIM also introduced unified flicker noise, gate tunneling leakage, intrinsic gate-input-resistance, nonquasi-static transient currents, RF substrate effect, self-heating, and many more models.

One example is grain-induced drain leakage (GIDL) [3]. A new leakage current mechanism was identified and explained as the band-to-band tunneling current induced by the gate-to-drain voltage Vgd. Once the mechanism was clearly understood, a simple analytical model became obvious and proved to be accurate. Another example is the flicker noise or 1/f noise. The unified flicker noise model includes fluctuations of both the number of inversion layer charge carriers (the number fluctuation) and the Coulombic scattering mobility (the mobility fluctuation). They are correlated because both are caused by the fluctuation in the number of trapped charges in the $SiO_2$ near the silicon/$SiO_2$ interface [4]. This model was characterized in detail using the random telegraphic noise measurements that can only be observed in transistors with very short lengths and narrow widths such that a transistor contains only one or two oxide charge traps [5]. These studies put the compact model of the MOSFET flicker noise on sound physics ground [6].

These new physical models were gradually added to the 1993 BSIM3 [7] and its updated versions. BSIM3v3 [8] won the R&D 100 Award from the R&D Magazine as one of the most significant R&D products of 1995.

The semiconductor foundry industry was beginning to grow. A foundry gives its customers, the circuit design companies, a process design kit (PDK) that contains the information needed to design ICs or develop IC design tools. One important part of the PDK is the compact models of the transistors. A foundry promises to deliver transistors that match the electrical characteristics predicted by compact models. A close match between the models' and the fabricated transistors' characteristics is of paramount importance to both the foundry and its customers.

Inspired by the accuracy and quality of BSIM3v3, several companies started to organize a movement to promoted the adoption of a standard compact model by the semiconductor industry. At that time, most large semiconductor companies developed their transistor models. Often a single company would have many models in active use at the same time. Smaller companies would license models from CAD tool companies. There were many dozens of MOSFET models in use.

In 1996 Sematech, a semiconductor industry consortium, organized a series of workshops that led to the formation of the Compact Model Council (CMC) that defined and executed a year-long process of selecting an industry-standard MOSFET model. Several models competed in the selection process. In 1997, BSIM3v3 was selected by CMC as the world's first standard transistor model for IC simulation. The industry rallied around BSIM. BSIM3v3 was used for the 0.5 μm 0.35 μm 0.25 μm 0.18 μm and 0.15 μm technology nodes.

With the release of BSIM4 [9] in 2000, BSIM was able to support the 100 nm CMOS technologies and the growth of high-speed analog, mixed-signal, and radio-frequency (RF) CMOS integrated circuits that enabled the new wireless lifestyle of the 21st century. A major new model was the holistic noise model for the channel thermal noise and the induced gate noise [10]. The induced gate noise and its correlation with the channel thermal noise are modeled. Accuracy at high frequencies close to the cut-off frequency of transistors was achieved with a simple intrinsic input resistance (Rii) model and a substrate resistance network. A novel quantum effect model was introduced to predict the inversion charge layer thickness that became no longer negligible compared to the decreasing gate oxide thickness [11]. The first gate direct-tunneling leakage current model was also introduced to anticipate the rise of gate leakage currents [12]. The self-heating model was also introduced. BSIM4 has been and continues to be used for the 0.18 μm, 0.13 μm 90 nm, 65 nm, 45/40 nm, 32/28 nm, and 22/20 nm technology nodes.

## 1.3 BSIM-BULK

BSIM1, 2, 3, and 4 used Vth-based current-voltage (IV) model equations that are explicit functions and do not require numerical iteration to calculate terminal currents. This made BSIM ideal for simulating large logic and memory circuits. However, analog and RF circuit designers prefer surface-potential or surface-charge-based model equations, which are implicit and require iterative solving of surface potential or charge but provide smooth derivatives of currents when Vds changes sign (positive to or from negative). With the advent of fast computers, SPICE simulators became more capable of solving implicit compact model equations, and surface potential/charge-based compact models became the clear favorite.

BSIM-BULK was initially released in 2013 as BSIM6 [13–15] and was renamed BSIM-BULK [16] when it was selected as a new standard compact model of bulk-substrate planar MOSFET as opposed to BSIM-SOI (for SOI MOSFETs) [17], BSIM-CMG (for common multi-gate transistors such as FinFET, GAA and nano-wire transistors) [18], and BSIM-IMG (for independent multi-gate transistors such as ultra-thin-body or FDSOI transistors with independent top and bottom gates) [19].

BSIM-BULK uses surface-charge/potential-based IV model equations and produces smooth IV and QV and their derivatives, even up to the 6th or order. BSIM-BULK includes all the models of advanced transistor behaviors, including those described in Section 1.3. All the models have been modified to ensure smoothness and differentiability without compromising their accuracy and flexibility. The names of the model parameters have been kept the same or similar to BSIM4 to help BSIM4 users to transition to BSIM-BULK. It is coded in Verilog-A, the presently preferred language of standard compact models, while BSM4 was coded in the C language.

All new improvements of the planar bulk MOSFET model made by the BSIM researchers have been introduced through BSIM-BULK. BSIM4 continues to be widely used as a legacy model. One example of the new models available only in BSIM-BULK is a more accurate QV or CV model. Another is a new HV (high-voltage) model LDMOS (laterally-diffused MOS) transistors that can operate at high voltages for applications such as microwave power amplifiers for 4G and 5G communications, displays and motor drives. Another example is a flicker noise model capturing nonuniform trap distribution and accurate for both subthreshold and above-threshold operating ranges Table 1.1 summarizes key model improvements that are a part of the BSIM-BULK model and are not available in the BSIM4 model.

TABLE 1.1 Key real-device effect models available in BSIM-BULK but not in BSIM4.

| Description | Status |
|---|---|
| Self-heating effect [20] | BSIM6 6.0.0 |
| Parasitic drain to source current due to edge effect [21] | BSIM6 6.1.1 |
| Improved 1/f noise model for pocket implanted MOSFETs [22] | BSIM-BULK 106.2.0 |
| Subsurface leakage model [23] | BSIM-BULK 106.2.0 |
| High-Voltage model [24] | BSIM-BULK 107.0.0 |
| Improved NQS model [25] | Model developed |
| Accurate gm in pocket implanted MOSFETs [26] | Model developed |

## 1.4 Summary

BSIM-BULK is the culmination of 25 years of BSIM compact model research. The authors believe that BSIM-BULK will serve the semiconductor industry longer than BSIM has already served, i.e., more than 25 more years. This book is intended to be a comprehensive and in-depth presentation of the BSIM-BULK model theory, techniques, and methodology that create a production-worthy compact model. It covers MOSFET device operation and physics, manufacturing process effects, model formulations, parameter extraction, SPICE implementation, and their implications for integrated circuit design. This book is written for all compact model users, model engineers, and students of IC design as either an introduction to or a reference for BSIM-BULK.

# References


[1] B. Sheu, D. Scharfetter, C. Hu, D. Pederson, A compact IGFET charge model, IEEE Transactions on Circuits and Systems 31 (8) (1984) 745–748.

[2] M. Jeng, P. Ko, C. Hu, A deep-submicrometer MOSFET model for analog/digital circuit simulations, in: Technical Digest., International Electron Devices Meeting, 1988, pp. 114–117.

[3] T. Chan, J. Chen, P. Ko, C. Hu, The impact of gate-induced drain leakage current on MOSFET scaling, in: 1987 International Electron Devices Meeting, 1987, pp. 718–721.

[4] K. Hung, P.-K. Ko, C. Hu, Y. Cheng, A unified model for the flicker noise in metal-oxide-semiconductor field-effect transistors, IEEE Transactions on Electron Devices 37 (3) (Mar 1990) 654–665.

[5] P. Fang, K. Hung, P. Ko, C. Hu, Characterizing a single hot-electron-induced trap in submicron MOSFET using random telegraph noise, in: Digest of Technical Papers. 1990 Symposium on VLSI Technology, 1990, pp. 37–38.

[6] K. Hung, P.K. Ko, C. Hu, Y.C. Cheng, A physics-based MOSFET noise model for circuit simulators, IEEE Transactions on Electron Devices 37 (1990) 1323–1333.

[7] J.H. Huang, Z.H. Liu, M.C. Jeng, K. Hui, M. Chan, P.K. Ko, C. Hu, BSIM3 manual, University of California, Berkeley, 1993.

[8] Y. Cheng, M.-C. Jeng, Z. Liu, J. Huang, M. Chan, K. Chen, P.K. Ko, C. Hu, A physical and scalable I-V model in BSIM3v3 for analog/digital circuit simulation, IEEE Transactions on Electron Devices 44 (2) (1997) 277–287.

[9] W. Liu, X. Jin, K.M. Cao, C. Hu, BSIM4.0.0 MOSFET model – user's manual, Memorandum No. UCB/ERL M00/38, Electronics Research Laboratory, College of Engineering, University of California, Berkeley, 2000.

[10] X. Jin, J.-J. Ou, C.-H. Chen, W. Liu, P. Gray, C. Hu, An effective gate resistance model for CMOS RF and noise modeling, in: Tech. Dig. of IEDM, 1998, pp. 961–964.

[11] W. Liu, X. Jin, Y. King, C. Hu, An efficient and accurate compact model for thin-oxide-MOSFET intrinsic capacitance considering the finite charge layer thickness, IEEE Transactions on Electron Devices 46 (5) (1999) 1070–1072.

[12] K. Cao, W.-C. Lee, W. Liu, X. Jin, P. Su, S. Fung, J. An, B. Yu, C. Hu, BSIM4 gate leakage model including source-drain partition, in: International Electron Devices Meeting 2000. Technical Digest. IEDM (Cat. No. 00CH37138), 2000, pp. 815–818.

[13] Y.S. Chauhan, M.A. Karim, S. Venugopalan, H. Agarwal, P. Thakur, A. Niknejad, C. Hu, BSIM6.0.0 technical manual, University of California, Berkeley, 2013.

[14] Y.S. Chauhan, M.A. Karim, S. Venugopalan, S. Khandelwal, P. Thakur, N. Paydavosi, A. Sachid, A. Niknejad, C. Hu, BSIM6: symmetric bulk MOSFET model, in: Workshop on Compact Modeling, 2012.

[15] Y.S. Chauhan, S. Venugopalan, M.A. Chalkiadaki, M.A.U. Karim, H. Agarwal, S. Khandelwal, N. Paydavosi, J.P. Duarte, C.C. Enz, A.M. Niknejad, C. Hu, BSIM6: analog and RF compact model for bulk MOSFET, IEEE Transactions on Electron Devices 61 (2) (Feb 2014) 234–244, https://doi.org/10.1109/TED.2013.2283084.

[16] C. Gupta, R. Goel, H. Agarwal, C. Hu, Y.S. Chauhan, BSIM-BULK: accurate compact model for analog and RF circuit design, in: 2019 IEEE Custom Integrated Circuits Conference (CICC), 2019, pp. 1–8.

[17] BSIM-SOI Technical Manual. [Online]. Available: http://www-device.eecs.berkeley.edu/bsim/?page=BSIMSOI.

[18] BSIM-CMG Technical Manual. [Online]. Available: http://www-device.eecs.berkeley.edu/bsim/?page=BSIMCMG.

[19] BSIM-IMG Technical Manual. [Online]. Available: http://www-device.eecs.berkeley.edu/bsim/?page=BSIMIMG.

[20] H. Agarwal, S. Venugopalan, M. Chalkiadaki, N. Paydavosi, J.P. Duarte, S. Agnihotri, C. Yadav, P. Kushwaha, Y.S. Chauhan, C.C. Enz, A. Niknejad, C. Hu, Recent enhancements in BSIM6 bulk MOSFET model, in: 2013 International Conference on Simulation of Semiconductor Processes and Devices (SISPAD), 2013, pp. 53–56.

[21] S. Khandelwal, H. Agarwal, J.P. Duarte, K. Chan, S. Dey, Y.S. Chauhan, C. Hu, Modeling STI edge parasitic current for accurate circuit simulations, IEEE Transactions on Computer-Aided Design of Integrated Circuits and Systems 34 (8) (April 2015) 1291–1294, https://doi.org/10.1109/TCAD.2015.2419626.

[22] H. Agarwal, S. Khandelwal, S. Dey, C. Hu, Y.S. Chauhan, Analytical modeling of flicker noise in halo implanted MOSFETs, IEEE Journal of the Electron Devices Society 3 (4) (2015) 355–360, https://doi.org/10.1109/JEDS.2015.2424686.

[23] Y.K. Lin, S. Khandelwal, A.S. Medury, H. Agarwal, H.L. Chang, Y.S. Chauhan, C. Hu, Modeling of subsurface leakage current in low VTH short channel MOSFET at accumulation bias, IEEE Transactions on Electron Devices 63 (5) (2016) 1840–1845, https://doi.org/10.1109/TED.2016.2544818.

[24] H. Agarwal, C. Gupta, R. Goel, P. Kushwaha, Y. Lin, M. Kao, J. Duarte, H. Chang, Y.S. Chauhan, S. Salahuddin, C. Hu, BSIM-HV: high-voltage MOSFET model including quasi-saturation and self-heating effect, IEEE Transactions on Electron Devices 66 (10) (Oct 2019) 4258–4263, https://doi.org/10.1109/TED.2019.2933611.

[25] C. Gupta, N. Mohamed, H. Agarwal, R. Goel, C. Hu, Y.S. Chauhan, Accurate and computationally efficient modeling of nonquasi static effects in MOSFETs for millimeter-wave applications, IEEE Transactions on Electron Devices 66 (1) (2019) 44–51, https://doi.org/10.1109/TED.2018.2854671.

[26] H. Agarwal, C. Gupta, S. Dey, S. Khandelwal, C. Hu, Y.S. Chauhan, Anomalous transconductance in long channel halo implanted MOSFETs: analysis and modeling, IEEE Transactions on Electron Devices 64 (2) (2017) 376–383, https://doi.org/10.1109/TED.2016.2640279.

CHAPTER

# 2

# BSIM-BULK Core Model

## 2.1 Basic formulation

In this and all the subsequent Chapters, model and instance parameters are given in capital fonts. For a metal-oxide-semiconductor (MOS) device shown in Fig. 2.1, Poisson's equation is given as

$$\frac{\partial^2 \psi}{\partial y^2} = -\frac{q}{\epsilon_{si}}\left(N_D + p_0 e^{-\psi/V_t} - N_A - n_0 e^{\psi/V_t}\right) \tag{2.1}$$

here $\psi$ is the surface potential, and $N_D$, $N_A$ are the donor and acceptor doping density, respectively. $V_t$ is the thermal voltage, $n_0$ and $p_0$ represent equilibrium electron and hole charge density, respectively. It is possible to show that the total semiconductor charge per unit area ($Q_s$) can be ex-

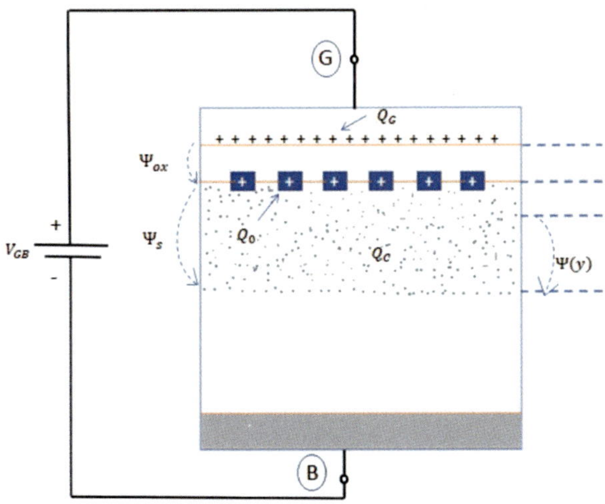


FIGURE 2.1 Illustration of two-terminal MOS system. The channel is along the x-axis, and the depth is along the y-axis (fig. not to scale).


https://doi.org/10.1016/B978-0-32-385677-5.00006-8

pressed in the following form after integrating the above equation [1]:

$$Q_s = -sgn(\psi_s)\sqrt{2q\epsilon_{si}V_t} \times \left( p_0(e^{-\psi_s/V_t} - 1) + (N_A - N_D)\psi_s/V_t + n_0(e^{\psi_s/V_t} - 1) \right) \tag{2.2}$$

The potential balance equation is given as [1]:

$$V_G = V_{FB} + \psi_S + \psi_{ox} = V_{FB} + \psi_S - \frac{Q_s}{C_{ox}} \tag{2.3}$$

here $V_G$, $V_{FB}$ and $\psi_{ox}$ are the applied gate voltage, flat-band voltage and voltage drop across the oxide layer. $\psi_S$ is the potential at the surface of the silicon, also known as surface-potential. $C_{ox}$ is the oxide capacitance and the total semiconductor charge expressed per unit area. $Q_s$ is made up of inversion charge ($Q_i$), depletion charge ($Q_{dep}$), and accumulation charge. We define the following variables that will be frequently used in the book [2]:

$$\text{Thermal voltage:} \quad V_t = \frac{kT}{q}; \quad \text{electron charge:} = q \tag{2.4}$$

$$\text{Bulk potential:} \quad \phi_b = \ln\left(\frac{\mathbf{NDEP}}{n_i}\right) \tag{2.5}$$

$$\text{Normalized body-factor:} \quad \gamma_0 = \frac{\sqrt{2 \cdot q \cdot \epsilon_{si} \cdot \mathbf{NDEP}}}{C_{ox}\sqrt{nV_t}} \tag{2.6}$$

$$\gamma_g = \frac{\sqrt{2 \cdot q \cdot \epsilon_{si} \cdot \mathbf{NGATE}}}{C_{ox}\sqrt{nV_t}} \tag{2.7}$$

$$\text{Body-factor:} \quad \gamma' = \gamma_0 \cdot \sqrt{nV_t} \tag{2.8}$$

$$\gamma_g' = \gamma_g \cdot \sqrt{nV_t} \tag{2.9}$$

$$\delta_{PD} = \frac{\mathbf{NDEP}}{\mathbf{NGATE}} \tag{2.10}$$

$$\left(\frac{\gamma_0}{\gamma_g}\right)^2 = \left(\frac{\frac{\sqrt{2 \cdot q \cdot \epsilon_{si} \cdot \mathbf{NDEP}}}{C_{ox}\sqrt{nV_t}}}{\frac{\sqrt{2 \cdot q \cdot \epsilon_{si} \cdot \mathbf{NGATE}}}{C_{ox}\sqrt{nV_t}}}\right)^2 = \frac{\mathbf{NDEP}}{\mathbf{NGATE}} = \delta_{PD} \tag{2.11}$$

$$\gamma = \frac{\gamma_0}{1 + \delta_{PD}} \tag{2.12}$$

In the above equations, **NDEP** is the substrate doping, $n_i$ is the intrinsic carrier density, $q$ is the electronic charge, $\epsilon_{si}$ is the permittivity of the silicon, $V_t$ is the thermal voltage and **NGATE** is the doping of the polysilicon gate.

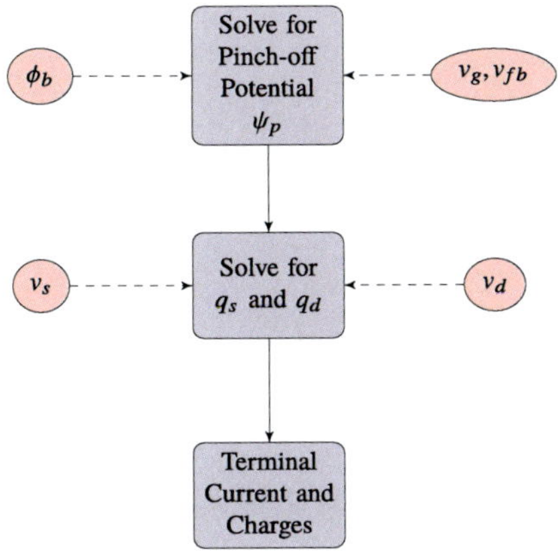


FIGURE 2.2 BSIM-BULK Core Model: Pinch-off potential ($\psi_P$) is first calculated, followed by inversion charge density at source and drain ends ($q_{s/d}$) and terminal currents/charges [4].

Fig. 2.2 shows the core model formulation. The pinch-off potential ($\psi_P$), which is defined as the value of surface potential when $\psi_P = \psi_S$, $Q_i = 0$ [3], is first calculated using the applied gate voltage. This is followed by the calculation of inversion charge densities at the source and drain ends. We will now discuss each of them in detail.

## 2.2 Inversion charge linearization and pinch-off potential

From (2.3), we can write

$$Q_S = -C_{ox}(V_G - V_{FB} - \psi_S) \tag{2.13}$$

Using the depletion approximation, the bulk charge is expressed as [1]

$$Q_b = -sign(\psi_s) \cdot \gamma' \cdot C_{ox} \cdot \sqrt{V_t.(e^{-\frac{\psi_s}{V_t}} - 1) + \psi_s} \tag{2.14}$$

If we include the poly-depletion effect, (2.3) gets modified to account for the drop in the depleted polysilicon layer. The new potential balance equation is then given by

$$V_G = V_{FB} + \psi_S - \frac{Q_i + Q_b}{C_{ox}} + \left(\frac{Q_i + Q_b}{\gamma_g'.C_{ox}}\right)^2 \tag{2.15}$$

Defining pinch-off potential, $\psi_P$, such that at $\psi_S = \psi_P$, $Q_i = 0$ [3]. $\psi_P$ is obtained by letting $Q_i = 0$ in (2.15). Using (2.14),

$$V_G - V_{FB} = \psi_P + \gamma' \cdot \sqrt{V_t.(e^{-\frac{\psi_P}{V_t}} - 1) + \psi_P + (\frac{\gamma'}{\gamma'_g})^2 \left( V_t.(e^{-\frac{\psi_P}{V_t}} - 1) + \psi_P \right)} \tag{2.16}$$

$$= \psi_P + \gamma' \cdot \sqrt{V_t.(e^{-\frac{\psi_P}{V_t}} - 1) + \psi_P + \delta_{PD} \left( V_t.(e^{-\frac{\psi_P}{V_t}} - 1) + \psi_P \right)} \tag{2.17}$$

Normalizing all the voltages by thermal voltage $V_t$, we get

$$v_g - v_{fb} = \psi_p + sgn(\psi_p)\gamma_0 \cdot \sqrt{e^{-\psi_p} + \psi_p - 1 + \delta_{PD} \left( e^{-\psi_p} - 1 + \psi_p \right)} \tag{2.18}$$

Note that the normalized quantities are represented by small-case letters. The above equation is implicit in terms of pinch-off potential $\psi_p$. We will first look at the analytical solution of (2.18).

**Analytical solution of pinch-off potential**

Explicit expression for $\psi_p$ can be derived in the asymptotic form by inspecting the behavior in three different regions. First, consider the depletion and inversion region of operation where $\psi_p >> 0$ so that $e^{-\psi_p}$ is very small. Let $\zeta_1 = e^{-\psi_p}$

$$v_g - v_{fb} = \psi_p + \gamma_0 \cdot \sqrt{\psi_p + \zeta_1 - 1 + \delta_{PD} \left( \zeta_1 - 1 + \psi_p \right)} \tag{2.19}$$

Let

$$\sqrt{\psi_p + \zeta_1 - 1} = x; \quad \psi_p = x^2 + 1 - \zeta_1 \tag{2.20}$$

from above two equations,

$$v_g - v_{fb} = x^2 + 1 - \zeta_1 + \gamma_0.x + \delta_{PD}.x^2 \tag{2.21}$$

or

$$x^2 + \frac{\gamma_0}{1 + \delta_{PD}} \cdot x + \frac{1 - \zeta_1}{1 + \delta_{PD}} - \frac{v_g - v_{fb}}{1 + \delta_{PD}} = 0 \tag{2.22}$$

This gives

$$x = \left[\sqrt{\frac{v_g - v_{fb} - 1 + \zeta_1}{1+\delta_{PD}} + \left(\frac{\gamma_0}{2\cdot(1+\delta_{PD})}\right)^2} - \frac{\gamma_0}{2\cdot(1+\delta_{PD})}\right] \tag{2.23}$$

$$\begin{aligned}\psi_p &= x^2 + 1 - \zeta_1 \\ &= \left[\sqrt{\frac{v_g - v_{fb} - 1 + \zeta_1}{1+\delta_{PD}} + \left(\frac{\gamma_0}{2\cdot(1+\delta_{PD})}\right)^2} - \frac{\gamma_0}{2\cdot(1+\delta_{PD})}\right]^2 + 1 - \zeta_1 \qquad (2.24) \\ &= \left[\sqrt{\frac{v_g - v_{fb} - 1 + \zeta_1}{1+\delta_{PD}} + \left(\frac{\gamma}{2}\right)^2} - \frac{\gamma}{2}\right]^2 + 1 - \zeta_1 \qquad (2.25)\end{aligned}$$

where $\gamma = \gamma_0/(1+\delta_{PD})$.

Now consider the case when $\psi_p$ is close to 0, represented by $\psi_{p0}$:

$$e^{-\psi_{p0}} \approx 1 - \psi_{p0} + \frac{\psi_{p0}^2}{2} - \frac{\psi_{p0}^3}{6} \tag{2.26}$$

$$e^{-\psi_{p0}} - 1 + \psi_{p0} = \frac{\psi_{p0}^2}{2}\left(1 - \frac{\psi_{p0}}{3}\right) \approx \frac{\psi_{p0}^2}{2\left(1 + \frac{\psi_{p0}}{3}\right)} \tag{2.27}$$

Using (2.18) and (2.27), we get

$$v_{gfb} = \psi_{p0} + \gamma_0 \cdot \sqrt{e^{-\psi_{p0}} + \psi_{p0} - 1} + \delta_{PD}\left(e^{-\psi_{p0}} - 1 + \psi_{p0}\right) \tag{2.28}$$

$$= \psi_{p0} + \gamma_0 \frac{\psi_{p0}}{\sqrt{2}} \cdot \sqrt{1 - \frac{\psi_{p0}}{3}} + \delta_{PD}\left[\frac{\psi_{p0}^2}{2}(1 - \frac{\psi_{p0}}{3}))\right] \tag{2.29}$$

here we have used $v_{gfb} = v_g - v_{fb}$. This is a quadratic equation, which is solved to give:

$$\psi_{p0} = \left[\frac{v_{gfb}}{2} - 3(1 + \frac{\gamma}{\sqrt{2}})\right] + \sqrt{\left[\frac{v_{gfb}}{2} - 3(1 + \frac{\gamma}{\sqrt{2}})\right]^2 + 6v_{gfb}} \tag{2.30}$$

In the accumulation region, $\psi_p << 0$. Let $\zeta_2 = \psi_p$,

$$\psi_p = -\ln\left[1 - \zeta_2 + \left(\frac{v_{gfb} - \zeta_2}{\gamma}\right)^2\right] \tag{2.31}$$

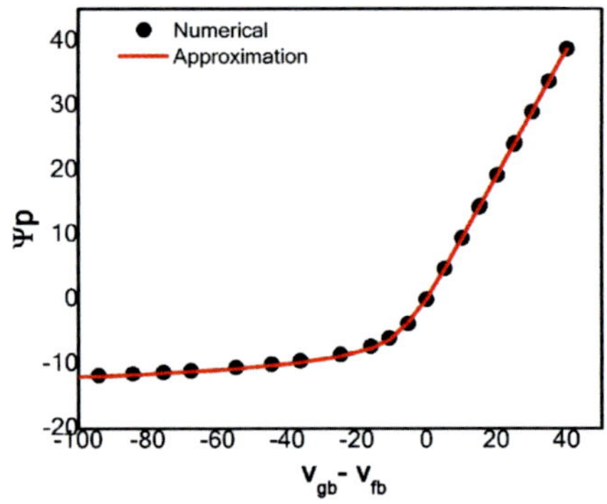


FIGURE 2.3 Comparison of exact solution of (2.18) with analytical approximation given in (2.32). $T_{ox} = 1\,\text{nm}$, **NDEP** $= 5e16/\text{cm}^3$. All the quantities in this figure are normalized to $V_t$.

Thus the pinch-off potential is expressed as

$$\psi_p = \begin{cases} -\ln\left[1 - \psi_{p0} + \left(\frac{v_{gfb} - \psi_{p0}}{\gamma}\right)^2\right] & \text{if } v_g - v_{fb} < 0 \\ 1 - e^{-\psi_{p0}} + \left[\sqrt{v_{gfb} - 1 + e^{-\psi_{p0}} + \left(\frac{\gamma}{2}\right)^2} - \frac{\gamma}{2}\right]^2 & \text{otherwise} \end{cases} \tag{2.32}$$

Fig. 2.3 compares the analytical solution given by the above equation with numerical solution of (2.18). Also, note that the derivatives of $\psi_p$ are continuous in all regions.

## 2.3 Normalized charge density

As shown in Fig. 2.2, the next step after analytically calculating pinch-off potential is to determine inversion charge density. The charge sheet model approximates inversion charge density as [1]

$$Q_i = -\gamma'.C_{ox}.\sqrt{V_t}\left[\sqrt{\frac{\psi_S}{V_t} + e^{\frac{\psi_S - 2.\phi_F - V_{ch}}{V_t}}} - \sqrt{\frac{\psi_S}{V_t}}\right] \tag{2.33}$$

Using inversion charge linearization [3],

$$Q_i = n_q.C_{ox} \cdot (\psi_S - \psi_P); \quad \psi_S = \psi_P + \frac{Q_i}{n_q.C_{ox}} \tag{2.34}$$

Substituting $\psi_S$ from (2.34) in (2.33),

$$-\frac{Q_i}{\gamma'.C_{ox}.\sqrt{V_t}} = \left[\sqrt{\frac{\psi_P + \frac{Q_i}{n_q.C_{ox}}}{V_t} + e^{\frac{\psi_P + \frac{Q_i}{n_q.C_{ox}} - 2.\phi_F - V_{ch}}{V_t}}} - \sqrt{\frac{\psi_P + \frac{Q_i}{n_q.C_{ox}}}{V_t}}\right] \tag{2.35}$$

rearranging,

$$\left[-\frac{Q_i}{\gamma'.C_{ox}.\sqrt{V_t}} + \sqrt{\frac{\psi_P + \frac{Q_i}{n_q.C_{ox}}}{V_t}}\right]^2 = \left[\sqrt{\frac{\psi_P + \frac{Q_i}{n_q.C_{ox}}}{V_t} + e^{\frac{\psi_P + \frac{Q_i}{n_q.C_{ox}} - 2.\phi_F - V_{ch}}{V_t}}}\right]^2 \tag{2.36}$$

$$e^{\frac{\psi_P + \frac{Q_i}{n_q.C_{ox}} - 2.\phi_F - V_{ch}}{V_t}} = \left(-\frac{Q_i}{\gamma'.C_{ox}.\sqrt{V_t}}\right)^2 - 2.\left(\frac{Q_i}{\gamma.C_{ox}.\sqrt{V_t}}\right)\cdot\sqrt{\frac{\psi_P + \frac{Q_i}{n_q.C_{ox}}}{V_t}} \tag{2.37}$$

This reduces to

$$\frac{\psi_P + \frac{Q_i}{n_q.C_{ox}} - 2.\phi_F - V_{ch}}{V_t}$$

$$= \ln\left[\left(-\frac{Q_i}{\gamma'.C_{ox}.\sqrt{V_t}}\right)^2 - 2.\left(\frac{Q_i}{\gamma'.C_{ox}.\sqrt{V_t}}\right)\cdot\sqrt{\frac{\psi_P + \frac{Q_i}{n_q.C_{ox}}}{V_t}}\right] \tag{2.38}$$

$$= \ln\left[-\frac{Q_i}{\gamma'.C_{ox}.\sqrt{V_t}}\left(-\frac{Q_i}{\gamma'.C_{ox}.\sqrt{V_t}} + 2\cdot\sqrt{\frac{\psi_P + \frac{Q_i}{n_q.C_{ox}}}{V_t}}\right)\right] \tag{2.39}$$

Normalizing inversion charge to $-2V_t.n_q.C_{ox}$ and all voltages to $V_t$,

$$\psi_p - 2.q_i - 2.\phi_f - v_{ch} = \ln\left[\frac{2n_q.q_i}{\gamma_0}\left(\frac{2.n_q.q_i}{\gamma_0} + 2\cdot\sqrt{\psi_p - 2q_i}\right)\right] \tag{2.40}$$

which gives

$$\ln(q_i) + \ln\left[\frac{2n_q}{\gamma_0}(q_i\frac{2n_q}{\gamma_0} + 2\sqrt{\psi_p - 2q_i})\right] + 2q_i = \psi_p - 2\phi_f - v_{ch} \tag{2.41}$$

This is a general equation that can be solved to give normalized inversion charge density. The procedure of obtaining an initial guess for the solution of the above equation for weak inversion is described below [5]. Note that to generalize the process, subscript "$i$" is dropped from the term $q_i$.

Let $v = \psi_p - 2\phi_f - v_{ch} - \ln(\frac{4n_q\sqrt{\psi_p}}{\gamma}) = \ln q + 2q$

$$v = \ln q + 2q = \ln q + 2e^{\ln q} = \ln q + \frac{1}{F(\ln q)} \tag{2.42}$$

Here, in the second term, $q$ has been used as $\ln(e^q)$. The function F is defined as

$$F = \frac{1}{2e^{\ln q}} = \frac{1}{2e^(\ln q + \ln q_t - \ln q_t)} \tag{2.43}$$

$$= \frac{1}{2q_t e^{\ln \frac{q}{q_t}}} = \frac{1}{2q_t} e^{-\Delta} \tag{2.44}$$

where $\Delta = \ln \frac{q}{q_t}$. Expanding (2.44) around $\Delta = 0$ and using Taylor series expansion (as $|2q| << |\ln q|$),

$$F = \frac{1}{2q_t}.[1 - e^{-0}.\Delta]] = \frac{1}{2q_t}(1 - \ln \frac{q}{q_t}) \tag{2.45}$$

substituting in (2.42),

$$v = \ln q + \frac{2q_t}{1 - \ln q + \ln q_t} \tag{2.46}$$

This equation is solved for $q$. Let ln(q)=x,

$$v = x + \frac{2q_t}{1 + \ln q_t - x} \tag{2.47}$$

$$v(1 + \ln q_t) - vx - x(1 + \ln q_t) + x^2 - 2q_t = 0 \tag{2.48}$$

$$x = \frac{v + (1 + \ln q_t) - \sqrt{(v + (1 + \ln q_t)^2 - 4v(1 + \ln q_t) + 8q_t}}{2} \tag{2.49}$$

For subthreshold region, normalized inversion charge density will be $|q| << 1$, and $|\ln q| >> |2q|$. The initial value is taken at a point where $|\ln q| = 2.|2q|$ that gives

$$q_t = 0.301; \quad 1 + \ln q_t = -0.201491 \tag{2.50}$$

substituting in (2.49),

$$x = \frac{v - 0.201491 - \sqrt{(v - 0.201491)^2 - 4v(-0.201491) + 8(0.301)}}{2} \tag{2.51}$$

$$x = \ln q = \frac{v - 0.201491 - \sqrt{(v + 0.402982)v + 2.446562}}{2} \tag{2.52}$$

Once the initial guess is known, the final value is obtained using analytical method as shown below

$$n_{q0} = 1 + \frac{\gamma}{2\sqrt{\psi_p}} \tag{2.53}$$

$$v = \psi_p - 2\phi - v_{ch} - ln\left(4.0 \cdot \frac{n_{q0}}{\gamma} \cdot \sqrt{\psi_p}\right) \tag{2.54}$$

$$ln_{q0} = \frac{1}{2}\left[v - 0.201491 - \sqrt{v \cdot (v + 0.402982) + 2.446562}\right] \tag{2.55}$$

$$q_0 = e^{ln_{q0}} \tag{2.56}$$

$$\text{if } ln_{q0} <= -80.0$$

$$q_{s/d} = f = q_0 \cdot \left[1 + \psi_p - 2\phi - v_{ch} - ln_{q0} - \ln\left(2 \cdot \frac{n_{q0}}{\gamma}\left(2 \cdot q_0 \cdot \frac{n_{q0}}{\gamma} + 2 \cdot \sqrt{\psi_p}\right)\right)\right] \tag{2.57}$$

In this equation, if $\ln(q_0)$ becomes very large and negative, then $q_0 = e^{\ln q_0}$ may be out of range of the precision limit of the simulator. Therefore it is approximated as follows

$$\text{if} \quad \ln(q_0) < -110, q_0 = e^{-100} \tag{2.58}$$

$$\text{if} \quad \ln(q_0) > -90, q_0 = e^{\ln(q_0)} \tag{2.59}$$

$$\text{else} \quad q_0 = exp\left[-100 + 20\left(\frac{5}{64} + \frac{z}{2} + z^2\left(\frac{15}{16} - z^2(1.25 - z^2)\right)\right)\right] \tag{2.60}$$

where $z = \frac{\ln q_0 + 100}{20}$.

The above polynomial provides smooth derivatives for $q$.

For $\ln q_0 > -80$,

$$f = 2q_0 + \ln\left(2q_0\frac{n_q}{\gamma}\left(2q_0\frac{n_q}{\gamma} + 2\sqrt{\psi_p}\right)\right) - (v_p - 2\phi_f - v_{ch}) \tag{2.61}$$

$$f' = 2 + \frac{1}{q_0} + \frac{\frac{n_{q0}}{\gamma} - \frac{1}{\sqrt{\psi_p}}}{\frac{n_{q0}}{\gamma} \cdot q_0 + \sqrt{\psi_p}} \tag{2.62}$$

$$q_1 = q_0 - \frac{f}{f'} \tag{2.63}$$

The accuracy of this initial guess is further improved by the following procedure

$$f = 2q_1 + \ln\left(2q_1\frac{n_q}{\gamma}\left(2q_1\frac{n_q}{\gamma} + 2\sqrt{\psi_p}\right)\right) - (v_p - 2\phi_f - v_{ch}) \tag{2.64}$$

$$f' = 2 + \frac{1}{q_1} + \frac{\frac{n_{q1}}{\gamma} - \frac{1}{\sqrt{\psi_p}}}{\frac{n_{q1}}{\gamma} \cdot q_1 + \sqrt{\psi_p}} \tag{2.65}$$

Applying Halley's method,

$$f'' = -\frac{1}{q_1^2} - \frac{1}{\left[(\psi_p)^{\frac{3}{2}}\right] \cdot \left[\frac{n_{q0}}{\gamma} \cdot q_1 + \sqrt{\psi_p}\right]} - \left[\frac{\frac{n_{q0}}{\gamma} - \frac{1}{\sqrt{\psi_p}}}{\frac{n_{q0}}{\gamma} \cdot q_1 + \sqrt{\psi_p}}\right]^2 \tag{2.66}$$

$$q_{s/d} = q_1 - \frac{f}{f'} \cdot \left(1 + \frac{f \cdot f''}{2 \cdot f'^2}\right) \tag{2.67}$$

## 2.4 Drain current model

To understand the expression for the drain to source channel current ($I_{ds}$), it is convenient to first consider the $I_{ds}$ without velocity saturation effect and then extend it to include the carrier velocity saturation effect.

### 2.4.1 Without velocity saturation

$I_{ds}$ can be expressed as a sum of drift and diffusion components of the current:

$$I_{ds} = I_{drift} + I_{diff} \tag{2.68}$$

$$I_{ds} = -W_{eff}.Q_i \cdot \mu_{eff} \frac{d\psi_s}{dx} + W_{eff} \cdot \mu_{eff} \cdot V_t \frac{dQ_i}{dx} \tag{2.69}$$

from charge linearization, $\psi_s = \psi_p + \frac{Q_i}{n_q.C_{ox}}$. Thus $\frac{d\psi_s}{dx} = \frac{1}{n_q.C_{ox}} \cdot \frac{dQ_i}{dx}$

$$I_{ds} = \mu_{eff}.W_{eff} \cdot \left[-\frac{Q_i}{n_q.C_{ox}} + V_t\right] \frac{dQ_i}{dx} \tag{2.70}$$

normalizing inversion charge to $-2n_q C_{ox}.nV_t$ and using $\xi = \frac{x}{L}$,

$$I_{ds} = \mu_{eff} \cdot \frac{W_{eff}}{L_{eff}} \cdot \left[-\frac{(-2.n_q.C_{ox}.nV_t.q)}{n_q.C_{ox}} + V_t\right] \frac{d(-2.n_q.C_{ox}.nV_t.q)}{d\xi} \tag{2.71}$$

$$= -2 \cdot n_q \cdot \mu_{eff} \cdot \frac{W_{eff}}{L_{eff}} \cdot C_{ox} \cdot nV_t^2 \cdot (2q+1) \frac{dq}{d\xi} \tag{2.72}$$

Total drain current,

$$I_{DS} = \int_0^1 I_{ds}\, d\xi = -2 \cdot n_q \cdot \mu_{eff} \cdot \frac{W_{eff}}{L_{eff}} \cdot C_{ox} \cdot nV_t^2 \cdot \int_{q_s}^{q_d} (2q+1) dq \tag{2.73}$$

which gives

$$I_{DS} = 2 \cdot n_q \cdot \mu_{eff} \cdot \frac{W_{eff}}{L_{eff}} \cdot C_{ox} \cdot nV_t^2 \cdot [(q_s - q_d)(q_s + q_d + 1)] \quad (2.74)$$

$n_q$ is the slope factor in the charge-based model and $nV_t$ is $n.\frac{KT}{q}$

$$I_{DS} = 2 \cdot n_q \cdot \mu_{eff} \cdot \frac{W_{eff}}{L_{eff}} \cdot C_{ox} \cdot nV_t^2 \cdot [(q_s - q_d)(q_s + q_d + 1)] \quad (2.75)$$

normalizing the drain current by $2 \cdot n_q \cdot \mu_{eff} \cdot \frac{W_{eff}}{L_{eff}} \cdot C_{ox} \cdot nV_t^2$. Thus

$$i_{ds} = (q_s - q_d)(q_s + q_d + 1) \quad (2.76)$$

## 2.4.2 With velocity saturation

As the device gets smaller and smaller, the lateral electric field strength and therefore the kinetic energy of the carriers increases. On reaching optical phonon energy levels, they release optical phonon by virtue of the reduction in kinetic energy and therefore lose velocity [6]. The effect of velocity on mobility is captured as follows

$$\mu = \frac{\mu_{eff}}{\sqrt{1 + (\frac{E}{E_c})^2}} = \frac{\mu_{eff}}{\sqrt{1 + (\frac{1}{E_c} \cdot \frac{d\psi_s}{dx})^2}} \quad (2.77)$$

from (2.69) and (2.77),

$$I_{ds} = W_{eff} \cdot \frac{\mu_{eff}}{\sqrt{1 + (\frac{1}{E_c} \cdot \frac{d\psi_s}{dx})^2}} \cdot \left[ -Q_i \cdot \frac{d\psi_s}{dx} + V_t \frac{dQ_i}{dx} \right] \quad (2.78)$$

Using charge linearization, we can get the following form:

$$I_{ds} = -2 \cdot n_q \cdot \frac{\mu_{eff}}{\sqrt{1 + (\frac{1}{E_c} \cdot \frac{d\psi_s}{dx})^2}} \cdot \frac{W_{eff}}{L_{eff}} \cdot C_{ox} \cdot nV_t^2 \cdot (2q + 1)\frac{dq}{d\xi} \quad (2.79)$$

$$= z \cdot \frac{(2q + 1)\frac{dq}{d\xi}}{\sqrt{1 + (\frac{1}{E_c} \cdot \frac{d\psi_s}{dx})^2}} \quad (2.80)$$

with $z = -2\mu_{eff} \cdot n_q \cdot \frac{W_{eff}}{L_{eff}} \cdot C_{ox} \cdot nV_t^2$.

Total current,

$$I_{DS} = \int_0^1 I_{ds}\, d\xi = z \cdot \int_{q_s}^{q_d} \frac{(2q + 1)}{\sqrt{1 + (\frac{1}{E_c} \cdot \frac{d\psi_s}{dx})^2}} dq \quad (2.81)$$

$$I_{DS}\int_0^1 \sqrt{1+\left(\frac{1}{E_c}\cdot\frac{d\psi_s}{dx}\right)^2}\,d\xi = z\cdot\int_{q_s}^{q_d}(2q+1)dq \tag{2.82}$$

Also

$$\int_{q_s}^{q_d}(2q+1)dq = -(q_s-q_d)(q_s+q_d+1) \tag{2.83}$$

From (2.82) and (2.83),

$$I_{DS}\int_0^1 \sqrt{1+\left(\frac{1}{E_c}\cdot\frac{d\psi_s}{dx}\right)^2}\,d\xi = z\cdot(-(q_s-q_d)(q_s+q_d+1)) \tag{2.84}$$

Let

$$D_{vsat} = \sqrt{1+\left(\frac{1}{E_c}\cdot\frac{d\psi_s}{dx}\right)^2}\,d\xi \tag{2.85}$$

It is evaluated by assuming that lateral electric field $(-\frac{d\psi_s}{d\xi})$ increases linearly from 0 at source to $2\cdot\left(\frac{\psi_{s,D}-\psi_{s,S}}{L}\right)$ at drain [7], i.e.,

$$-\frac{d\psi_s}{dx} = 2\cdot\frac{\psi_{s,D}-\psi_{s,S}}{L}\cdot\frac{x}{L} = 2\cdot\frac{\psi_{s,D}-\psi_{s,S}}{L}\cdot\xi \tag{2.86}$$

from charge linearization, $\psi_s = \psi_p + \frac{Q_i}{n_q.C_{ox}}$. Thus

$$\psi_{s,S} = \psi_P + \frac{Q_S}{n_q.C_{ox}} = \psi_P - 2V_t.q_s \tag{2.87}$$

$$\psi_{s,D} = \psi_P + \frac{Q_D}{n_q.C_{ox}} = \psi_P - 2V_t.q_d \tag{2.88}$$

$$\psi_{s,D} - \psi_{s,S} = 2.V_t(q_s-q_d) \tag{2.89}$$

substituting in (2.86),

$$-\frac{d\psi_s}{dx} = 2\cdot\frac{2.V_t(q_s-q_d)}{L^2}\cdot x = 2\cdot\frac{2.V_t(q_s-q_d)}{L}\cdot\xi \tag{2.90}$$

$$-\frac{1}{E_c}\frac{d\psi_s}{dx} = 2\cdot\frac{2.V_t}{E_cL}\cdot(q_s-q_d)\xi = 2\lambda_c(q_s-q_d)\xi \tag{2.91}$$

where $\lambda_c = \frac{2V_t}{E_c.L}$. Thus $D_{vsat}$ can be given as

$$D_{vsat} = \int \sqrt{1 + \left(\frac{1}{E_c} \cdot \frac{d\psi_s}{dx}\right)^2} d\xi \tag{2.92}$$

$$= \int \sqrt{1 + (2\lambda_c(q_s - q_d)\xi)^2 d\xi} = \int \sqrt{1 + (2\lambda_c.\Delta q \cdot \xi)^2 d\xi} \tag{2.93}$$

$$= \frac{1}{2}\left[\sqrt{1 + (2.\lambda_c.\Delta q)^2} + \frac{1}{2.\lambda_c.\Delta q}.\ln\left(2.\lambda_c.\Delta q + \sqrt{1 + (2.\lambda_c.\Delta q)^2}\right)\right] \tag{2.94}$$

$$= 0.5 * (T2 + (1.0/T1) * asinh(T1)) \tag{2.95}$$

with $\Delta q = q_s - q_d$, $T1 = 2.\lambda_c.\Delta q$, and $T2 = \sqrt{1 + T1^2}$.

$$I_{DS} = 2 \cdot n_q \cdot \mu_{eff} \cdot \frac{W_{eff}}{L_{eff}} \cdot C_{ox} \cdot nV_t^2 \cdot \frac{[(q_s - q_d)(q_s + q_d + 1)]}{D_{vsat}} \tag{2.96}$$

normalizing the drain current by $2 \cdot n_q \cdot \mu_{eff} \cdot \frac{W_{eff}}{L_{eff}} \cdot C_{ox} \cdot nV_t^2$. Thus

$$i_{ds} = \frac{(q_s - q_d)(q_s + q_d + 1)}{D_{vsat}} \tag{2.97}$$

### 2.4.3 Drain saturation voltage ($V_{dsat}$) calculations

Starting from

$$\ln(q_i) + \ln\left[\frac{2n_q}{\gamma_0}(q_i \frac{2n_q}{\gamma_0} + 2\sqrt{\psi_p - 2q_i})\right] + 2q_i = \psi_p - 2\phi_f - v_{ch} \tag{2.98}$$

Using (2.98), we can take out the expression of $V_{dsat}$ by replacing $q_i$ with $q_{dsat}$ and $v_{ch}$ by $v_{dsat}$

$$v_{dsat} = \psi_p - 2\phi_f - 2q_{dsat} - \ln\left[\frac{2n_q}{\gamma_0}(q_{dsat} \frac{2n_q}{\gamma_0} + 2\sqrt{\psi_p - 2q_{dsat}})\right] \tag{2.99}$$

**Charge density at the drain side when $v_{ch} = v_{dsat}$**

The velocity-field model is given by [8]

$$V_{Drift} = VSAT \cdot \frac{\frac{E}{E_c}}{1 + \frac{E}{2 \cdot E_c}} \tag{2.100}$$

or in normalized form

$$v = \frac{V_{Drift}}{VSAT} = \frac{e}{1 + \frac{e}{2}} \tag{2.101}$$

where, $e = E/E_c$.

The corresponding effective mobility-field model is then given by

$$\mu_{eff} = \frac{V_{Drift}}{E} = \frac{\mu_0}{1 + \frac{E}{2E_c}} \tag{2.102}$$

or in normalized form

$$u = \frac{\mu_{eff}}{\mu_0} = \frac{1}{1 + \frac{e}{2}} \tag{2.103}$$

The longitudinal field can be expressed in terms of the inversion charge density gradient using [8]

$$E = \frac{d\psi_s}{dx} = \frac{1}{n_q \cdot C_{ox}} \frac{dQ_i}{dx} \tag{2.104}$$

or in a normalized form

$$e = -\lambda_C \frac{dq_i}{d\xi} \tag{2.105}$$

Therefore the normalized velocity and mobility can then be written as [8]

$$v = -\frac{\lambda_C \frac{dq_i}{d\xi}}{1 - \frac{\lambda_C}{2}\frac{dq_i}{d\xi}} \tag{2.106}$$

$$u = \frac{1}{1 - \frac{\lambda_C}{2}\frac{dq_i}{d\xi}} \tag{2.107}$$

From (2.72),

$$i_{ds} = -u \cdot (2q_i + 1)\frac{dq_i}{d\xi} \tag{2.108}$$

From (2.107) and (2.108),

$$i_{ds} = -\frac{(2q_i + 1)\frac{dq_i}{d\xi}}{1 - \frac{\lambda_C}{2}\frac{dq_i}{d\xi}} \tag{2.109}$$

By simplification

$$i_{ds} = \frac{(q_s^2 - q_d^2) + (q_s - q_d)}{1 + \frac{\lambda_C}{2} \cdot (q_s - q_d)} \tag{2.110}$$

And therefore

$$i_{dsat} = \frac{(q_s^2 - q_{dsat}^2) + (q_s - q_{dsat})}{1 + \frac{\lambda_C}{2} \cdot (q_s - q_{dsat})} \tag{2.111}$$

Also, the drift component of current in the normalized form is given by

$$i_{ds} = 2 \cdot q_i \cdot \frac{v}{\lambda_C} \tag{2.112}$$

When the drift velocity saturates right at the drain, the drain current cannot increase anymore and is then limited to the saturation value $i_{dsat}$ given by (2.105) evaluated at $\xi = 1$ with $v = 1$ and $q_i = q_{dsat}$. The expression is given by

$$i_{dsat} = 2 \cdot q_{dsat} \cdot \frac{1}{\lambda_c} \tag{2.113}$$

Comparing (2.111) and (2.113) results into

$$q_{dsat} = \frac{\lambda_C}{2} \cdot \frac{q_s^2 + q_s}{1 + \frac{\lambda_C}{2} \cdot (1 + q_s)} \tag{2.114}$$

The expression of $q_{dsat}$ calculated above will be used as an initial guess for the calculation of $v_{dsat}$. From (2.95), we can write

$$D_{vsat} = \frac{1}{2}\left[\sqrt{1 + T_1^2} + \frac{1}{T_1}.\ln\left(T_1 + \sqrt{1 + T_1^2}\right)\right] \tag{2.115}$$

where $T_1 = 2.\lambda_c.(q_s - q_d)$

From (2.115), we can write

$$\frac{dD_{vsat}}{dT_1} = \frac{1}{2} \cdot \left[\frac{T_1}{T_2} + \frac{1}{T_1 \cdot T_2} - \frac{1}{T_1^2} ln(T_1 + T_2)\right] \tag{2.116}$$

where $T_2 = \sqrt{1 + T_1^2}$

Also, we can write

$$\frac{dT_1}{dq_d} = -2 \cdot \lambda_c \tag{2.117}$$

Therefore (2.116) and (2.117) lead to

$$\frac{dD_{vsat}}{dq_d} = -\lambda_c \left[\frac{T_1}{T_2} + \frac{1}{T_1.T_2} - \frac{1}{T_1^2}.ln(T_1 + T_2)\right] \tag{2.118}$$

Comparing (2.97) and (2.113) and replacing $q_d$ with $q_{dsat}$ and $i_{ds}$ with $i_{dsat}$ in (2.97) results into

$$2 \cdot q_{dsat} \cdot \frac{1}{\lambda_c} = \frac{(q_s - q_d)(q_s + q_d + 1)}{D_{vsat}} \tag{2.119}$$

Let

$$f(q_{dsat}) = q_{dsat} \cdot D_{vsat} - \frac{\lambda_c}{2}[q_s^2 + q_s - q_{dsat}^2 - q_{dsat}] = 0 \tag{2.120}$$

Then

$$f'(q_{dsat}) = D_{vsat} + q_{dsat} \cdot D'_{vsat} + \frac{\lambda_c}{2}[2 \cdot q_{dsat} + 1] \tag{2.121}$$

Now, using the Newton-Raphson numerical method

$$q_{dsat,1} = q_{dsat,0} - \frac{f(q_{dsat,0})}{f'(q_{dsat,0})} \tag{2.122}$$

Similarly, we can also calculate the second iteration for the $q_{dsat}$. Once, we get the final expression of $q_{dsat}$, then (2.99) can be modified into

$$v_{dsat} = \psi_p - 2\phi_f - 2q_{dsat} - \ln\left[\frac{2n_q \cdot q_{dsat}}{\gamma_0}\left[\frac{2n_q \cdot q_{dsat}}{\gamma_0} + \frac{\gamma_0}{n_q - 1}\right]\right] \tag{2.123}$$

where, the term $2.\sqrt{\psi_p - q_{dsat}}$ of (2.99) is approximated by $\gamma_0/(n_q - 1)$

$$V_{dsat} = v_{dsat} \cdot nV_t \tag{2.124}$$

$$V_{dssat} = V_{dsat} - V_s \tag{2.125}$$

$$V_{dseff} = \frac{V_{ds}}{\left[1 + \left(\frac{V_{ds}}{V_{dssat}}\right)^{1/DELTA}\right]^{DELTA}} \tag{2.126}$$

$$v_{deff} = \frac{V_{dseff} + V_s}{nV_t} \tag{2.127}$$

## 2.5 Comparison of BSIM4 and BSIM-BULK models

BSIM4 is a threshold voltage model which is widely used in the industry. Researchers and engineers are familiar with BSIM4 model equations, and the core IV model of BSIM-BULK may appear very different from BSIM4. However, we should not forget that both BSIM4 and BSIM-BULK are compact models of the same device and capture the same physics.

Therefore their equations should be similar too. In this section, we will discuss how BSIM-BULK model equations behave as compared to the BSIM4 model. For clarity, we divide the discussion into strong and weak inversion parts.

### 2.5.1 Strong inversion region

As we know, the simplified expression of $I_{DS}$ as a function of inversion charge densities in BSIM-BULK is given by

$$I_{DS} = 2 \cdot n_q \cdot \mu \cdot \frac{W}{L} \cdot C'_{ox} \cdot V_t^2 \cdot \Delta q \cdot (1 + q_s + q_d) \tag{2.128}$$

where, $\Delta q = q_s - q_d$

Also, (2.98) can be approximated to

$$\ln(q_i) + 2q_i = \psi_p - 2\phi_f - v_{ch} \tag{2.129}$$

Further, (2.129) can be written as

$$\ln(q_s) + 2q_s = \psi_p - 2\phi_f - v_s \quad \text{at source end} \tag{2.130}$$

$$\ln(q_d) + 2q_d = \psi_p - 2\phi_f - v_d \quad \text{at drain end} \tag{2.131}$$

In the strong inversion region $2.q_i >> ln(q_i)$. Therefore, from (2.130) and (2.131), we can have $\Delta q = v_{ds}/2$.

As a result, in the strong inversion and linear region, $I_{DS}$ can be approximated by

$$I_{DS} \approx 2 \cdot n_q \cdot \mu \cdot \frac{W}{L} \cdot C'_{ox} \cdot V_t^2 \cdot \frac{v_{ds}}{2} \cdot \left(q_s + q_s - \frac{v_{ds}}{2}\right) \tag{2.132}$$

with $Q_s = q_s \cdot (-2 \cdot n_q \cdot V_t \cdot C'_{ox})$ and $V_{DS} = V_t \cdot v_{ds}$, (2.132) gets the following form:

$$I_{DS} \approx \mu \cdot \frac{W}{L} \cdot C'_{ox} \cdot V_{DS} \cdot \left(-\frac{Q_s}{C'_{ox}} - n_q \cdot \frac{V_{DS}}{2}\right) \tag{2.133}$$

using $Q_s = -C'_{ox}(V_{GS} - V_{TH})$ in (2.133)

$$I_{DS} \approx \mu \cdot \frac{W}{L} \cdot C'_{ox} \cdot V_{DS} \cdot \left(V_{GS} - V_{TH} - n_q \cdot \frac{V_{DS}}{2}\right) \tag{2.134}$$

where, $V_{GS}$ and $V_{TH}$ represent the effective gate voltage and threshold voltages, respectively.

Further, the strong inversion saturation region of operation is characterized by pinch-off, i.e., $q_d \approx 0$ and $q_s >> 1$. Hence, the $I_{DS}$ is given by

$$I_{DS} \approx 2 \cdot n_q \cdot C'_{ox} \cdot V_t^2 \cdot \mu \cdot \frac{W}{L} \cdot \frac{Q_s^2}{(2n_q C'_{ox} V_t)^2}$$

or

$$I_{DS} \approx \mu \cdot \frac{W}{L} \cdot \frac{Q_s^2}{2 \cdot n_q \cdot C'_{ox}} \tag{2.135}$$

Now, put $Q_s = -C'_{ox}(V_{GS} - V_{TH})$ in (2.135), then we get

$$I_{DS} \approx \mu \cdot C'_{ox} \cdot \frac{W}{L} \cdot \frac{(V_{GS} - V_{TH})^2}{2 \cdot n_q} \tag{2.136}$$

If we compare (2.134) and (2.136) with the drain current expressions of BSIM4 [9,10] in linear and saturation regions, respectively, then one can easily notice the similarities between these $V_{TH}$ and charge-based models. The role of $Abulk$ in BSIM3/BSIM4 is played by $n_q$ in BSIM-BULK [11]. The bulk charge coefficient $Abulk = f(A0, AGS)$, where $A0$ and $AGS$ are parameters, provides flexibility in data fitting, especially in the saturation region. In BSIM-BULK, $n_q$ is calculated from basic equations and does not have any parameter associated with it [11].

### 2.5.2 Weak inversion

In the weak inversion region, $\ln(q_i) >> 2.q_i$ and $q_i << 1$. As a result (2.129) gets the following form:

$$\ln(q_i) = v_p - v_{ch} \tag{2.137}$$

$$\text{where, } v_p = \psi_p - 2\phi_f = \frac{(v_{gs} - v_{th})}{n_q}$$

Here, $v_p$ indicates the pinch-off voltage [8]. Using (2.137), the (2.128) can be written as

$$\begin{aligned} I_{DS} &= 2 \cdot n_q \cdot \mu \cdot \frac{W}{L} \cdot C'_{ox} \cdot V_t^2 \cdot (q_s - q_d) \\ &= 2 \cdot n_q \cdot \mu \cdot \frac{W}{L} \cdot C'_{ox} \cdot V_t^2 \cdot \left[\exp\left(\frac{v_{gs} - v_{th}}{n_q}\right) - \exp\left(\frac{v_{gs} - v_{th} - v_{ds}}{n_q}\right)\right] \\ &= 2 \cdot n_q \cdot \mu \cdot \frac{W}{L} \cdot C'_{ox} \cdot V_t^2 \cdot \exp\left(\frac{V_{GS} - V_{TH}}{n_q.V_t}\right) \cdot \left[1 - \exp\left(\frac{-V_{DS}}{n_q.V_t}\right)\right] \end{aligned} \tag{2.138}$$

If we compare (2.138) with the drain current expressions of BSIM4 [9,10] in the subthreshold region, then one would notice that charge-based formulation takes the same form as BSIM4 once we simplify the BSIM-BULK expressions.

## References


[1] Y. Tsividis, Operation and Modeling of MOS Transistors, Oxford Publications, 2010.

[2] Y.S. Chauhan, S. Venugopalan, M.A. Chalkiadaki, M.A.U. Karim, H. Agarwal, S. Khandelwal, N. Paydavosi, J.P. Duarte, C.C. Enz, A.M. Niknejad, C. Hu, BSIM6: analog and RF compact model for bulk MOSFET, IEEE Transactions on Electron Devices 61 (2) (Feb 2014) 234–244.

[3] J.-M. Sallese, M. Bucher, F. Krummenacher, P. Fazan, Inversion charge linearization in MOSFET modeling and rigorous derivation of the EKV compact model, Solid-State Electronics 47 (2003) 677–683.

[4] H. Agarwal, S. Venugopalan, M. Chalkiadaki, N. Paydavosi, J.P. Duarte, S. Agnihotri, C. Yadav, P. Kushwaha, Y.S. Chauhan, C.C. Enz, A. Niknejad, C. Hu, Recent enhancements in BSIM6 bulk MOSFET model, in: 2013 International Conference on Simulation of Semiconductor Processes and Devices (SISPAD), 2013, pp. 53–56.

[5] F. Pregaldiny, F. Krummenacher, B. Diagne, F. Pecheux, J.-M. Sallese, C. Lallement, Explicit modeling of the double-gate MOSFET with VHDL-AMS, International Journal of Numerical Modelling: Electronic Networks, Devices and Fields 19 (3) (May 2006) 239–256.

[6] C. Hu, Modern Semiconductor Devices for Integrated Circuits, vol. 2, Prentice Hall, Upper Saddle River, NJ, 2010.

[7] R. Van Langevelde, A. Scholten, D. Klaassen, Physical background of MOS Model 11, Measurements 6 (2003) 9.

[8] C.C. Enz, E.A. Vittoz, Charge-Based MOS Transistor Modeling, vol. 68, John Wiley & Sons Inc, 2006.

[9] Y. Cheng, M. Chan, K. Hui, M. Jeng, Z. Liu, J. Huang, K. Chen, J. Chen, R. Tu, P. Ko, et al., BSIM3 Version 3.1 Manual, [Online]. Available: http://wwwdevice.eecs.berkeley.edu/bsim/?page=BSIM3.

[10] C.K. Dhabhi, S.S. Parihar, H. Agrawal, N. Paydavosi, T.H. Morshed, D. Lu, et al., BSIM4 Technical Manual, [Online]. Available: https://bsim.berkeley.edu/models/bsim4/.

[11] H. Agarwal, C. Gupta, H.-L. Chang, S. Khandelwal, J.P. Duarte, Y.S. Chauhan, S. Salahuddin, C. Hu, BSIM-BULK106.2.0 Technical Manual, [Online]. Available: http://www-device.eecs.berkeley.edu/bsim/.

CHAPTER

# 3

# BSIM-BULK Real Device Effect Models

## 3.1 Analytical modeling of bulk charge effect

When the channel length is long and/or $V_{DS} > 0$, the depletion region of the channel is not uniform along the channel length, and it leads to the changing threshold voltage along the channel. As a result, $V_{TH}$ will be a function of the position, that is, a function of $V(y)$ along the channel, where, $V(y)$ is the channel potential voltage [1,2].

Fig. 3.1(a) shows the drain current ($I_{DS}$) - with drain to source voltage ($V_{DS}$) simulation for different values of $VSAT$ with default parameter set of BSIM-BULK (without the Abulk model). Here $VSAT$ is a parameter signifying saturation velocity, which controls the drain saturation voltage ($V_{d,sat}$) [3]. For long-channel MOSFETs, the effect of velocity saturation effect on $V_{d,sat}$ is negligible, even if $VSAT$ is set to nonphysical large values ($> 10^7$ m/s). In this case, the user has no other way to adjust the drain saturation current. Fig. 3.1(b) shows the measured $I_{DS}$ - $V_{DS}$ for a long-channel MOSFET. It is evident from Table 3.1 that the BSIM-BULK model, without including the Abulk model, overestimates the drain saturation voltage for a particular $V_{GS}$ value of a long-channel MOSFET as compared to $V_{d,sat}$ values extracted from the measured data of Fig. 3.1(b) [3]. And the extracted values of $n_q = 1 + \gamma/(2\sqrt{2\phi_F})$ and $V_{TH}$ are approximately given by 1.18 and 0.5 V, respectively [3], while all the symbols used have been described in Table 3.2. Hence, a bias-dependent parameter similar to $Abulk$ in BSIM3/BSIM4 is needed as a modifier to $n_q$.

Now, we will explain the derivation of bulk charge effect or Abulk model in the BSIM-BULK model, which is required for the following reasons: (i) it provides the fitting flexibility for $I_{DS}$ - $V_{DS}$ in the saturation region, even at very high $VSAT$, as shown in Fig. 3.1(a). (ii) By doing so, it also provides an accurate value of $V_{d,sat}$, as already explained in Fig. 3.1(b). (iii) Also, this bulk charge effect model is incorporated into the expressions of total source ($Q_S$), drain ($Q_D$), and body ($Q_B$) charges to enhance the fitting flexibility of capacitances in the saturation region.

BSIM-Bulk MOSFET Model for IC Design - Digital, Analog, RF and High-Voltage
https://doi.org/10.1016/B978-0-32-385677-5.00007-X

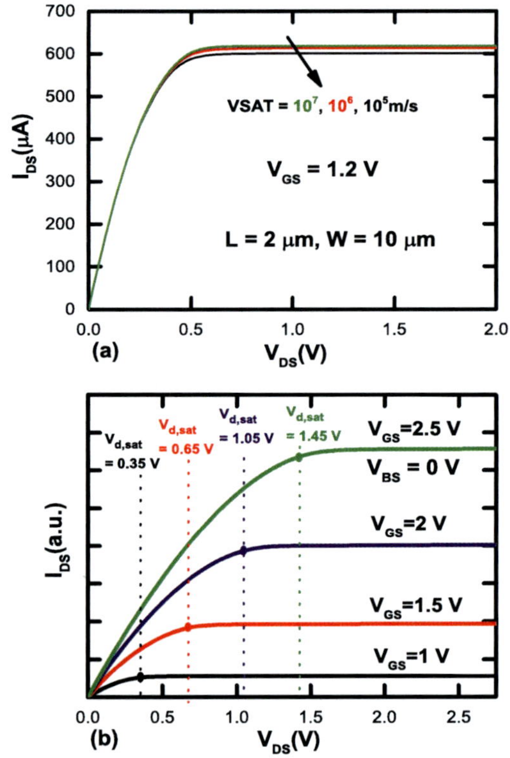


FIGURE 3.1 (a) Saturation characteristics from the existing BSIM-BULK model for different $VSAT$: $I_{DS}$ vs. $V_{DS}$ simulation for L=2 μm and W=10 μm. (b) The measured $I_{DS}$ vs. $V_{DS}$ curves for the long-channel nMOSFET device.

TABLE 3.1 Extraction of $V_{d,sat}$ values.

| $V_{GS}$ | $V_{d,sat}$ from measured data | $V_{d,sat} \approx (V_{GS} - V_{TH})/n_q$ of BSIM-BULK |
|---|---|---|
| 1 V | 0.35 V | 0.42 V |
| 1.5 V | 0.65 V | 0.84 V |
| 2 V | 1.05 V | 1.27 V |
| 2.5 V | 1.45 V | 1.70 V |

## 3.1.1 Bulk charge model for channel current

### *3.1.1.1 Comparison of BSIM3/BSIM4 and BSIM-BULK models*

The expression of $I_{DS}$ as a function of inversion charge densities in BSIM-BULK is given by [3,4]

$$I_{DS} = 2 \cdot n_q \cdot \mu \cdot \frac{W}{L} \cdot C'_{ox} \cdot V_t^2 \cdot \Delta q \cdot (1 + q_s + q_d) \tag{3.1}$$

TABLE 3.2 Description of the symbols used in this work.

| Symbol | Description | Symbol | Description |
|---|---|---|---|
| $L$ | Channel length | $W$ | Channel width |
| $V_{TH}$ | Threshold voltage | $V_t$ | Thermal voltage |
| $\gamma$ | Body coefficient | $n_q$ | Slope factor |
| $\phi_F$ | $V_t.ln(N_A/n_i)$ | $\mu$ | Effective mobility |
| $\Delta q$ | $q_s - q_d$ | $q_{ia}$ | $(q_s + q_d)/2$ |
| $V_{GS}$ | Gate-to-source voltage | $V_{gsteff}$ | $V_{GS} - V_{TH}$ |
| $m$ | $(-WLC'_{ox}V_t)$ | $N_A$ | Impurity concentration |
| $I_{DS}$ | Drain current | $V_{DS}$ | Drain-to-source voltage |
| $VSAT$ | Saturation velocity | $V_{d,sat}$ | Drain saturation voltage |
| $t_{ox}$ | Interfacial oxide ($SiO_2$) thickness | | |
| $n_i$ | Intrinsic carrier concentration | | |
| $C'_{ox}$ | Oxide capacitance per unit area | | |
| $q_s, q_d$ | Normalized inversion charge densities at source/drain side | | |
| $Q_s$ | Inversion charge density at source side | | |
| $V_{dseff}$ | Effective drain-to-source voltage | | |
| $v_{ds}$ | Normalized drain-to-source voltage | | |
| $v_{ds,sat}$ | Normalized drain-to-source saturation voltage | | |
| $\psi_P$ | Normalized pinch-off potential | | |
| $v_P$ | Normalized pinch-off voltage | | |

where all the symbols used have been defined in Table 3.2. In the strong inversion and linear region, the $I_{DS}$ (without including the Abulk model) can be approximated by [3,4]

$$I_{DS} \approx \mu \cdot \frac{W}{L} \cdot C'_{ox} \cdot V_{DS} \cdot \left( \frac{Q_s}{C'_{ox}} - n_q . \frac{V_{DS}}{2} \right) \tag{3.2}$$

The saturation region of operation is characterized by pinch-off, i.e., $q_d \approx 0$. Hence, the $I_{DS}$ (without including the Abulk model) is given by [3,4]

$$I_{DS} = 2 \cdot n_q \cdot C'_{ox} \cdot V_t^2 \cdot \mu \cdot \frac{W}{L} \cdot \frac{Q_s^2}{(2n_q C'_{ox} V_t)^2} \tag{3.3}$$

If we compare (3.2) and (3.3) with the drain current expressions of BSIM3/BSIM4 [5,6] in linear and saturation regions, respectively, then one can easily notice the similarities between these $V_{TH}$ and charge-based models. The role of *Abulk* in BSIM3/BSIM4 is played by $n_q$ in BSIM-BULK [3]. The bulk charge coefficient $Abulk = f(A0, AGS)$, where $A0$ and $AGS$ are parameters, provides accuracy in data fitting, especially in saturation region since the drain saturation voltage in the absence of velocity saturation effect is given as $V_{d,sat} = \frac{V_{gs} - V_{TH}}{Abulk}$ [3,4,7]. In BSIM-BULK, $n_q$ is calculated from basic equations and does not have any parameter associated with it [3].

#### 3.1.1.2 Expression of bulk charge effect ($Abulk$)

Similar to BSIM3/BSIM4 [5,6], $Abulk$ expression for BSIM-BULK can be shown by [1,7]

$$Abulk \approx \frac{1+\left[A0 \cdot T_1 \left(1 - AGS \cdot q_s \cdot V_t \cdot T_1^2\right)\right]}{(1+KETA \cdot V_{BS})} \tag{3.4}$$
$$\text{where, } T_1 = \frac{L}{L+2\sqrt{X_j \cdot X_{DEP}}}$$

where, $A0$, $AGS$, and $KETA$ are the coefficients of channel length, gate voltage, and body-bias dependence, respectively [1,7]. $X_j$ and $X_{DEP}$ represent the source/drain junction depth and depletion region width, respectively.

#### 3.1.1.3 Model implementation (strong inversion)

In-line with BSIM4, we can write the following formulation in BSIM-BULK

$$V_{DS,new} = V_{DS} \cdot Abulk, \quad I_{DS,new} = \frac{I_{DS}}{Abulk} \tag{3.5}$$

In strong inversion linear region ($q_s, q_d >> 1$), $v_{dseff} = (V_{dseff}/V_t) = v_{ds,new} = Abulk \cdot v_{ds}$, where $v_{dseff}$ represents the normalized effective drain to source voltage [3,7]. Hence $2q_d = 2q_s - Abulk.v_{ds}$ giving $\Delta q = \frac{Abulk.v_{ds}}{2}$ [3]. From (3.1),

$$\begin{aligned} I_{DS,new} &= \frac{I_{DS}}{Abulk} \\ &= \mu \cdot \frac{W}{L} \cdot C'_{ox} V_{DS} \left( \frac{Q_s}{C'_{ox}} - \frac{Abulk}{2} V_{DS} \right) \end{aligned} \tag{3.6}$$

One would easily notice that the effect of Abulk model in the linear region is insignificant. Whereas, in strong inversion saturation region ($q_s >> 1$ and $q_d \approx 0$) in the absence of velocity saturation effect, $v_{dseff} = v_{ds,sat} = v_p \approx \psi_p - 2\phi_f$. Therefore $\Delta q \approx q_s \approx v_{ds,sat}/2$ [3,7]. Finally, in saturation, from (3.1)

$$I_{DS,new} = \mu \cdot \frac{W}{2 \cdot (n_q \cdot Abulk) \cdot L \cdot C'_{ox}} \cdot Q_s^2 \tag{3.7}$$

This is also similar to the $I_{DS}$ expression of BSIM3/BSIM4 in saturation [1,5]. Eqs. (3.6) and (3.7) show the similar representation of bulk charge effect in BSIM-BULK as of BSIM4, which fulfills our requirements of getting the accurate values of $V_{d,sat}$, and, in addition to that, it also gives sufficient fitting flexibility to $g_m$ and $g_{ds}$ while fitting a long-channel device at high $V_{DS}$ [3,7].

#### 3.1.1.4 *Model implementation (weak inversion)*

Since in sub-threshold region $q_s(q_d) << 1$, $I_{DS}$ becomes

$$I_{DS,new} \approx 2 \cdot n_q \cdot \mu \cdot \frac{W}{L.Abulk} \cdot C'_{ox} \cdot V_t^2 (q_s - q_d) \tag{3.8}$$

Although the presence of $Abulk$ in $I_{DS}$ slightly impacts the drain current in sub-threshold region, its effect can be overcome with the help of other BSIM-BULK parameters.

### 3.1.2 Bulk charge model for capacitances

The bulk charge effect is also added in the $Q_S$, $Q_D$, and $Q_B$ expressions of the BSIM-BULK model, to get the similar capacitances trend as of "CAPMOD=2" mode of BSIM4 [1]. The modified charge expressions (without CLM and Velocity Saturation effect) inside the CVMOD = 1 [3] of BSIM-BULK, including the bulk charge effect are given below [3,7]:

$$\ln(q_s) + 2q_s = v_p - v_s \quad \text{at source end}$$

$$\ln(q_d) + 2q_d = v_p - v_s - Abulk \cdot v_{dseff,cv} \quad \text{at drain end}$$

$$\text{where, } V_{dseff,cv} = \frac{V_{DS}}{\left[1 + \left(\frac{V_{DS}}{V_{ds,satcv\_new}}\right)^{1/DELTA}\right]^{DELTA}}$$

$$v_{dseff,cv} = \frac{V_{dseff,CV}}{V_t}. \text{ Also, } V_{ds,satcv\_new} = \frac{V_{ds,satcv}}{Abulk} \tag{3.9}$$

From the above equations, we can write

$$q_s + q_{deff} = 2 \cdot \left(v_p - \frac{Abulk \cdot v_{deff,cv}}{2}\right) \tag{3.10}$$

$$q_s - q_{deff} = Abulk \cdot v_{deff,cv} \tag{3.11}$$

The $V_{TH}$ based notation of (3.10) and (3.11) can be given by

$$Q_s + Q_d = 2 \cdot C'_{ox} \cdot \left(V_{gsteff} - \frac{Abulk \cdot V_{deff,CV}}{2}\right) \tag{3.12}$$

$$Q_s - Q_d = C'_{ox}.Abulk \cdot V_{deff,CV} \tag{3.13}$$

$$Q_I = W \cdot \int_0^L Q_i \, dx$$

$$= -WL \cdot C'_{ox} \cdot V_t \int_0^1 2.Abulk.q_i \, d\xi$$

where, $\xi = x/L$. And here, we have normalized the charge density by the factor of $(-2.Abulk.V_t.C'_{ox})$. To simplify the charge expressions, we have assumed the following notations: $\Delta q = q_s - q_d$, $q_{ia} = (q_s + q_d)/2$ and $m = (-WLC'_{ox}V_t)$ [3,7]. After rearranging of terms, the modified expression of total inversion charge as a function of $Abulk$ is given by [3,7]

$$Q_I = m.Abulk.\left[2.q_{ia} + \frac{1}{3}.\frac{\Delta q^2}{(1+2.q_{ia})}\right] \tag{3.14}$$

Similarly, the total drain charge as a function of $Abulk$ is given by

$$\begin{aligned} Q_D &= W \cdot \int_0^L \frac{x}{L}.Q_i \, dx \\ &= m.\int_0^1 2.Abulk.q_i.\xi.d\xi \end{aligned}$$

After rearranging of terms, the modified expression of $Q_D$, containing the effect of bulk charge, is given by [3,7]

$$Q_D = m.Abulk.\left[q_{ia} - \frac{\Delta q}{6}.\left(1 - \frac{\Delta q}{(1+2.q_{ia})} - \frac{1}{5}.\frac{\Delta q^2}{(1+2.q_{ia})^2}\right)\right] \tag{3.15}$$

Likewise, the total source charge as a function of $Abulk$ is given by

$$\begin{aligned} Q_S &= W \cdot \int_0^L \left(1 - \frac{x}{L}\right).Q_i \, dx \\ &= m.\int_0^1 2.Abulk.q_i.(1-\xi).d\xi \\ &= m.Abulk.\left[\left(2.q_{ia} + \frac{1}{3}.\frac{\Delta q^2}{(1+2.q_{ia})}\right) - \right. \\ &\left.\left(q_{ia} - \frac{\Delta q}{6}.\left(1 - \frac{\Delta q}{(1+2.q_{ia})} - \frac{1}{5}.\frac{\Delta q^2}{(1+2.q_{ia})^2}\right)\right)\right] \end{aligned} \tag{3.16}$$

Finally, the expression of total bulk charge, containing the effect of bulk charge, is given by [3,7]

$$\begin{aligned} Q_B &= W \cdot \int_0^L Q_b \, dx \\ &= -WL \cdot C'_{ox} \cdot \left[V_G - V_{FB} - \psi_P + \left(1 - \frac{1}{Abulk}\right) \cdot \int_0^1 \frac{Q_i}{C'_{ox}} \, d\xi\right] \end{aligned}$$

where $Q_b$ is the bulk charge density. After rearranging of terms, the modified expression of $Q_B$ containing the effect of bulk charge is given by

$$Q_B = m.\left[v_g - v_{fb} - \psi_p - (Abulk - 1)\cdot\left(2.q_{ia} - \frac{1}{3}\cdot\frac{\Delta q^2}{1+2.q_{ia}}\right)\right] \quad (3.17)$$

where, $v_g$, $v_{fb}$, and $\psi_p$ indicate the normalized gate voltage, normalized flatband voltage, and normalized pinch-off potential, respectively. Using Eqs. (3.9), (3.12), (3.13), (3.14), (3.15), (3.16), and (3.17), $V_{TH}$-based formulation for $Q_I$, $Q_D$, $Q_S$ and $Q_B$ in the linear region, which also contain the effect of bulk charge effect, are given by (3.18) to (3.21). In addition to that, (3.22) to (3.25) represent the $V_{TH}$-based, modified expressions of $Q_I$, $Q_D$, $Q_S$ and $Q_B$ in the saturation region. The unified expression of $Q_I$, $Q_D$, $Q_S$ and $Q_B$, which used a single set of equations for the charges, is function of $V_{dseff}$ term and is applicable for both linear and saturation given by (3.26) to (3.29). These expressions of total charges are very similar to the $V_{TH}$-based expression of charges in BSIM4 for "CAPMOD=2" mode [3,7], [1].

**Linear Region**

$$Q_I = -W.L.C'_{ox}.\left[V_{gsteff} - \frac{1}{2}.Abulk.V_{DS} + \frac{Abulk^2 V_{DS}^2}{12.\left(V_{gsteff} - \frac{Abulk.V_{DS}}{2}\right)}\right] \quad (3.18)$$

$$Q_D = -\frac{W.L.C'_{ox}}{2.\left(V_{gsteff} - \frac{Abulk.V_{DS}}{2}\right)}.\left[V^3_{gsteff} - \frac{5}{3}.V^2_{gsteff}.Abulk.V_{DS} + V_{gsteff}.(Abulk.V_{DS})^2 - \frac{1}{5}.(Abulk.V_{DS})^3\right] \quad (3.19)$$

$$Q_S = -\frac{W.L.C'_{ox}}{2.\left(V_{gsteff} - \frac{Abulk.V_{DS}}{2}\right)}.\left[V^3_{gsteff} - \frac{4}{3}.V^2_{gsteff}.Abulk.V_{DS} + \frac{2}{3}.V_{gsteff}.(Abulk.V_{DS})^2 - \frac{2}{15}.(Abulk.V_{DS})^3\right] \quad (3.20)$$

$$Q_B = W.L.C'_{ox}.\left[V_{FB} - V_{TH} + 2\phi_F - 2.(1 - Abulk)\left(V_{gsteff} - \frac{Abulk.V_{DS}}{2}\right) + \frac{(1 - Abulk).Abulk.V^2_{DS}}{12.\left(V_{gsteff} - \frac{Abulk.V_{DS}}{2}\right)}\right] \quad (3.21)$$

### Saturation Region

$$Q_I = -W.L.C'_{ox}.\left[V_{gsteff} - \frac{1}{2}.V_{ds,satcv} + \frac{V_{ds,satcv}^2}{12.\left(V_{gsteff} - \frac{V_{ds,satcv}}{2}\right)}\right] \tag{3.22}$$

$$Q_D = -\frac{W.L.C'_{ox}}{2.\left(V_{gsteff} - \frac{V_{ds,satcv}}{2}\right)}.\left[V_{gsteff}^3 - \frac{5}{3}.V_{gsteff}^2.V_{ds,satcv} + V_{gsteff}.V_{ds,satcv}^2 - \frac{1}{5}.V_{ds,satcv}^3\right] \tag{3.23}$$

$$Q_S = -\frac{W.L.C'_{ox}}{2.\left(V_{gsteff} - \frac{V_{ds,satcv}}{2}\right)}.\left[V_{gsteff}^3 - \frac{4}{3}.V_{gsteff}^2.V_{ds,satcv} + \frac{2}{3}.V_{gsteff}.V_{ds,satcv}^2 - \frac{2}{15}.V_{ds,satcv}^3\right] \tag{3.24}$$

$$Q_B = W.L.C'_{ox}.\left[V_{FB} - V_{TH} + 2\phi_F - 2.(1 - Abulk)\left(V_{gsteff} - \frac{V_{ds,satcv}}{2}\right) + \frac{(1 - Abulk).\frac{V_{ds,satcv}^2}{Abulk}}{12.\left(V_{gsteff} - \frac{V_{ds,satcv}}{2}\right)}\right] \tag{3.25}$$

### Unified Equations

$$Q_I = -W.L.C'_{ox}.\left[V_{gsteff} - \frac{1}{2}.Abulk.V_{dseff} + \frac{Abulk^2 V_{dseff}^2}{12.\left(V_{gsteff} - \frac{Abulk.V_{dseff}}{2}\right)}\right] \tag{3.26}$$

$$Q_D = -\frac{W.L.C'_{ox}}{2.\left(V_{gsteff} - \frac{Abulk.V_{dseff}}{2}\right)^2}.\left[V_{gsteff}^3 - \frac{5}{3}.V_{gsteff}^2.Abulk.V_{dseff} + V_{gsteff}.(Abulk.V_{dseff})^2 - \frac{1}{5}.(Abulk.V_{dseff})^3\right] \tag{3.27}$$

$$Q_S = -\frac{W.L.C'_{ox}}{2.\left(V_{gsteff} - \frac{Abulk.V_{dseff}}{2}\right)}.\left[V_{gsteff}^3 - \frac{4}{3}.V_{gsteff}^2.Abulk.V_{dseff} + \frac{2}{3}.V_{gsteff}.(Abulk.V_{dseff})^2 - \frac{2}{15}.(Abulk.V_{dseff})^3\right] \tag{3.28}$$

$$Q_B = W.L.C'_{ox}.\left[V_{FB} - V_{TH} + 2\phi_F - 2.(1 - Abulk)\left(V_{gsteff} - \frac{Abulk.V_{dseff}}{2}\right) + \frac{(1 - Abulk).Abulk.V_{dseff}^2}{12.\left(V_{gsteff} - \frac{Abulk.V_{dseff}}{2}\right)}\right] \tag{3.29}$$

### 3.1.3 Model for $I_{DS} - V_{DS}$

Fig. 3.2 shows the impact of bulk charge parameters ($A0$ and $AGS$) on $I_{DS} - V_{DS}$ for high values of $VSAT$. We can see from Fig. 3.2(a) that for $VSAT = 10^6$ m/s, $I_{DS}$ is quite sensitive to $A0$. Hence, Abulk model provides significant fitting flexibility for the long MOSFET device, where the velocity saturation effect is almost negligible. Fig. 3.2(b) shows the additional gate voltage-dependent flexibility, obtained from the $AGS$, in the $I_{DS} - V_{GS}$ characteristics at $V_{DS} = 1.5$ V. Further, the Abulk model also passes the GST, even for high values of $A0$ and $AGS$.

Fig. 3.3 shows that the new model improves the $g_m$ fitting for a long nMOSFET for both $V_{BS} = 0$ and $V_{BS} \neq 0$. Fig. 3.4(a) and Fig. 3.4(b) show the model validation with the measured data, for $I_{DS}$ and $g_{ds}$ as a function of $V_{DS}$ for $V_{BS} = 0$ and $V_{BS} \neq 0$. Fig. 3.5(a) and Fig. 3.5(b), on the other hand, show the model validation for $I_{DS}$ and $g_{ds}$ as a function of $V_{DS}$ for $V_{BS} \neq 0$. This clearly indicates that the model without Abulk, does not give a good fit, whereas our proposed Abulk model shows the considerable improvement in the $g_{ds}$ fitting. It may be noticed here that for $V_{BS} \neq 0$, the body bias-dependent parameter $KETA$ of Abulk model has been used to further improve the fitting. Fig. 3.6, Fig. 3.7, and Fig. 3.8 show the model validation for different technology node.

### 3.1.4 Model for capacitances

Fig. 3.9(a) shows the $C_{gg}$ versus $V_{GS}$ trend of a long-channel MOSFET for $V_{DS} = 1.5$ V, $V_{BS} = 0$ V, for different values of $A0$. In spite of using a large value of $VSAT$, we are getting the significant shift in the $C_{GG}$ in both up and down directions, depending on the sign of $A0$. Fig. 3.9(a) shows that for high values of gate voltage, $C_{GG}$ curves exhibit a crisscross for different values of $A0$. This is due to the fact that for higher $V_{GS}$, device switches from the saturation region to the linear region, i.e., $q_s \approx q_d$, and, hence, the effect of Abulk model becomes almost insignificant. We can also observe that the effect of Abulk in the weak inversion region is negligible, as we required. Fig. 3.9(b) shows the $C_{GG}$ behavior as a function of $V_{DS}$ at $V_{GS} = 2$ V for different $A0$ values. Due to the same reason, as mentioned above, we also see a crisscross in $C_{GG}$ versus $V_{DS}$ characteris-

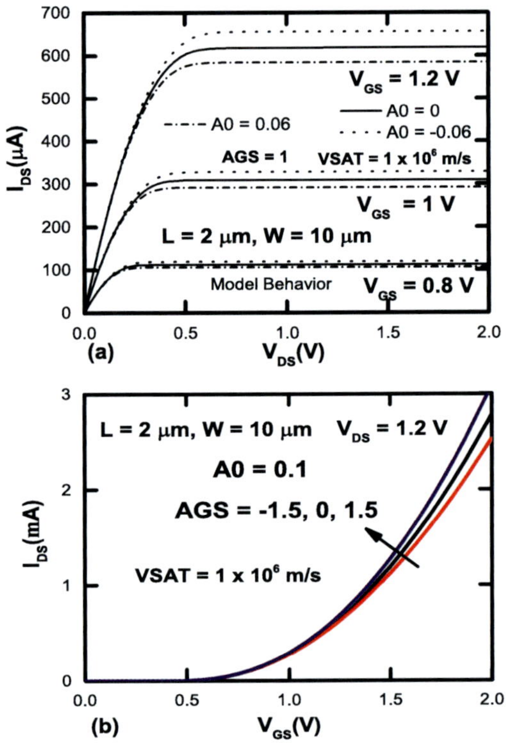


FIGURE 3.2 Effect of the Abulk model: (a) $I_{DS}$ vs. $V_{DS}$ for different values of the parameter $A0$. $A0$ provides additional tuning flexibility to vary the drain saturation voltage, and hence drain current (b) $I_{DS}$ vs. $V_{GS}$ in the saturation region for different values of the parameter $AGS$.

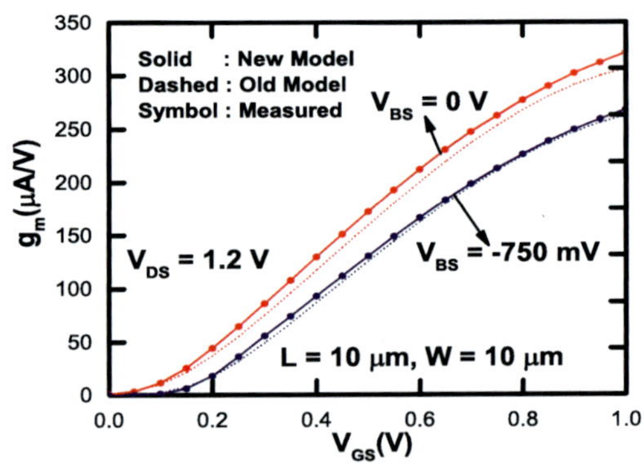


FIGURE 3.3 $g_m$ vs. $V_{GS}$ validation against another silicon data for L = 10 μm and W=10 μm nMOSFET at $V_{BS}$ = 0, -750 mV and $V_{DS}$ = 1.2 V. Lines: Model including Abulk, Dashed Lines: Model without Abulk, and Symbol: Data.

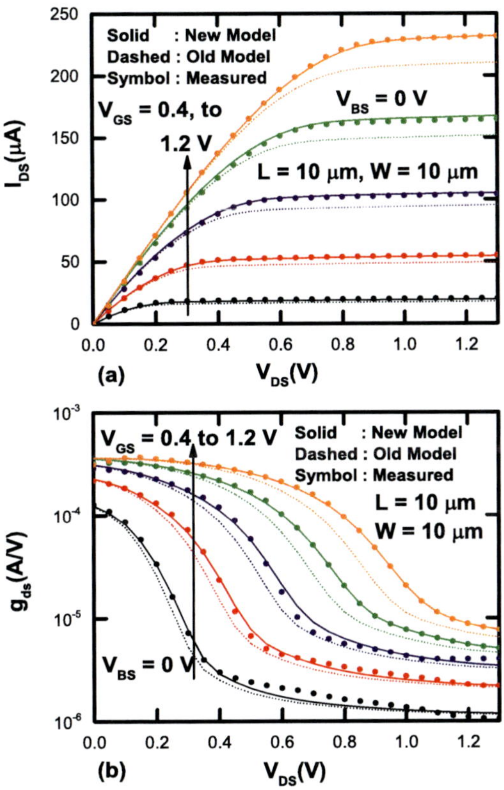


FIGURE 3.4 (a) $I_{DS}$ (b) $g_{ds}$ vs. $V_{DS}$ validation against experimental data for nMOSFET at $V_{BS} = 0\,\text{V}$ and $V_{GS} = 0.4$ to $1.2\,\text{V}$. L = 10 μm and W=10 μm. Lines: Model including Abulk, Dashed Lines: Model without Abulk, and Symbol: Data.

tics, as we move from low $V_{DS}$ to high $V_{DS}$. Fig. 3.9(c) shows the $C_{GS}$ and $C_{GD}$ characteristics as a function of drain voltage at $V_{GS} = 2\,\text{V}$ for different $A0$ values. Here, it is important to note that the capacitances' behavior in Fig. 3.9(a) - Fig. 3.9(c) shows trends similar to "CAPMOD=2" of BSIM4 [1]. The new Abulk model of capacitances also passes the AC symmetry test.

Fig. 3.10(a) - Fig. 3.10(c) compare the models (with and without the bulk charge effect) with the Sentaurus Technology Computer-Aided Design (TCAD) [8] data of NMOS. $C_{GG}$ versus $V_{GS}$ plot, in Fig. 3.10(a), indicates that for high gate and drain biases, the model without Abulk does not give a good fit. However, the BSIM-BULK, with Abulk model, exhibits large improvement in the $C_{gg}$ fitting, especially for large $V_{DS}$ values. Similarly, Fig. 3.10(b) and Fig. 3.10(c) show significant improvement in the $C_{GS}$ and $C_{GD}$ fitting when the effect of new Abulk model is included in the BSIM-BULK model [3].

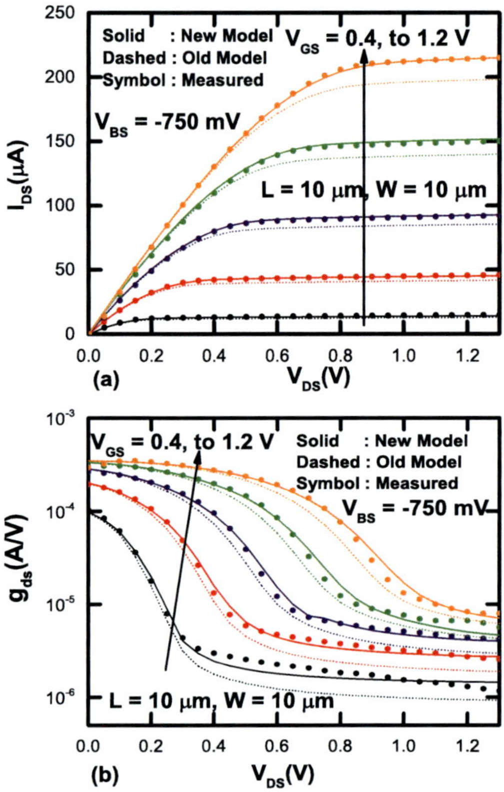


FIGURE 3.5 (a) $I_{DS}$ (b) $g_{ds}$ vs. $V_{DS}$ validation for $V_{BS}$ = -750 mV and $V_{GS}$ = 0.4 V to 12 V. L = 10 µm and W=10 µm. Lines: Model including Abulk, Dashed Lines: Model without Abulk, and Symbol: Data.

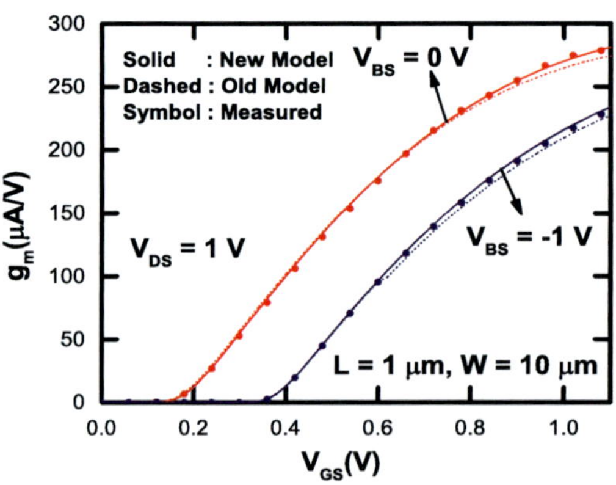


FIGURE 3.6 $g_m$ vs. $V_{GS}$ validation against another silicon data for L = 1 µm and W=10 µm nMOSFET at $V_{BS}$ = 0, -1 V and $V_{DS}$ = 1 V. Lines: Model including Abulk, Dashed Lines: Model without Abulk, and Symbol: Data.

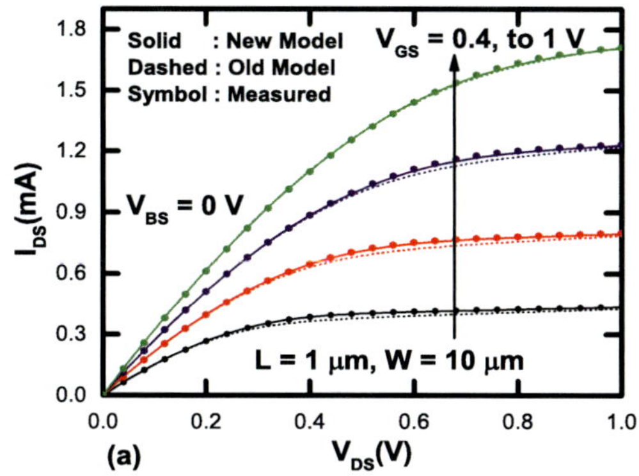


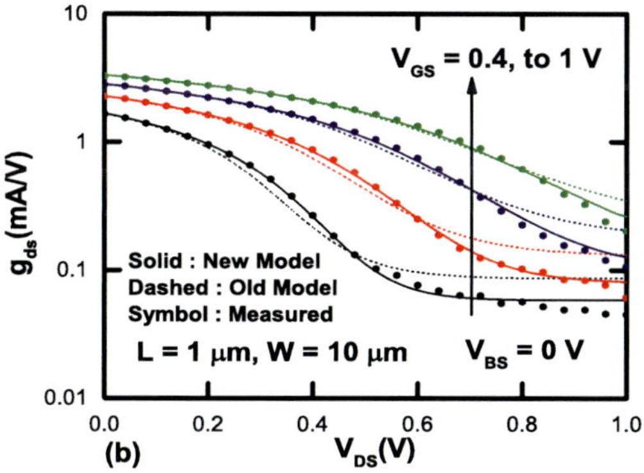


FIGURE 3.7 (a) $I_{DS}$ (b) $g_{ds}$ vs. $V_{DS}$ validation against experimental data for nMOSFET at $V_{BS} = 0\,\text{V}$ and $V_{GS} = 0.4, 0.6, 0.8, 1.0\,\text{V}$. L = 1 μm and W=10 μm. Lines: Model including Abulk, Dashed Lines: Model without Abulk, and Symbol: Data.

## 3.2 Enumeration of various real device effects in threshold voltage

Unlike the ideal transistor assumption, the threshold voltage ($V_{TH}$) of a real MOSFET gets impacted by multiple physical effects: $V_{TH}$ Roll-off, Reverse-short-channel effect (RSCE), drain-induced barrier lowering (DIBL), narrow-width effects, drain-induced threshold shift (DITS), Vertical Nonuniform Doping (NUD), etc. To reap the optimum performance of any circuit design, it is important to incorporate all these effects in the $V_{TH}$ expression of an ideal transistor.

### 3.2.1 Threshold voltage roll-off

In a short-channel device, since the source is in close proximity to the drain, the drain starts to interfere to the potential barrier at the source side [9–12]. The threshold voltage of a long-channel device is independent of

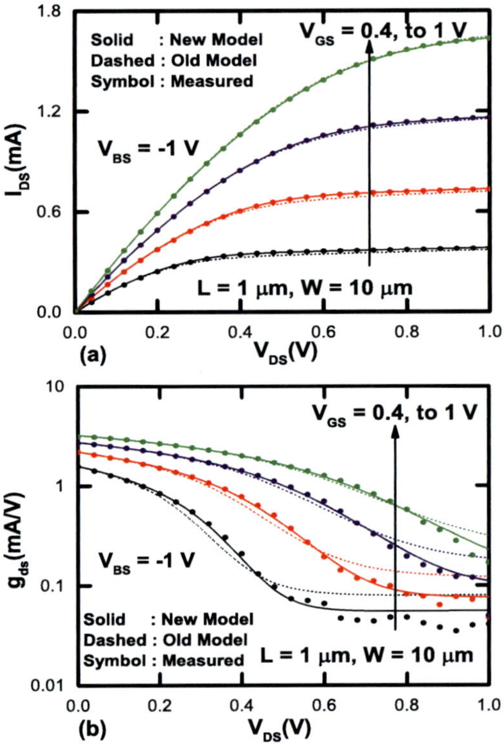


FIGURE 3.8 (a) $I_{DS}$ (b) $g_{ds}$ vs. $V_{DS}$ validation for $V_{BS}$ = -1 V and $V_{GS}$ = 0.4, 0.6, 0.8, 1.0 V. L = 1 μm and W=10 μm. Lines: Model including Abulk, Dashed Lines: Model without Abulk, and Symbol: Data.

the channel length $L$ and the drain voltage $V_{DS}$ [9–12]. However, the dependence of $V_{TH}$ on $L$ and $V_{DS}$ in short-channel devices cannot be ignored. It has been observed experimentally that $V_{TH}$ of the device decreases as $L$ decreases or $V_{DS}$ increases. This effect is called the short-channel effect or $V_{TH}$ roll-off, where $V_{TH}$ decreases monotonically with the decrease in the channel length $L$, and it leads to a shift in $V_{TH}$ represented by $\Delta V_{TH,SCE}$ [9–12].

However, in devices using halo or pocket implantation, it has been noticed that $V_{TH}$ initially increases with decreasing channel length. This is called the reverse short channel effect (RSCE), or $V_{TH}$ roll-up, represented by $\Delta V_{TH,RSCE}$. $V_{TH}$ reaches a maximum value at a certain channel length, and as $L$ decreases further, $V_{TH}$ starts to decrease. The latter is the $V_{TH}$ roll-off. The combined RSCE and $V_{TH}$ roll-off effects result in a 'hump' in the characteristics of $V_{TH}$ versus $L$ [9–12]. The cause of RSCE is nonuniform lateral doping. In fact, for some technologies, such as pocket implantation, the channel doping concentration near the source and drain

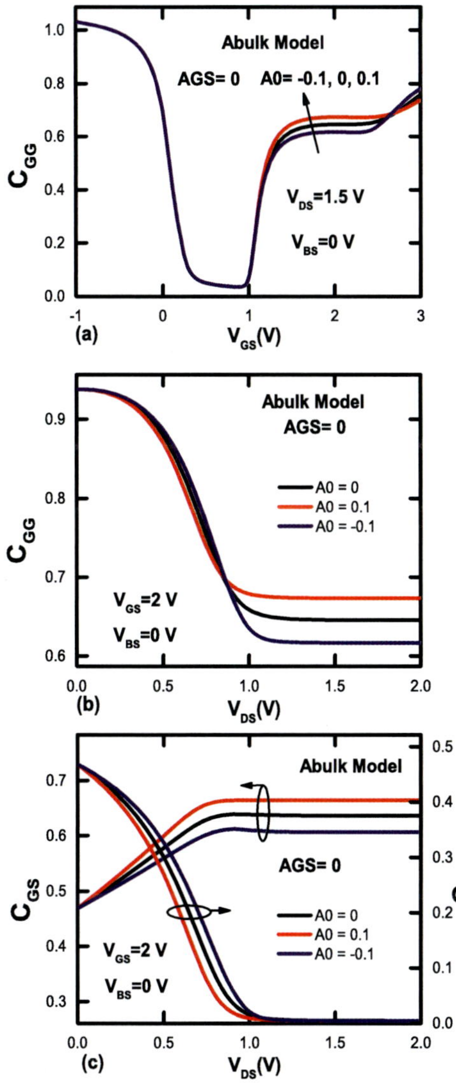


FIGURE 3.9 Normalized capacitances behavior of BSIM-BULK in the presence of the proposed Abulk model. (a) $C_{GG}$ vs. $V_{GS}$ with $V_{DS} = 1.5$ V (b) $C_{GG}$ vs. $V_{DS}$ with $V_{GS} = 2$ V (c) $C_{GS}$ vs. $V_{DS}$ (Left Y-axis), and $C_{GD}$ vs. $V_{DS}$ (Right Y-axis) with $V_{GS} = 2$ V. In all the above plots, A0 is varied from -0.1 to 0.1. The proposed bulk charge model gives similar trends for all the combinations of capacitances as that of "CAPMOD=2" in BSIM4. L = 2 μm, W = 10 μm, $V_{BS} = 0$ and $VSAT = 10^6$ m/s.

is higher than in the middle of the channel. The increased doping concentration in these regions can result in an increase in $V_{TH}$ if the channel length becomes small.

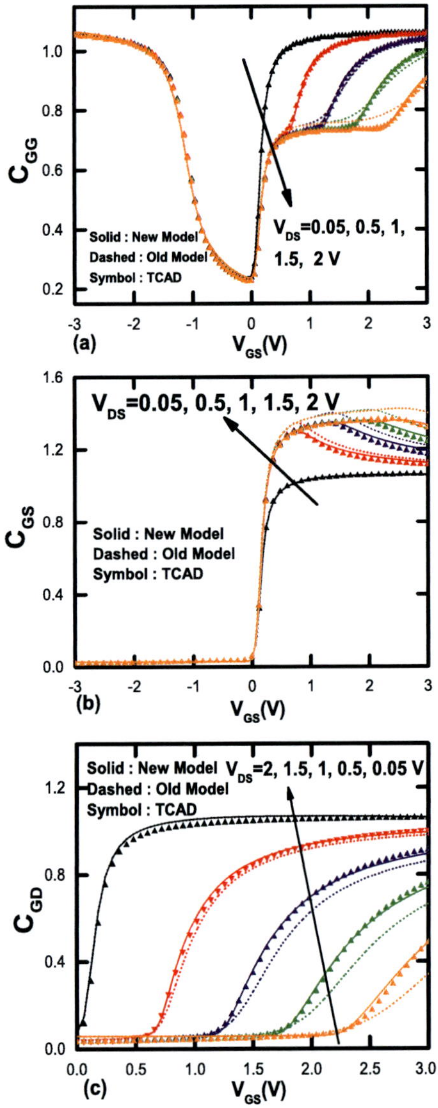


FIGURE 3.10 Model validation with the TCAD data for (a) Normalized $C_{GG}$ vs. $V_{GS}$ (b) Normalized $C_{GS}$ vs. $V_{GS}$ (c) Normalized $C_{GD}$ vs. $V_{GS}$. nMOSFET with L = 2 µm, W=10 µm and $T_{OX} = 3\,\text{nm}$ at $V_{BS} = 0\,\text{V}$ and $V_{DS} = 0.05$, 0.5, 1, 1.5 V. Solid Lines: Model including Abulk, Dashed Lines: Model without Abulk, and Symbol: TCAD Data.

Although there are an abundant number of models for modeling threshold voltage roll-off and RSCE, BSIM-BULK uses a physical-based geometry scaling of the channel doping parameter ($NDEP$) such that it is computationally efficient and implementable as shown below [3]:

$$NDEP_{eff} = NDEP.[1+ \\ NDEPL1.\left(\frac{1}{L_{eff}^{NDEPLEXP1}} - \frac{1}{LLONG^{NDEPLEXP1}}\right)+ \\ NDEPL2.\left(\frac{1}{L_{eff}^{NDEPLEXP2}} - \frac{1}{LLONG^{NDEPLEXP2}}\right)+ \\ NDEPW.\left(\frac{1}{W_{eff}^{NDEPWEXP}} - \frac{1}{WWIDE^{NDEPWEXP}}\right)+ \\ NDEPWL.\left(\frac{1}{(W_{eff}.L_{eff})^{NDEPWL}}\right) \quad (3.30)$$

Here, the parameters $NDEPL1$ and $NDEPLEXP1$ are responsible for the reverse short-channel effects in the $V_{TH}$ versus $L_{eff}$ trends, and $\Delta V_{TH,RSCE}$ will be taken care of by these parameters. On the other hand, $NDEPL2$ and $NDEPLEXP2$ determine the short-channel effects in the same $V_{TH}$ versus $L_{eff}$ plot, and $\Delta V_{TH,SCE}$ is automatically taken care of by these two parameters.

To match the VTCC versus $L$ trends of the model with the data, the length scaling parameters associated with the channel doping have to be extracted. First, the parameter $NDEP$ has to be extracted with the help of the longest channel device. Further, to capture the intermediate region length trends, which refer to the reverse short-channel effects, the length scaling parameters $NDEPL1$ and $NDEPLEXP1$ will be used. Finally, to match the short-channel effects region, the scaling parameters, $NDEPL2$ and $NDEPLEXPL2$, will be extracted as shown in Fig. 3.11. Fig. 3.12, on the other hand, illustrates the corresponding VTCC versus $L$ trends for W=3 µm and $V_{ds} = 0.1$ V. It is clearly evident from Fig. 3.12 that VTCC exhibits the RSCE for intermediate lengths, and for further shorter lengths, the $V_{TH}$ decreases.

### 3.2.2 DIBL effect on the threshold voltage

In short-channel MOSFET devices, the $V_{TH}$ of the transistor starts to decrease due to the decrease in the channel length $L$ and the increase in $V_{DS}$. As $L$ decreases and $V_{DS}$ increases, the potential barrier at the source side is lowered, as a result, less gate voltage is needed to built the inversion charge layer in the channel region, hence, we can also say that the $V_{TH}$ of the transistor reduces. This physical phenomenon is also known as a drain-induced threshold shift (DIBL) [9–12]. Another impact of DIBL is seen on the $I_{DS} - V_{DS}$ characteristics, where it causes a finite slope in $I_{DS} - V_{DS}$ and increases output conductance.

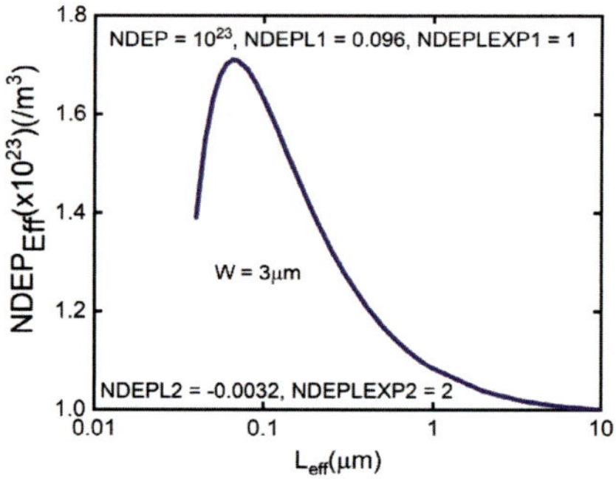


FIGURE 3.11 Length scaling of the BSIM-BULK model: Effective value of channel doping ($NDEP$) as a function of channel length.

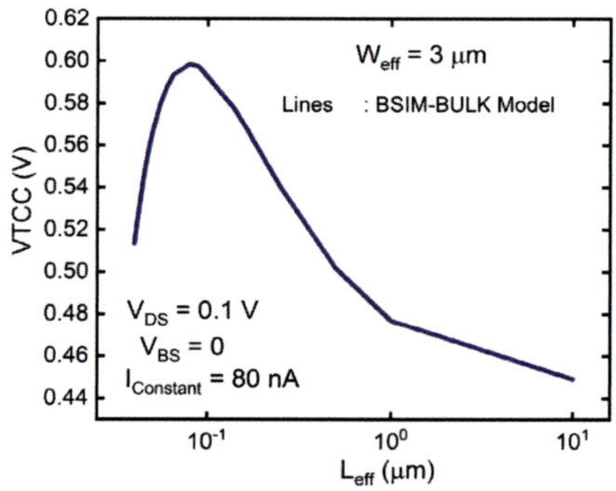


FIGURE 3.12 Length scaling of the BSIM-BULK model: VTCC versus $L_{eff}$ trends for $V_{DS}$= 0.1 V and $I_{Constant}$= 80 nA.

Using a quasi-2-D numerical approach to solve the Poisson equation in the channel of the MOSFET, analytical models for the threshold voltage can be derived. It has been shown that the threshold voltage model developed using the above-mentioned approach can accurately predict the short-channel effects. Thus the threshold voltage shift $\Delta V_{TH}$ can be expressed as [2,12]

$$\Delta V_{TH,DIBL} = -[3(\phi_{bi} - \phi_0) + V_{DS}]e^{-L/\lambda} \tag{3.31}$$

where $\phi_{bi}$ is the source (or drain) to channel junction built-in potential, and $\lambda$ is the characteristic length, which is given by

$$\lambda = \sqrt{\frac{\epsilon_S.T_{OX}.d_B}{\epsilon_{OX}}} \tag{3.32}$$

here, $d_B$ represents the depletion region below the channel region, near the source. Also, (3.31) is derived by considering the assumption that $L >> d_B$.

Eqs. (3.31) and (3.32) show that the following factors can influence DIBL [2,12,13]:

(a.) The channel length $L$: shorter $L$ leads to stronger DIBL effects because $d_B$ and hence $\lambda$ increase.
(b.) The gate oxide thickness $T_{OX}$: thicker $T_{OX}$ leads to stronger DIBL effects, as $T_{OX}$ increases then $\lambda$ also increases.
(c.) The source/drain junction depth $X_j$: deeper $X_j$ leads to stronger DIBL effects. Although (3.32) does not show the $X_j$ dependence, it has been suggested to add it empirically.
(d.) The channel doping concentration $N_{ch}$: lower $N_{ch}$ leads to stronger DIBL effects, since lower $N_{ch}$ leads to large value of $d_B$ and hence $\lambda$.
(e.) The body bias $V_{BS}$: higher $V_{BS}$ leads to stronger DIBL effects because with increasing $V_{BS}$ the $d_B$ and hence $\lambda$ increase.

BSIM-BULK model uses the empirical expression to capture the effect of DIBL on $V_{TH}$, because of the difficulty in obtaining the accurate results from (3.31) as shown below:

$$\Delta V_{TH,DIBL} = -(ETA0 + ETAB.V_{BS}).V_{DS} \tag{3.33}$$

where, $ETA0$ and $ETAB$ are the DIBL related parameters containing the physical geometry scaling for length and width.

### 3.2.3 Normal narrow-width effects

Here we are going to consider two different cases that will introduce a shift in the $V_{TH}$ expression represented by $\Delta V_{TH,NWE}$:

**STI Process**:

Due to the presence of shallow-trench isolation (STI) walls, the depletion charge gets confined near the region just below the gate. The field lines coming from the gate get terminated to the acceptor ions of the depletion region just immediately to the gate, leading to the thin oxide capacitance. However, some of the additional field lines coming from the gate, so-called the fringing field, terminate on the side wall of the channel. Since these fringing field lines increase the surface potential, which helps the inversion charge start early, it reduces the $V_{TH}$ of the transistor [2,13]. If $W$ is wide enough, then these fringing field lines constitute a small percentage of the total and can be neglected, but these extra field lines cannot be ignored for the narrow $W$ devices. Many different approaches have been proposed in the literature to model $V_{TH}$ changes caused by the narrow width effect, but the BSIM-BULK model uses the physical width scaling of $NDEP$, as shown in Fig. 3.13 [3]. Here the negative value of the parameter

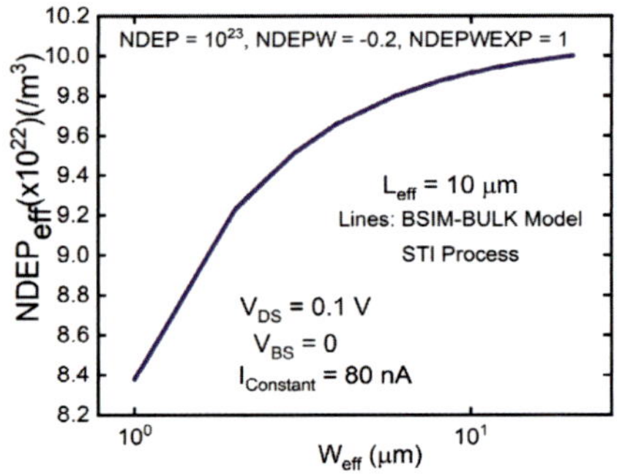


**FIGURE 3.13** Width scaling of the BSIM-BULK model: Effective value of channel doping ($NDEP$) as a function of channel width for the STI process.

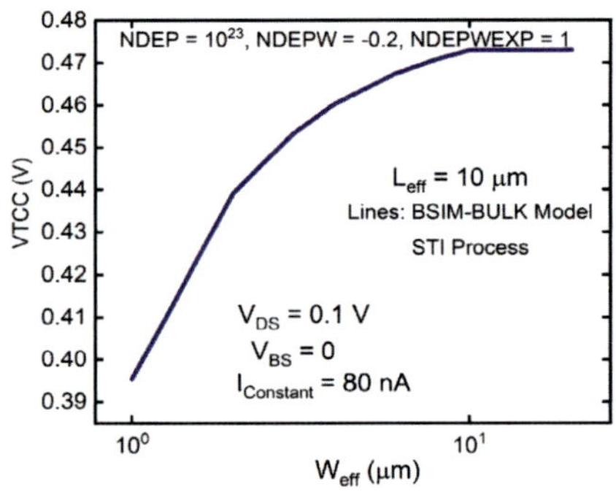


**FIGURE 3.14** Width scaling of the BSIM-BULK model: VTCC versus $W_{eff}$ trends for the STI process with $V_{DS}$= 0.1 V and $I_{Constant}$= 80 nA.

$NDEPW$ is used so that the overall channel doping reduces as the channel width decreases, and $\Delta V_{TH,NWE}$ is calculated. Fig. 3.14, on the other hand, illustrates the corresponding VTCC versus $W_{eff}$ trends for L=10 μm and $V_{DS} = 0.1$ V. It is clearly evident from Fig. 3.14 that for $NDEPW = -0.2$ and $NDEPWEXP = 1$, VTCC starts to reduce with the width.

**LOCOS Process:**

Now we consider the devices that use the local oxidation of silicon (LOCOS) process. This process has been used for high-voltage integrated circuits. Here, the depletion charge is not just limited to just below the thin oxide; rather some field lines from the gate get terminated to the depletion ions staying near the side area. Since some of the field lines from the gate are wasted to induce the depletion charge on the sides rather than directly below the gate, it increases the overall $V_{TH}$ of the device [2,13]. If $W$ is wide enough, then these fringing field lines constitute a small percentage of the total and can be neglected, but for the narrow $W$ devices, these extra field lines cannot be ignored. To capture the desired $V_{TH}$ trends as a function

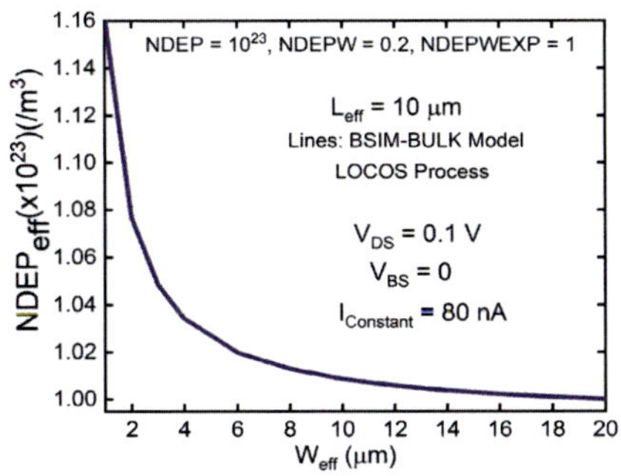


FIGURE 3.15 Width scaling of the BSIM-BULK model: Effective value of channel doping ($NDEP$) as a function of channel width for the LOCOS process.

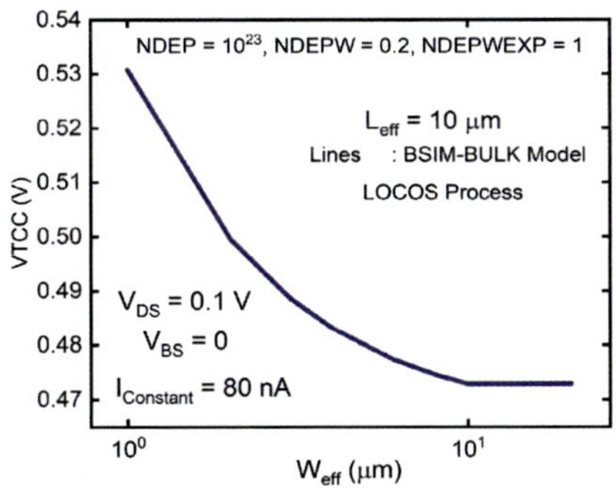


FIGURE 3.16 Width scaling of the BSIM-BULK model: VTCC versus $W_{eff}$ trends for the LOCOS process with $V_{DS}$= 0.1 V and $I_{Constant}$= 80 nA.

of width, the BSIM-BULK model uses the physical geometry scaling of $NDEP$ as shown in Fig. 3.15 [3]. However, here, we have to use a positive value of $NDEPW$ so that the overall channel doping increases as the channel width decreases, and $\Delta V_{TH,NWE}$ is calculated. Fig. 3.16, on the other hand, illustrates the corresponding VTCC versus $W_{eff}$ trends for L=10 μm and $V_{DS} = 0.1$ V. It is clearly evident from Fig. 3.16 that for $NDEPW = 0.2$ and $NDEPWEXP = 1$, VTCC starts to increase with the decrease in the channel width.

### 3.2.4 Effect of drain-induced threshold shift (DITS) on threshold voltage

Pocket implant is widely used in deep sub-micron CMOS technologies to reduce $V_{TH}$ roll-off and punch-through [2,13]. This technique, however, produces large drain-induced $V_{TH}$ shift and low $R_{out}$ in long-channel devices [2,13] greatly affecting analog circuit design and performance. The threshold voltage shift due to DITS is given by [1,14]:

$$\Delta V_{TH,DITS} = -n.V_t.ln\left(\frac{(1-e^{-V_{DS}/V_t}).L_{eff}}{L_{eff}+DVTP0.(1+e^{-DVTP1.V_{DS}})}\right) \quad (3.34)$$

In the BSIM-BULK model, the above equation is simplified and implemented as [3]

$$\Delta V_{TH,DITS} = -n.V_t.ln\left(\frac{L_{eff}}{L_{eff}+DVTP0.(1+e^{-DVTP1.V_{DS}})}\right) \quad (3.35)$$

It was observed that $V_{TH}$ shift dependence of $V_{DS}$ shows asymmetric behavior that can be accurately captured by introducing an $tanh()$ function in (3.35). The complete expression for $\Delta V_{TH,DITS}$ is shown below [3]:

$$\Delta V_{TH,DITS} = -n.V_t.ln\left(\frac{L_{eff}}{L_{eff}+DVTP0.(1+e^{-DVTP1.V_{DS}})}\right) - \left(DVTP5+\frac{DVTP2}{L_{eff}^{DVTP3}}\right).tanh(DVTP4.V_{DS}) \quad (3.36)$$

## 3.2.5 Modeling of the vertical nonuniform doping effects [13]

By solving the one-dimensional poison equation in the vertical direction for a device with a long and wide channel geometry and uniformly doped substrate, $V_{TH}$ can be derived easily, as shown below [2,13]

$$V_{TH} = V_{TH0} + \gamma.(\sqrt{\phi_s + V_{SB}} - \sqrt{\phi_s}) \quad (3.37)$$

where,

$$V_{TH0} = V_{FB} + \phi_s + \gamma.\sqrt{\phi_s}$$

and,

$$\gamma = \sqrt{\frac{2.\epsilon_s.q.N_A}{C'_{OX}}} \quad (3.38)$$

where $\phi_s$ is the surface potential at threshold, $V_{FB}$ is the flat band voltage, $V_{TH0}$ is the threshold voltage of a long-channel device at zero body bias, $\gamma$ is the coefficient of the body-bias effect, $N_A$ is the substrate doping concentration, and $C'_{OX}$ is the gate oxide capacitance per unit area.

From (3.37), we see that the $V_{TH}$ should depend linearly on $\sqrt{\phi_s + V_{SB}}$, with a slope of $\gamma$. However, experimental data in general displays a nonlinear dependence. The slope $\gamma$ becomes smaller as the body bias becomes

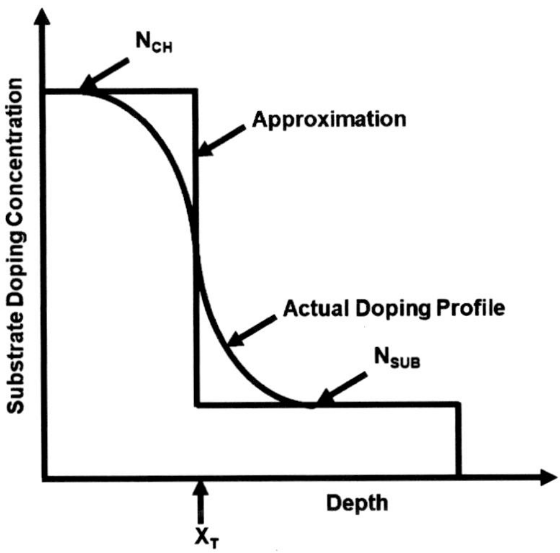


FIGURE 3.17 One possible body doping profile and approximation.

more negative (for NMOS). This nonlinearity comes from the nonuniform substrate doping in the vertical direction of the body [2,13].

The nonuniform body doping along the vertical direction makes the $\gamma$ in (3.37) a function of the substrate bias. As shown in Fig. 3.17, an approximate step doping profile can be used to obtain an analytical expression of $V_{TH}$ as a function of $V_{BS}$. If the depletion width is less than $X_T$ as shown in Fig. 3.17, $N_A$ in (3.37) is equal to $N_{CH}$; otherwise, it is equal to $N_{SUB}$. Therefore we have [2,13]

$$V_{TH} = V_{TH0} + \gamma_1.(\sqrt{\phi_s + V_{SB}} - \sqrt{\phi_s}), |V_{BS}| < |V_{BX}| \quad (3.39)$$

$$V_{TH} = V_{TH0} + \gamma_1.(\sqrt{\phi_s + V_{SB}} - \sqrt{\phi_s}) + \gamma_2.(\sqrt{\phi_s + V_{SB}} - \sqrt{\phi_s - V_{BX}}), |V_{BS}| \geq |V_{BX}| \quad (3.40)$$

Here, $\gamma_1$ and $\gamma_2$ are given by:

$$\gamma_1 = \sqrt{\frac{2.\epsilon_s.q.N_{CH}}{C'_{OX}}} \quad (3.41)$$

$$\gamma_2 = \sqrt{\frac{2.\epsilon_s.q.N_{SUB}}{C'_{OX}}} \quad (3.42)$$

where $V_{BX}$ is the body bias at which the depletion width is equal to $X_T$. Therefore $V_{BX}$ satisfies

$$\frac{q.N_{CH}.X_T^2}{2.\epsilon_s} = \phi_S - V_{BX} \quad (3.43)$$

The unified expression of $V_{TH}$ shift in the BSIM-BULK model due to vertical nonuniform doping by merging (3.41) and (3.42) is proposed as

$$\Delta V_{TH,VNUD} = K_1.(\sqrt{\phi_s + V_{SB}} - \sqrt{\phi_s}) - K_2.V_{BS} \tag{3.44}$$

### 3.2.6 Complete threshold voltage expression

Finally, all effects are combined in a single threshold voltage shift as follows:

$$\begin{aligned}\Delta V_{TH,All} = &\Delta V_{TH,SCE} + \Delta V_{TH,RSCE} + \Delta V_{TH,DIBL} \\ &+ \Delta V_{TH,NWE} + \Delta V_{TH,DITS} + \Delta V_{TH,NUD}\end{aligned} \tag{3.45}$$

## 3.3 Vertical-field mobility degradation

In the BSIM-BULK model, the effect of mobility degradation on the channel current is incorporated through the $D_{mobs}$ term [3,15]. The vertical-field-dependent mobility model of BSIM-BULK includes the mobility degradation due to different scattering mechanisms, including phonons, coulomb scattering, and surface roughness scattering. The model is taken from BSIM4 model but uses charges instead of the threshold voltage. The unified mobility model used in BSIM-BULK can be expressed as [3,15]

$$\mu_{eff} = \frac{U_0}{D_{mobs}} \tag{3.46}$$

where

$$D_{mobs} = 1 + (UA + UC.V_{bseff}).(E_{eff})^{EU} + \frac{UD}{\left[\frac{1}{2}.\left(1 + \frac{q_{ia}}{q_{bs}}\right)\right]^{UCS}} \tag{3.47}$$

where

$$\begin{aligned}q_{ia} &= \frac{q_{is} + q_{id}}{2} \\ q_{bs} &= nV_t.(vgfb - \psi_P - 2.q_{is}.(n_q - 1)) \\ E_{eff} &= 10^{-8}.\left(\frac{q_{bs} + \eta.q_{ia}}{\epsilon_{ratio}.EOT}\right)\end{aligned} \tag{3.48}$$

$UA$, $UC$, and $UD$ are used for phonon-surface roughness scattering, the body effect coefficient for mobility, and coulomb scattering, respectively.

## 3.4 Drain saturation voltage

Current saturation attributed to the pinch-off or velocity saturation effects can occur. Drain saturation voltage ($V_{dsat}$) is needed to get the effective drain voltage ($V_{dseff}$), which, in turn, is used to calculate the normalized charge density at the drain side ($q_d$). Since $V_{dsat}$ depends only on the normalized charge density at the source side ($q_s$) and not on $V_{DS}$, one can say that $V_{dsat}$ is a function of applied $V_{GS}$ only. Also, the presence of source/drain resistance in the transistor affects $V_{dsat}$. The $V_{dsat}$ expression in BSIM-BULK is adopted from BSIM4 with the required modification. For the convenience of the reader, we present the detailed derivation of $V_{dsat}$ for RDSMOD = 0 and RDSMOD = 1,2 [3,15,16].

### 3.4.1 Extrinsic case (RDSMOD = 0 and 1)

Due to the parasitic resistance, the saturation voltage $V_{dsat}$ will be larger than the no parasitic resistance case. When sufficiently high drain voltage is applied, the lateral electric field is large enough to cause carrier velocity to saturate near the drain end. Thus the expression of $q_{dsat}$ in the absence of parasitic resistances is calculated below [3,15,16]

$$T_0 = PSAT.V_{BS}$$
$$T_1 = \sqrt{0.1 + T_0.T_0}$$
$$T_2 = 0.5.(1 - T_0 + \sqrt{(1 - T_0).(1 - T_0) + T_1})$$
$$\lambda_C = \frac{2.U_0.nV_t}{(D_{mobs})^{PSAT}.VSAT.L_{eff}}.\left[1 + PTWG.\frac{10.PSATX.q_s.T_0}{10.PSATX + q_s.T_0}\right] \tag{3.49}$$

We use Eq. (2.114) given in Chapter 2, as the initial guess for the $q_{dsat}$ calculations, to calculate its final expression, and, finally, the same expression is used to calculate the $V_{dsat}$ and hence $V_{dseff}$, from which the normalized charge density at the drain side can be calculated. Therefore

$$q_{dsat}(0) = \frac{\lambda_C}{2}.\frac{q_s^2 + q_s}{1 + \frac{\lambda_C}{2}.(1 + q_s)} \tag{3.50}$$
$$T_3 = 2.\lambda_C.(q_s - q_{dsat}(0))$$
$$T_4 = \sqrt{1 + T_3.T_3}$$
$$T_5 = ln(T_3 + T_4)$$
$$T_6 = \begin{cases} T_4 + (1/T_3).T_5 & \text{if } T_3 \neq 0 \\ T_4 + (1/T_4) & \text{otherwise} \end{cases}$$
$$f(q_{dsat}(0)) = q_{dsat}(0).T_6 - \lambda_C.((q_s.q_s + q_s) - (q_{dsat}(0) * q_{dsat}(0) + q_{dsat}(0))) \tag{3.51}$$

$$T_7 = \begin{cases} (-2.0.\lambda_C).(T_3.T_4 - T_5)/(T_3.T_3) & \text{if } T_3 \neq 0 \\ (-2.0.\lambda_C).(T_3/T_4).(T_3/T_4).(T_3/T_4) & \text{otherwise} \end{cases}$$

$$f'(q_{dsat}(0)) = q_{dsat}(0).T_7 + T_6 + \lambda_C.(2.q_{dsat}(0) + 1) \tag{3.52}$$

$$q_{dsat}(1) = q_{dsat}(0) - f(q_{dsat}(0))/f'(q_{dsat}(0)) \tag{3.53}$$

The above expression is evaluated using (3.50), (3.51), and (3.52).

$$T_3 = 2.\lambda_C.(q_s - q_{dsat}(1))$$

$$T_4 = \sqrt{1 + T_3.T_3}$$

$$T_5 = ln(T_3 + T_4)$$

$$T_6 = \begin{cases} T_4 + (1/T_3).T_5 & \text{if } T_3 \neq 0 \\ T_4 + (1/T_4) & \text{otherwise} \end{cases}$$

$$f(q_{dsat}(1) = q_{dsat}(1).T_6 - \lambda_C.((q_s.q_s + q_s) - (q_{dsat}(1) * q_{dsat}(1) + q_{dsat}(1))) \tag{3.54}$$

$$T_7 = \begin{cases} (-2.0.\lambda_C).(T_3.T_4 - T_5)/(T_3.T_3) & \text{if } T_3 \neq 0 \\ (-2.0.\lambda_C).(T_3/T_4).(T_3/T_4).(T_3/T_4) & \text{otherwise} \end{cases}$$

$$f'(q_{dsat}(1)) = q_{dsat}(1).T_7 + T_6 + \lambda_C.(2.q_{dsat}(1) + 1) \tag{3.55}$$

$$q_{dsat}(final) = q_{dsat}(1) - f(q_{dsat}(1))/f'(q_{dsat}(1)) \tag{3.56}$$

The above expression is evaluated using (3.53), (3.54), and (3.55).

### 3.4.2 Intrinsic case (RDSMOD = 0 and 2)

In the intrinsic case, the part of source/drain resistances is modeled using an internal resistor – that is, RDSMOD = 0. If the drain voltage is significantly high, the carrier velocity near the drain saturates. At the boundary between the two portions, the channel voltage is the saturation voltage ($V_{dsat}$), and the lateral electric field is equal to $E_{sat}$. The expression of internal channel resistance is given by [3,15,16]:

$$T_1 = 1 + PRWG.q_{ia}$$

$$T_2 = PRWB.(\sqrt{\phi_s - V_{BS}} - \sqrt{\phi_s})$$

$$T_3 = \frac{1}{T_1} + T_2;\; T_4 = \frac{1}{2}.[T_3 + \sqrt{T_3{}^2 + 0.01}]$$

$$R_{ds}(V) = R_{s,geo} + NF.\left(W_{eff}^{WR}\left[RDSWMIN + RDSW.T_4\right]\right) + R_{d,geo} \tag{3.57}$$

The expression of $q_{dsat}$ in the absence of parasitic resistances is calculated below

$$T_1 = W_{eff}.n_q.C_{ox}.nV_t.VSAT$$
$$T_2 = T_1.\lambda_C.R_{ds}(V)/(2.nV_t)$$
$$q_{dsat}(0) = \frac{\lambda_C}{2}.\frac{q_s^2+q_s}{1+\frac{\lambda_C}{2}.(1+q_s)} \tag{3.58}$$
$$T_3 = 2.\lambda_C.(q_s - q_{dsat}(0))$$
$$T_4 = \sqrt{1+T_3.T_3}$$
$$T_5 = ln(T_3+T_4)$$
$$T_6 = \begin{cases} T_4 + (1/T_3).T_5 & \text{if } T_3 \neq 0 \\ T_4 + (1/T_4) & \text{otherwise} \end{cases}$$
$$f(q_{dsat}(0)) = q_{dsat}(0).T_6 + T_2.q_{dsat}(0).(q_s + q_{dsat}(0) + 1) - \lambda_C.((q_s.q_s + q_s) - (q_{dsat}(0) * q_{dsat}(0) + q_{dsat}(0))) \tag{3.59}$$
$$T_7 = \begin{cases} (-2.0.\lambda_C).(T_3.T_4 - T_5)/(T_3.T_3) & \text{if } T_3 \neq 0 \\ (-2.0.\lambda_C).(T_3/T_4).(T_3/T_4).(T_3/T_4) & \text{otherwise} \end{cases}$$
$$f'(q_{dsat}(0)) = q_{dsat}(0).T_7 + T_6 + T_2.(q_s + 2 * q_{dsat}(0) + 1) + \lambda_C.(2.q_{dsat}(0) + 1) \tag{3.60}$$
$$q_{dsat}(1) = q_{dsat}(0) - f(q_{dsat}(0))/f'(q_{dsat}(0)) \tag{3.61}$$

From (3.58), (3.59), and (3.60), (3.61) will be calculated

$$T_3 = 2.\lambda_C.(q_s - q_{dsat}(1))$$
$$T_4 = \sqrt{1+T_3.T_3}$$
$$T_5 = ln(T_3+T_4)$$
$$T_6 = \begin{cases} T_4 + (1/T_3).T_5 & \text{if } T_3 \neq 0 \\ T_4 + (1/T_4) & \text{otherwise} \end{cases}$$
$$f(q_{dsat}(1) = q_{dsat}(1).T_6 + T_2.q_{dsat}(1).(q_s + q_{dsat}(1) + 1) - \lambda_C.((q_s.q_s + q_s) - (q_{dsat}(1) * q_{dsat}(1) + q_{dsat}(1))) \tag{3.62}$$
$$T_7 = \begin{cases} (-2.0.\lambda_C).(T_3.T_4 - T_5)/(T_3.T_3) & \text{if } T_3 \neq 0 \\ (-2.0.\lambda_C).(T_3/T_4).(T_3/T_4).(T_3/T_4) & \text{otherwise} \end{cases}$$
$$f'(q_{dsat}(1)) = q_{dsat}(1).T_7 + T_6 + T_2.(q_s + 2 * q_{dsat}(1) + 1) + \lambda_C.(2.q_{dsat}(1) + 1) \tag{3.63}$$
$$q_{dsat}(final) = q_{dsat}(1) - f(q_{dsat}(1))/f'(q_{dsat}(1)) \tag{3.64}$$

From (3.61), (3.62), and (3.63), (3.64) will be calculated.

### 3.4.3 Drain saturation voltage calculations

The expressions of $q_{dsat}$ calculated for extrinsic and intrinsic cases ((3.56) and (3.64)) are used to calculate $V_{dsat}$, as given below

$$v_{dsat} = \psi_P - \frac{2.\phi_b}{n} - 2q_{dsat} - ln\left[\frac{2q_{dsat}.n_q}{\gamma}.\left(\frac{2q_{dsat}.n_q}{\gamma} + \frac{\gamma}{n_q - 1}\right)\right] \quad (3.65)$$

where $\gamma$ is the normalized body factor.

Also,

$$V_{dsat} = v_{dsat}.nV_t, \; V_{dssat} = V_{dsat} - V_s$$
$$V_{dseff} = \frac{V_{ds}}{\left[1 + \left(\frac{V_{ds}}{V_{dssat}}\right)^{1/DELTA}\right]^{DELTA}} \quad (3.66)$$

and,

$$v_{deff} = \frac{V_{dseff} + V_s}{nV_t} \quad (3.67)$$

Finally, (3.67) is used to calculate the normalized charge density at the drain side ($q_d$).

## 3.5 Output conductance

Unlike the ideal transistor assumption, drain current of a real MOSFET shows a finite slope w.r.t. $V_{ds}$. There are multiple physical effects that causes such behavior: channel length modulation (CLM), drain-induced barrier lowering (DIBL), drain-induced threshold shift (DITS), substrate current induced body effect (SCBE), etc. Almost all of these are prominent in a shorter channel device.

The finite slope of $I_{ds} - V_{ds}$ characteristics is nothing but the output conductance $g_{ds}$. The logic designer can not worry about $g_{ds}$ accuracy if the on-current, off-current, and others are well modeled. However, for the analog and RF designs, accurate modeling of derivatives is a necessity since amplifier gain, distortion, and others depend on that. In the BSIM-BULK model, output conductance in the saturation region is modeled as [1]:

$$I_{ds} = I_{ds,sat}\cdot\left[1 + \frac{\Delta V_{ds}}{V_A}\right] \quad (3.68)$$

where $V_A$ represents early voltage and

$$\Delta V_{ds} = V_{ds} - V_{dseff} \tag{3.69}$$

We will now discuss output conductance modeling in the BSIM-BULK model.

### 3.5.1 Channel length modulation (CLM)

For a given gate voltage, as $V_{DS}$ increases from zero, the current increases almost linearly (triode region in Fig. 3.18). The inversion charge carrier concentration decreases along the channel from source to drain while carrier velocity increases to support the current. At sufficiently large $V_{DS}$, carriers near the drain undergo velocity saturation, and their concentration is minimum. On further increase in $V_{DS}$, the carrier velocity saturation point moves towards the source, thus leading to channel length modulation. In BSIM-BULK, it is modeled similar to BSIM4 [1], and the model equations are given as:

$$I_{ds} = I_{ds} * M_{CLM} \tag{3.70}$$

where

$$M_{CLM} = 1 + C_{clm} \ln\left[1 + \frac{\Delta V_{ds}}{V_{asat}} \cdot \frac{1}{C_{clm}}\right] \tag{3.71}$$

$$V_{asat} = V_{dssat} + E_{sat} \cdot L_{eff} \tag{3.72}$$

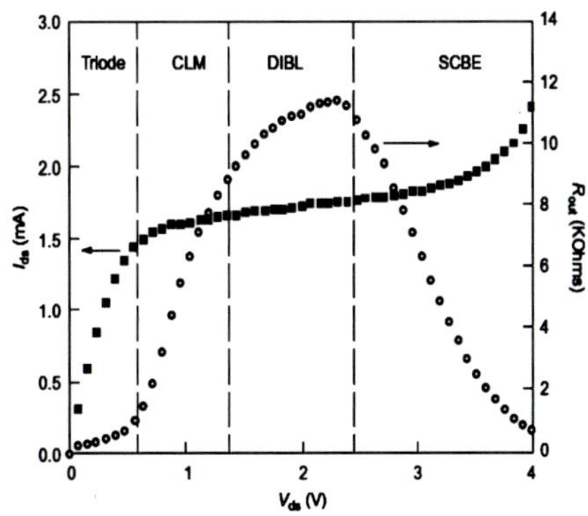


FIGURE 3.18 Illustration of the output conductance model [17].

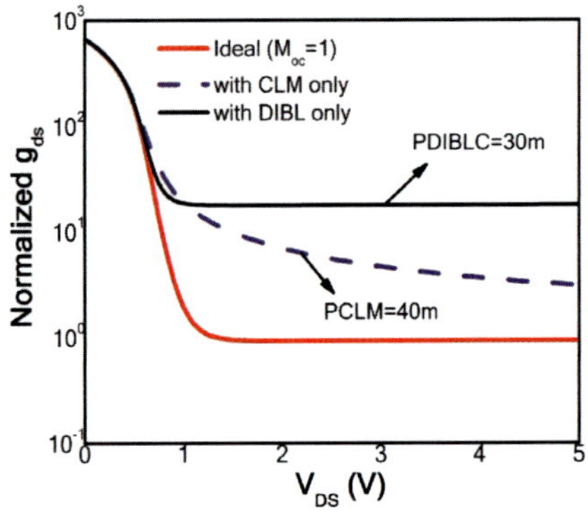


FIGURE 3.19 Impact of CLM and DIBL on the output conductance.

and

$$C_{clm} = \begin{cases} PCLM \cdot \left(1 + PCLMG \cdot \frac{q_{ia}}{E_{sat} \cdot L_{eff}}\right) \frac{1}{F} & \text{for } PCLMG > 0 \\ \frac{PCLM}{\left(1 - PCLMG \cdot \frac{q_{ia}}{E_{sat} \cdot L_{eff}}\right)} \frac{1}{F} & \text{for } PCLMG < 0 \end{cases} \tag{3.73}$$

$$F = \begin{cases} 1 & \text{for } FPROUT \leq 0 \\ \frac{1}{1 + \frac{FPROUT \cdot \sqrt{L_{eff}}}{q_{ia} + 2 \cdot n V_t}} & \text{for } FPROUT > 0 \end{cases} \tag{3.74}$$

Critical electric field for velocity-saturation is defined as:

$$E_{sat} = \frac{2 \cdot VSAT}{\frac{U0}{D_{mob}}} \tag{3.75}$$

PCLM is a key parameter for the CLM model, and PCLMG provides gate-bias dependence. Fig. 3.19 shows the impact of PCLM on the output conductance ($g_{ds}$).

## 3.5.2 Drain induced barrier lowering (DIBL)

We have seen that DIBL affects threshold voltage of a transistor at high $V_{ds}$. In fact, it also has a significant impact on output conductance. The BSIM-BULK model equations are given below:

$$M_{DIBL} = \left(1 + \frac{\Delta V_{ds}}{V_{ADIBL}}\right) \tag{3.76}$$

where

$$V_{ADIBL} = \frac{1}{PDIBLC} \cdot \frac{(q_{ia} + 2V_t)^2}{V_{dssat} + q_{ia} + 2V_t} \cdot PVAGfactor \cdot f(V_{bs}) \quad (3.77)$$

$$f(V_{bs}) = \frac{1}{1 + PDIBLCB \cdot V_{bsx}} \quad (3.78)$$

$$PVAGfactor = \begin{cases} 1 + PVAG \cdot \frac{q_{ia}}{E_{sat}L_{eff}} & \text{for } PVAG > 0 \\ \frac{1}{1 - PVAG \cdot \frac{q_{ia}}{E_{sat}L_{eff}}} & \text{for } PVAG < 0 \end{cases} \quad (3.79)$$

Though the model is similar to the BSIM4, model equations are suitably updated to ensure continuity and symmetry. Parameters PDIBLCB and PVAG provide body-bias and gate-bias dependency to the DIBL model and are very useful. Fig. 3.19 shows the impact of DIBL and CLM models on $g_{ds}$. For an ideal transistor, $g_{ds}$ is constant and does not change with $g_{ds}$ (simulated by setting ETA0=0 and choosing parameters such that output conductance model is not active).

### 3.5.3 Drain induced threshold shift (DITS)

DITS is a long-channel phenomenon observed in transistors using pocket implants. Pocket implants introduce energy barriers at the source and the drain ends. At high $V_{ds}$, the barrier at the drain side is lowered first, causing threshold voltage shift. In BSIM-BULK, this effect is modeled as:

$$M_{DITS} = \left(1 + \frac{\Delta V_{ds}}{V_{ADITS}}\right) \quad (3.80)$$

$$V_{ADITS} = \frac{F}{PDITS} \cdot \left[1 + (1 + PDITSL \cdot L_{eff}) \exp(PDITSD \cdot V_{ds})\right] \quad (3.81)$$

F in the above equation is defined in (3.74). PDITS and PDITSD are the model parameters, representing coefficient for drain-induced threshold voltage shift and $V_{ds}$ dependence of drain-induced threshold voltage shift, respectively. Fig. 3.20(a) shows the impact of DITS model on $g_{ds}$. The increase in $g_{ds}$ is more initial when the transistor enters into saturation. However, as $V_{ds}$ increases, $V_{ADITS}$ also increases, and $g_{ds}$ reduces (as $V_{ADITS}$ approaches $\infty$, $M_{ADITS}$ approaches 1).

### 3.5.4 Substrate current induced body effect (SCBE)

At high drain biases, channel carriers gain sufficient energy to cause impact-ionization. This results in the generation of electron-hole pair. For n-channel MOS, holes flow out of the body terminal through the Si body

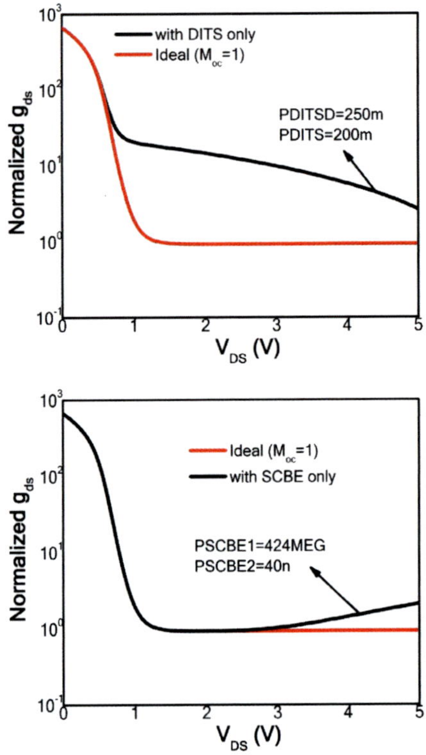


FIGURE 3.20 Illustration of DITS and SCBE model parameters effect on the output conductance $g_{ds}$.

resistance and create a positive voltage drop across the body and source junction. This impacts $g_{ds}$ since positive voltage drop (for n-MOS) causes threshold voltage reduction. It is modeled as:

$$M_{SCBE} = \left(1 + \frac{\Delta V_{ds}}{V_{ASCBE}}\right) \tag{3.82}$$

$$V_{ASCBE} = \frac{L_{eff}}{PSCBE2} \cdot \exp\left(\frac{PSCBE1 \cdot litl}{\Delta V_{ds}}\right) \tag{3.83}$$

where the characteristic length is given as:

$$litl = \sqrt{(\epsilon_{sub}/\epsilon_{ox}) \cdot TOXE \cdot XJ} \tag{3.84}$$

PSCBE2 is the substrate current body-effect parameter with default value of $10^{-8}$ m/V. If this parameter is set to zero, above $V_{ASCBE}$ is not evaluated, and it is set to a large value ($5.540622384 * 10^{34}$ in BSIM-BULK 107.1.0).

### 3.5.5 Complete output conductance model

$M_{oc}$ is multiplied to the drain current $I_{ds}$ in the final drain current expression.

$$M_{oc} = M_{DIBL} \cdot M_{CLM} \cdot M_{DITS} \cdot M_{SCBE} \tag{3.85}$$

$$I_{ds} = I_{ds} \cdot M_{oc} \tag{3.86}$$

## 3.6 Sub-threshold hump model

Shallow Trench Isolation (STI) is a widely used technique for device isolation. However, it has been reported that it can cause the parasitic drain to source current to the edge effects [18].

One way of modeling the total current of a device with $I_{edge}$ is to use two full MOSFET compact models: one for the main channel current ($I_{main}$) and another for the edge parasitic current. These two models can then be connected in parallel via a sub-circuit. However, this solution is not attractive from a simulation speed point of view. The simulation speed will be reduced not only by using the sub-circuit, which may create additional nodes, but also by a large number of extra model equation evaluations required by the full compact model used for $I_{edge}$. Additionally, it introduces complexity for model parameter extraction. BSIM-BULK avoids the use of a sub-circuit approach, and the edge effect is modeled by a single compact model. Following key differences between the main channel current and $I_{edge}$ are considered:

1. $I_{edge}$ has a different sub-threshold slope (SS). This can be due to a different effective oxide thickness. There might also be a different depletion charge density due to the dopant loss around the edges. Also, the interface traps at the edge can be different as compared to the main channel resulting in a different SS.
2. $I_{edge}$ has a different body-bias coefficient ($\gamma$) than $I_{main}$. The main reason for this is a higher effective oxide capacitance due to the field crowding at the edge. $\gamma$ being the ratio of depletion and oxide capacitance is thus lower for $I_{edge}$.
3. $I_{edge}$ has a different short-channel behavior. As the channel length is scaled down, $I_{main}$ will suffer from short-channel effects (SCEs). $I_{edge}$ also gets affected by SCEs but not in exactly the same way. Typically, $I_{edge}$ has a better immunity to SCEs than $I_{main}$ due to a thinner effective oxide thickness, but it depends on exact characteristics of the STI edge.

Fig. 3.21 shows the schematic representation of the BSIM-BULK model with EDGEFET model. This model is activated by setting EDGEFET=1, and sub-threshold parasitic current is evaluated in a same way as BSIM-BULK main model current [19]. It uses new parameters for doping, DIBL

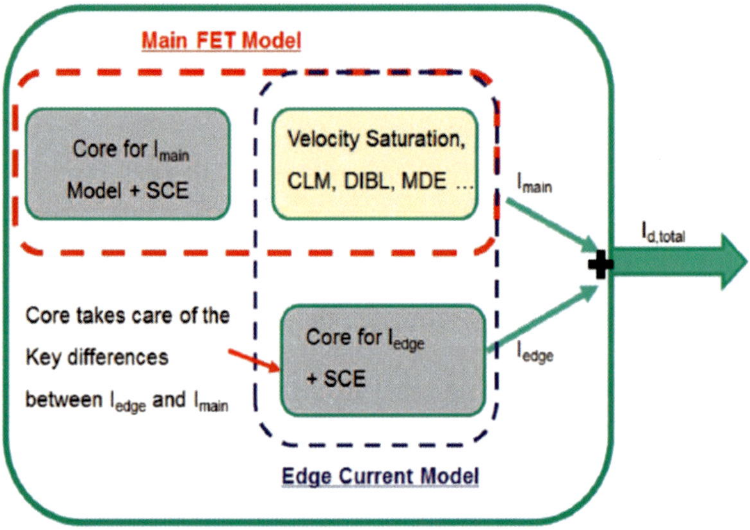


FIGURE 3.21 Schematic representation of the BSIM-BULK compact model with the EDGEFET model [19].

TABLE 3.3 EDGEFET main parameters.

| Name | Default |
|---|---|
| EDGEFET | 0 (0:disable, 1: enable) |
| WEDGE | 10 nm |
| DGAMMAEDGE | 0 |
| DVTEDGE | 0 |
| CITEDGE | 0 |
| NFACTOREDGE | 0 |
| CDSCDEDGE | 1n |
| CDSCBEDGE | 0 |
| ETA0EDGE | 0.08 |
| ETABEDGE | -0.07 |
| KT1EDGE | -0.11 |
| KT2EDGE | 0.022 |
| TNFACTOREDGE | 0 |
| TETA0EDGE | 0 |
| DVT0EDGE | 2.2 |
| DVT1EDGE | 0.53 |
| DVT2EDGE | 0 |
| TNFACTOREDGE | 0 |

and others, instead of main transistor parameters. Table 3.3 summarizes the list of parameters for EDGEFET. The model equations are given be-

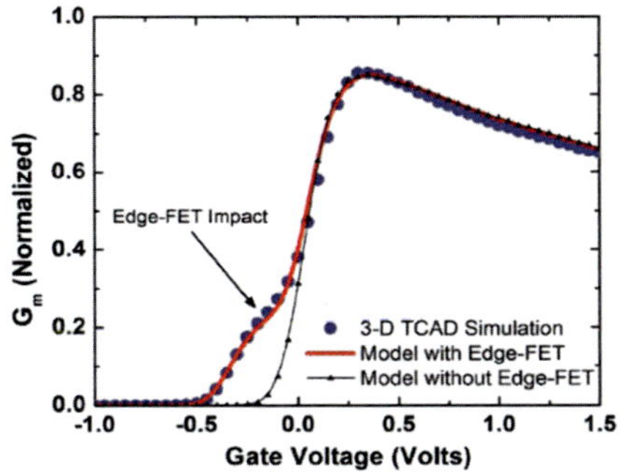


FIGURE 3.22 EDGEFET model validation: transconductance vs. gate voltage of a MOS transistor with STI. This highlights the importance of modeling parasitic drain to source current [19].

low:

$$I_{ds,EDGE} = 2 \cdot NF \cdot n_q \cdot \mu_{eff} \cdot WEDGE/L_{eff} \cdot C_{ox} \cdot \\ nV_t \cdot (q_s - q_{deff}) \cdot (1 + q_s + q_{deff}) \cdot M_{oc} \quad (3.87)$$

here $q_s$ and $q_{d,eff}$ are calculated in the same way as in the BSIM-BULK main channel model. The total current is then given as

$$I_{Total} = I_{ds,EDGE} + I_{ds} \quad (3.88)$$

The Edge FET Model current is added to the main current $I_{ds}$ only when the parameter EDGEFET is 1. Fig. 3.22 and Fig. 3.23 show the model validation results with the 3D TCAD simulation data and experimental data [19], respectively. The figures clearly highlight the importance of the STI model.

## 3.7 Series resistance model

The series resistance of a MOSFET is composed of different parts like contact resistance, LDD resistance, etc. Broadly, it can be classified into bias-dependent and bias-independent parts. For modeling, it is straightforward to instantiate resistance between the external drain (source) and the internal drain (source). However, this leads to the creation of two more nodes in the circuit and may cost higher simulation time. BSIM-BULK offers three ways to model the series resistance effect as shown in Fig. 3.24 and summarized below. A suitable model should be enabled based on accuracy and speed requirements.

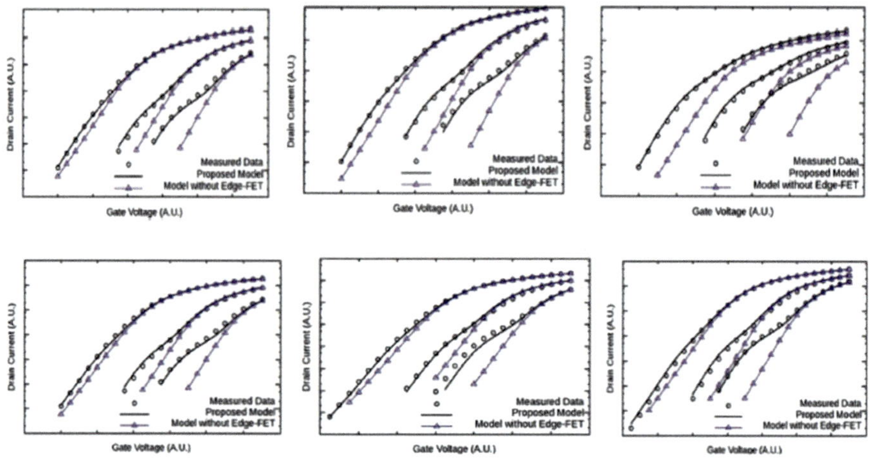


FIGURE 3.23 EDGEFET model validation: Drain current vs. gate voltage (a) for different body biases and multiple channel widths (b) for different body biases and multiple channel lengths [19].

(a) **RDSMOD=0**, External resistance is bias-independent, while internal resistance is bias-dependent.
(b) **RDSMOD=1**, No internal resistance. Both bias-dependent and independent resistors are kept externally.
(c) **RDSMOD=2**, No external resistance. Both bias-dependent and independent resistors are kept internally. Internal nodes are not created.

We will now discuss detailed model equations for all these modes.

### *3.7.0.1 RDSMOD=1: bias dependent external series resistance*

The bias-dependent external resistance model is adopted from BSIM4 and can be invoked by setting model selector RDSMOD=1. BSIM4 and BSIM-BULK allow the source extension resistance $R_s(V)$ and the drain extension resistance $R_d(V)$ to be external and asymmetric (i.e., $R_s(V)$ and $R_d(V)$ can be connected between the external and internal source and drain nodes, respectively; furthermore, $R_s(V)$ does not have to be equal to $R_d(V)$). This feature makes accurate RF CMOS simulation possible.

The source/drain series resistance is the sum of a bias-independent ($R^{bias\_indep}$) component and a bias-dependent ($R^{bias\_dep}$) component:

$$R_{source} = R_s^{bias\_dep} + R_s^{bias\_indep} \tag{3.89}$$

$$R_{drain} = R_d^{bias\_dep} + R_d^{bias\_indep} \tag{3.90}$$

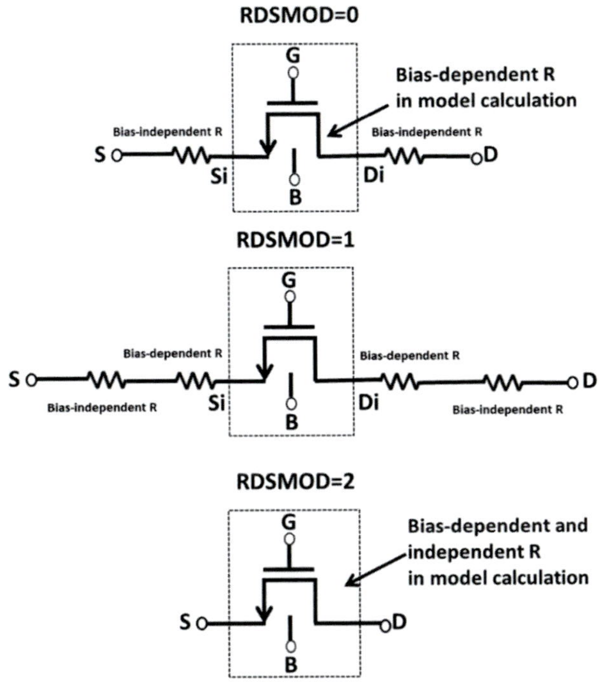


FIGURE 3.24 BSIM-BULK offers three models for modeling the series resistance effect. RDSMOD is the model selector. For RDSMOD=0, the bias-dependent part is internal and bias-dependent is external. RDSMOD=1, both bias-independent and bias-dependent parts are external. In RDSMOD=2, the resistance is internal, and no external node is created.

where the bias-dependent part is given as follows:

$$R_s^{bias\_dep} = \frac{1}{W_{eff}^{WR} \cdot NF} \cdot \Bigg( RSWMIN + RSW \cdot \left[ -PRWB \cdot V_{bs} + \frac{1}{1 + PRWG \cdot V_{gs,eff}} \right] \Bigg) \tag{3.91}$$

$$R_d^{bias\_dep} = \frac{1}{W_{eff}^{WR} \cdot NF} \cdot \Bigg( RDWMIN + RDW \cdot \left[ -PRWB \cdot V_{bd} + \frac{1}{1 + PRWG \cdot V_{gd,eff}} \right] \Bigg) \tag{3.92}$$

$W_{eff}$ in the above equations is expressed in μm. The effective gate-to-source (drain) voltage in above model equations are the no swap voltage,

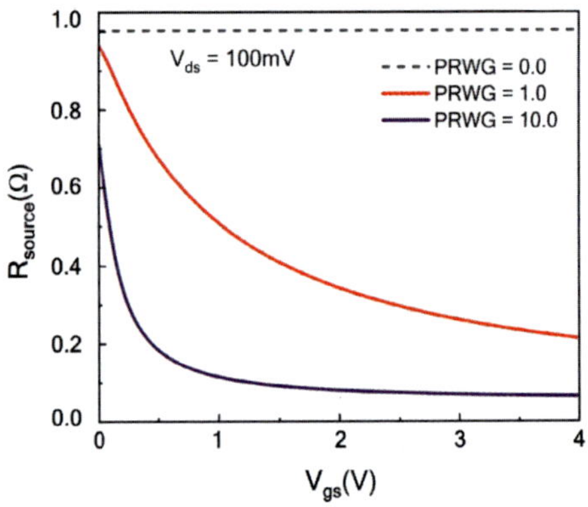


FIGURE 3.25 RDSMOD=1 model behavior: plot of $R_{source}$ vs. gate voltage $V_g$ for $V_d = V_s = V_b = 0$ V. Default model parameters are used.

i.e., they do not get swapped when negative $V_{ds}$ is applied. It is given as:

$$V_{gs,eff} = \frac{1}{2}\left[V_{gs} - V_{fbsdr} + \sqrt{(V_{gs} - V_{fbsdr})^2 + 10^{-2}}\right] \quad (3.93)$$

$$V_{gd,eff} = \frac{1}{2}\left[V_{gd} - V_{fbsdr} + \sqrt{(V_{gd} - V_{fbsdr})^2 + 10^{-2}}\right]$$

The bias-independent part of the resistance is the geometrical diffusion resistance or sheet resistance simply calculated as the sheet resistances ($RSHS$, $RSHD$) times the number of squares ($NRS$, $NRD$):

$$R_s^{bias\_indep} = R_{s,geo} = NRS \cdot RSHS \quad (3.94)$$

$$R_d^{bias\_indep} = R_{d,geo} = NRD \cdot RSHS \quad (3.95)$$

The total resistance is thus given by the sum of (3.91) and (3.94). Note that this model creates two additional nodes, as shown in Fig. 3.24. Fig. 3.25 shows the typical behavior of the series resistance model. When PRWG is 0, the resistance shows no gate-bias dependency. For positive PRWG, it decreases with the gate bias. The minimum value of the resistance is governed by the parameter RSWMIN.

#### *3.7.0.2 RDSMOD=0: bias independent external and bias dependent internal resistance*

In this RDSMOD, the external resistance is bias-independent. A degradation factor is added to the drain current model to account for the bias-dependent part of the series resistance. The model equations are summarized below:

$$R_{source} = R_s^{bias\_indep} = R_{s,geo} \tag{3.96}$$

$$R_{drain} = R_d^{bias\_indep} = R_{d,geo} \tag{3.97}$$

Current degradation factor ($D_r$) is determined as:

$$T_0 = 1 + PRWG \cdot q_{ia} \tag{3.98}$$

$$T_1 = PRWB \cdot (\sqrt{\phi_s - V_{bs}} - \sqrt{\phi_s}) \tag{3.99}$$

$$T_2 = \frac{1}{T_0} + T_1; \quad T_3 = \frac{1}{2}[T_2 + \sqrt{T_2^2 + 0.01}] \tag{3.100}$$

$$R_{ds}(V) = NF \cdot \left( W_{eff}^{WR} \left[ RDSWMIN + RDSW \cdot T_3 \right] \right) \tag{3.101}$$

$$D_r = 1.0 + \frac{\mu_0}{D_{mob}.D_{vsat}} \cdot C_{ox} \cdot \frac{W_{eff}}{L_{eff}} \cdot q_{ia} \cdot R_{ds} \tag{3.102}$$

$D_r$ goes into the denominator of the final $I_{ds}$ expression. The advantage of this mode is improved speed as compared to RDSMOD=1 since the external resistances are not bias-dependent. Note that two additional nodes are also created in this mode.

#### 3.7.0.3 Bias dependent internal resistance (RDSMOD=2)

This is mode where no external nodes are created. Both bias-dependent and bias-independent parts of the resistance are modeled through the current degradation factor. Model equations are given as:

$$R_{ds}(V) = R_{s,geo} + NF \cdot \left( W_{eff}^{WR} \left[ RDSWMIN + RDSW \cdot T_3 \right] \right) + R_{d,geo} \tag{3.103}$$

$$D_r = 1.0 + \frac{\mu_0}{D_{mob}.D_{vsat}} \cdot C_{ox} \cdot \frac{W_{eff}}{L_{eff}} \cdot q_{ia} \cdot R_{ds} \tag{3.104}$$

where T3 is given by (3.100).

**Note on Node Collapsing:** External resistance in RDSMOD=0 is bias-independent (i.e., $R_{source} = R_{s,geo}$). If $R_{s,geo} < minr$, $R_{source}$ is set to 0, and no additional nodes are created. If provided in the netlist, $minr$ takes the value from there. Otherwise, it assumes 1 m Ω by default.

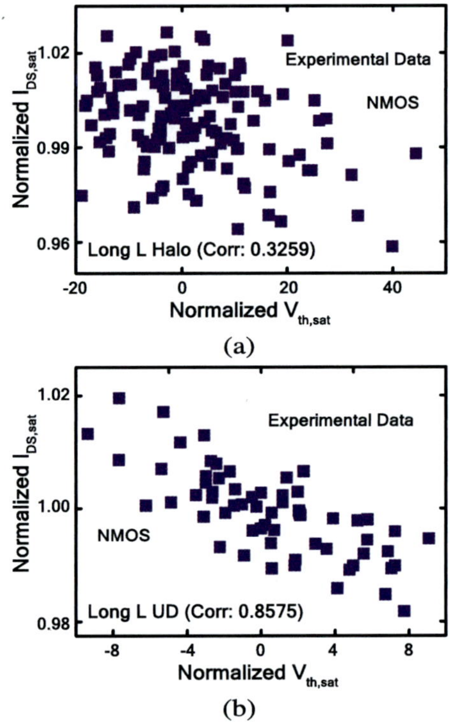


FIGURE 3.26 Measured correlation between $I_{D,sat}$ with $V_{th,sat}$ of long-channel (a) Halo implanted (b) Uniformly doped MOSFET. The long-channel halo device shows low correlation, indicating that different mechanisms are controlling the channel in weak and strong inversion, which cannot be modeled by conventional MOS models. Measurement data is from the same process in [20,21].

## 3.8 Impact of strong halo implants on current and transconductance

Halo or pocket implants are essential for the BULK transistors to combat short-channel effects. Halo devices evince peculiar terminal characteristics, especially in the case of strong pocket dose. In this chapter, we will discuss the impact on current and transconductance. In Chapter 6, the impact on noise behavior is discussed.

Fig. 3.26 shows correlation between the drain current and threshold voltage in saturation ($I_{D,sat}$ and $V_{th,sat}$) for long-channel device with and without halo. For the long-channel halo device, $I_{D,sat}$ and $V_{th,sat}$ show a small correlation coefficient. This contradicts conventional long-channel theory and understanding that $I_{D,sat}$ is proportional to $(V_{GS} - V_{th,sat})^2$ and therefore is expected to be fully correlated, as seen for the UD devices in

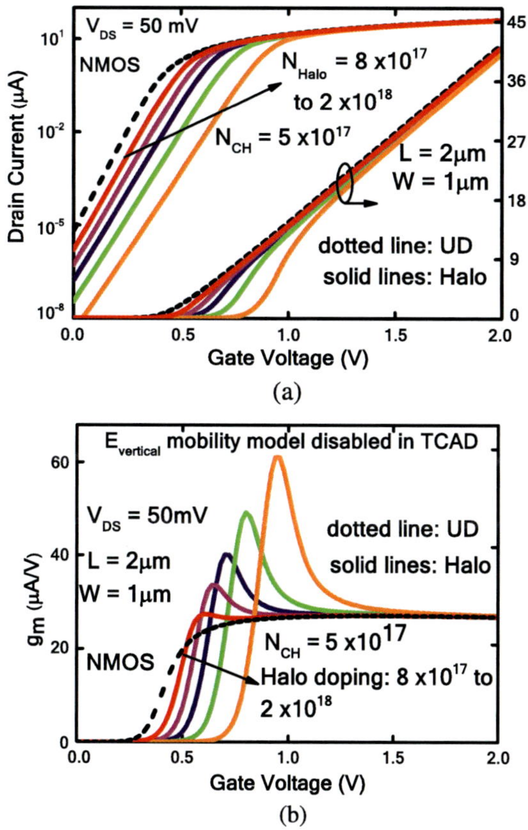


FIGURE 3.27 TCAD simulations of Halo and UD devices in linear region at $V_{DS} = 50\,\text{mV}$. Channel doping is the same in all the devices, and L=2 μm, W=1 μm. (a) Drain current (b) $g_m$ at different doping conditions. In (b), while the threshold voltage is different, drain current in SI is the same for all the cases, causing $g_m$ to peak in the vicinity of weak-strong inversion transition region. Here, we have intentionally turned off the mobility degradation model at high field to demonstrate that the $g_m$ peak in halo devices is not related to mobility degradation at high field as in conventional UD. Similar trends are also seen in the real device measurements [22].

Fig. 3.26(b). In the following sections, we discuss the impact of halo regions in different regions of operation.

### 3.8.1 Linear region

Fig. 3.27 compares the current and $g_m$ of uniformly doped transistor and halo transistor for different halo doping levels ($N_H$). Threshold voltage increases with $N_H$, as seen in Fig. 3.27(a), which is expected since the barrier height increases with doping. Interestingly, the strong inversion current is of the similar magnitude [23] due to the following reason: in strong inversion (SI) and at small drain voltages, MOSFET behaves like a resistor, with effective resistance given by the sum of the resistances of halo regions ($R_h$) and channel region ($R_{ch}$). In SI, both the halo and channel regions are strongly inverted, and therefore their effective resistances are determined by the respective lengths and doping of the regions. For long-channel devices, since the length of the channel region is much larger than the halo region length, the effective resistance of the overall device is determined by the resistance of the channel region. Since channel resistance ($R_{ch}$) is similar in all the cases, the strong inversion current remains the same. Note that for the device with the highest threshold voltage (highest halo doping), although $I_D$ remains low till a higher gate voltage, yet it still attains the same $I_D$ as the UD device with the lowest threshold voltage at $V_{GS} = V_{DD}$. Therefore there is a region in $I_D - V_G$ characteristics where current sharply increases to the current level of UD. Higher the halo doping, sharper is this increase. This leads to peak in the derivative $g_m$, which increases with the halo doping as shown in Fig. 3.27(b). For these TCAD simulations, high field mobility degradation models are intentionally turned off to demonstrate that the $g_m$ peak in halo devices is a result of the above-mentioned artifact and not related to mobility degradation at high field as in conventional UD. Note that the UD device, as expected, does not show a peak in $g_m$. Similar trends are also seen in the real device measurements [22].

### 3.8.2 Saturation region

Saturation region behavior of the halo transistor is also interesting. To study the same, we investigated Uniformly Doped (UD), Source Halo (SH), Drain Halo (DH), and both side halo (Halo) transistors. Fig. 3.28 shows characteristics of UD, SH, DH, and Halo devices at $V_D = 2\,\mathrm{V}$. Threshold voltage of DH is significantly smaller than that of SH and Halo as seen in Fig. 3.28(a). This is because the former experiences barrier lowering at the drain side, which essentially brings DH at the same threshold level as that of UD (UD and DH has similar $V_{th,sat}$). It is important to note that Halo also undergoes barrier lowering at the drain side, but high barrier at the source side prevents significant threshold voltage drop. Fig. 3.28(b) shows $g_m$ of all the four devices. In the case of SH and Halo, $g_m$ has very distinct behavior as it initially increases, then decreases, and then again increases with the gate voltage. At smaller gate voltages, the channel region is strongly inverted while halo regions are still in weak inversion. This makes the effective length of the device the same as the length of the halo regions [27]. Since $g_m \propto (\frac{1}{L})$, it continues to rise sharply till the halo regions also get inverted. Once the halo regions are inverted, the control of conduction is taken over by the channel, and $g_m$ behaves more conventionally. In Fig. 3.28(b), one can also see that in strong inversion, $g_m$ of the UD and SH devices are the same, while $g_m$ of DH and Halo

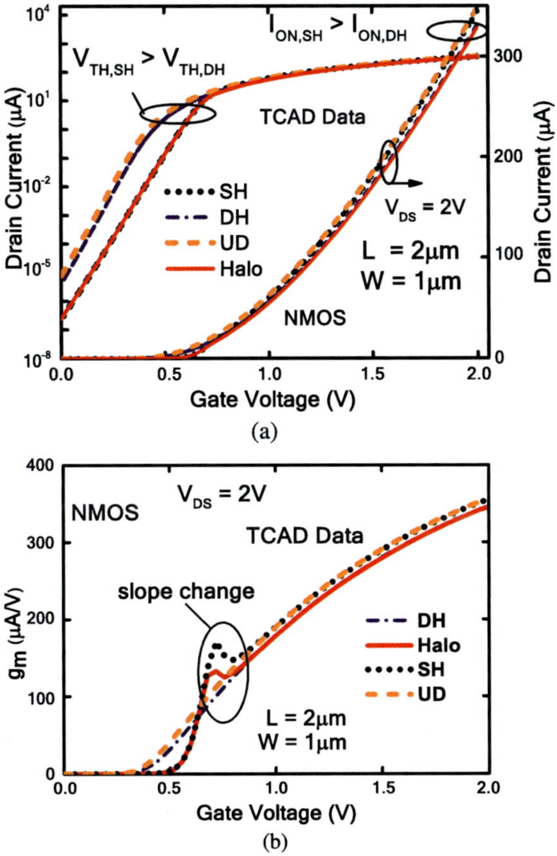


FIGURE 3.28 Comparison of (a) current (b) transconductance of SH, DH, Halo, and UD devices in the saturation at $V_{DS} = 2$ V. $g_m$ of DH is very different from that of SH and Halo, as it does not undergo change of slope around the threshold voltage. Most of the compact models model the saturation characteristics of halo implanted MOSFET by two transistor equivalent model [24,25] (proposed in [26]), one representing the drain side halo region and other representing channel. By neglecting the source side halo region, $g_m$ characteristics cannot be modeled accurately. L=2 μm, W=1 μm.

device asymptomatically follows $g_m$ of UD with an offset. This can be explained as follows: in the case of pinch-off, $g_m \propto (V_{GS} - V_{th,sat}|_D)$, where $V_{th,sat}|_D$ represents the threshold voltage of the drain region [28]. Since SH and UD do not have drain side halo implants, their $V_{th,sat}|_D$ is the same, and hence the same $g_m$ (and $I_{on}$). Similarly, DH and Halo have the same $V_{th,sat}|_D$ as both have drain side halo implants. Also, since $V_{th,sat}|_D$ for Halo (and DH) > $V_{th,sat}|_D$ for UD (and SH), their $g_m$ in SI as well as $I_{on}$ is smaller ($g_m \propto (V_{GS} - V_{th,sat}|_D)$). Note that threshold voltage of DH is smaller than that of SH (see Fig. 3.28(a)), but it carries smaller $I_{on}$ due to

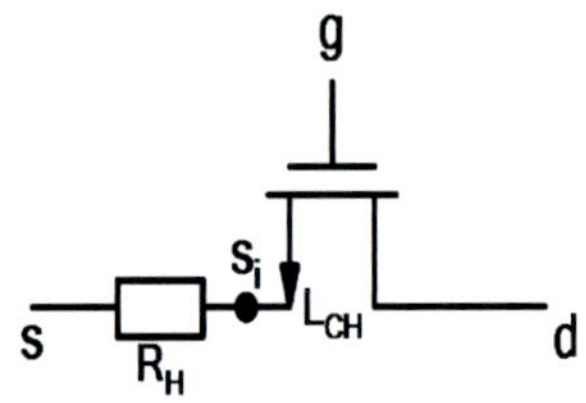


FIGURE 3.29 Modeling of Halo implanted MOSFET. The source side halo region is modeled using the bias-dependent resistor. The impact of drain side halo region on the threshold voltage and output conductance is analytically modeled in BSIM-BULK through the DITS model [26].

the same reason. It can, therefore, be inferred that $I_{on}$ and threshold voltage are not correlated in the long-channel halo devices, and hence poor correlation is observed in real device measurements in Fig. 3.26.

### 3.8.3 Compact modeling and validation

Saturation characteristics of the Halo implanted devices are traditionally modeled by two transistor equivalent sub-circuit: one representing the drain side halo and the other representing the channel region [26,28]. This, however, is insufficient to capture peculiar attributes of $g_m$ in the saturation region. As seen in Fig. 3.28(b), unlike SH, DH does not show unconventional $g_m$ behavior in saturation, and it can be inferred that source side electrostatics is responsible for strange $g_m$ characteristics. We model halo device using two element sub-circuit model as shown in Fig. 3.29, where BSIM-BULK is used for modeling the drain side halo and the channel region, and other bias-dependent elements for the source side halo region, which is characterized as

$$I_{R_H} = U0H.C_{ox}\frac{W}{L_H}.V_{s_i s,eff}.V_{gsteff} \tag{3.105}$$

where

$$V_{gsteff} = \frac{2N_H V_t.ln\left[1 + exp(\frac{V_{GS}-VTH}{2N_H V_t})\right]}{1 + 2N_H.exp(-\frac{V_{GS}-V_{TH}}{2N_H V_t})} \tag{3.106}$$

$$V_{s_i s,eff} = \frac{V_{s_i s}}{\left[1+\left(\frac{V_{s_i s}}{V_{dsat,H}}\right)^{\frac{1}{DELTAH}}\right]^{DELTAH}} \tag{3.107}$$

$$V_{dsat,H} = V_{gsteff} + 2V_t \tag{3.108}$$

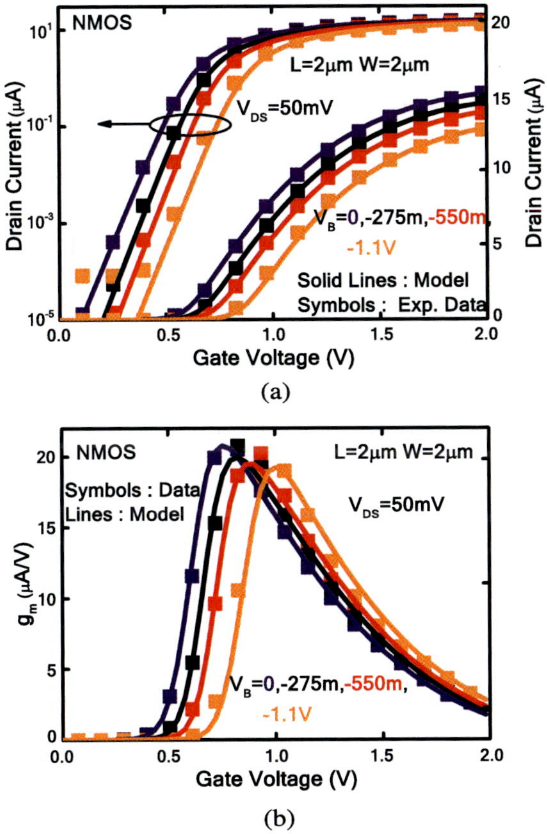


FIGURE 3.30 NMOS - Model validation with experimental data in linear region at $V_{DS} = 50\,\text{mV}$: (a) $I_D - V_G$ (b) $g_m - V_G$ at different body biases. L=2 µm, W=2 µm.

Here **U0H**, **LH**, **VTH**, and **NH** are the model parameters representing mobility, length, threshold voltage, and sub-threshold slope of the halo region. **DELTAH** is the smoothing parameter. It is important to note that the halo device can also be modeled using the transistor for both the source side halo region and the channel region; however, it will increase computational time.

**Model validation with NMOS:** The model is validated with low power 45 nm CMOS technology node measurements [20,21]. Fig. 3.30(a) and Fig. 3.30(b) show $I_D$ and $g_m$ vs $V_G$ for different body-bias conditions, respectively, at $V_{DS} = 50\,\text{mV}$. Fig. 3.31 shows $I_D$ and $g_m$ in saturation region $V_{DS} = 1.21\,\text{V}$. A compact model ignoring the source side halo region cannot model $g_{m,sat}$ as evident from Fig. 3.31(b), which also compares $g'_{m,sat}$ fitting result as obtained from proposed and conventional models. The

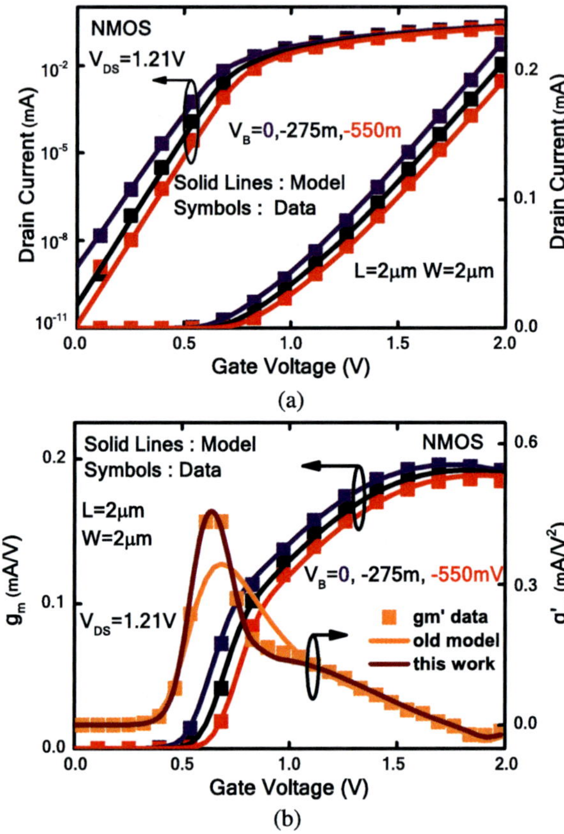


FIGURE 3.31 NMOS - Model validation with experimental data in saturation region at $V_{DS} = 1.21$ V: (a) $I_D - V_G$ (b) $g_m$ at different body biases. The proposed model accurately captures $g_m$ behavior across body biases. Compact model ignoring source side halo region electrostatic do not capture typical $g_m$ behavior. L=2 µm, W=2 µm.

proposed model accurately captures peculiar $g_{m,sat}$ trends across different body biases.

**Model validation with PMOS:** Fig. 3.32(a) shows the model-data overlay for a p-channel transistor with $L = W = 2\,\mu\text{m}$ at $|V_{DS}| = 50\,\text{mV}$ and $V_B = 0, 0.45, 0.9\,\text{V}$. It is interesting to note that peak value of $g_m$ in Fig. 3.32(a) initially increases and then decreases with $V_B$. It is known that peak $g_m$ of UD devices decreases with $V_B$ due to increase in scattering. To the best of our knowledge, such $g_m$ behavior is reported for the first time [29]. To understand this, TCAD simulations with $N_{CH}$=5X10$^{17}$/cm$^3$ and $N_H$=1.2X10$^{18}$/cm$^3$ at different body bias is done, and $g_m$ result is shown in Fig. 3.32(b). The TCAD result also shows nonmonotonous trend of max. $g_m$ with body bias as shown in the inset figure. The maximum $g_m$ is the strong function of the relative threshold voltage difference between the

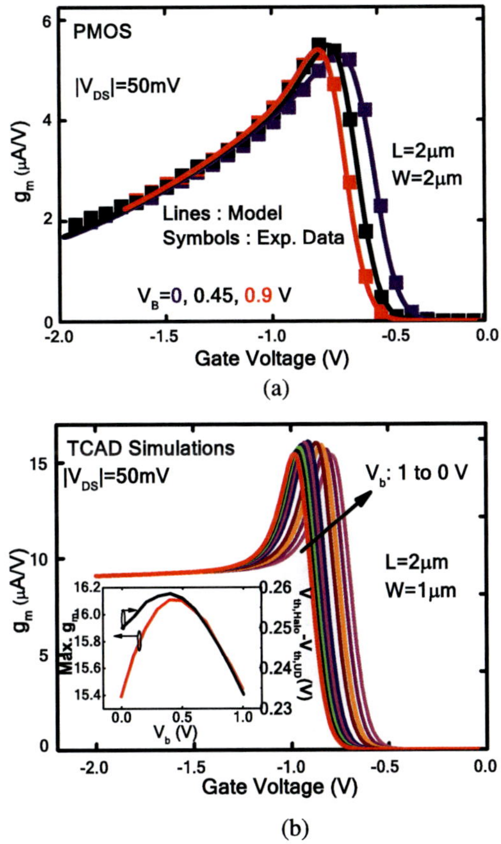


FIGURE 3.32 Impact of body bias: (a) $g_m - V_G$ of long channel PMOS in linear region for different $V_B$. Peak of $g_m$ initially increases and then decreases with $V_B$. Similar trends are also seen in TCAD simulation in (b). Inset figure in (b) shows that threshold voltage difference between the halo region and the channel region has nonmonotonous trend responsible for such behavior of max. $g_m$ with $V_B$.

channel and halo region ($\Delta V_{th}$). As body bias is swept, one would expect the halo region to experience higher increase in the threshold voltage as compared to the channel region since $\gamma_H > \gamma_{CH}$ ($N_H > N_{CH}$). This is indeed true, and therefore initially $\Delta V_{th}$ increases with $V_B$ as seen in the inset of Fig. 3.32(b). At larger $V_B$, the depletion width in the halo region increases, and for some $V_B$, it reaches the boundary between the halo and the well region. If $V_B$ is increased beyond this, the additional depletion charges are obtained by uncovering of the lightly doped substrate region, and thus $\gamma_H$ approaches $\gamma_{CH}$ [30,31]. As a result, $\Delta V_{th}$, and hence max. $g_m$, reduces at higher $V_B$. Such $V_B$ dependency on $V_{th,H}$ is modeled as

$$V_{th,H}(V_b) = V_{th,H}(V_b = 0) + K1.V_{BS} + K2.V_{BS}^2 \tag{3.109}$$

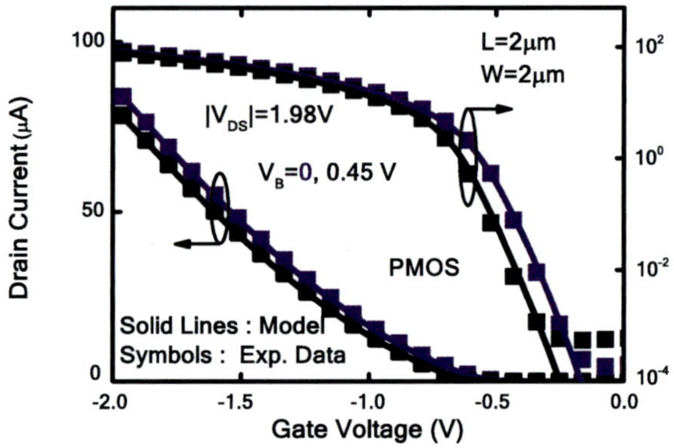


(a)

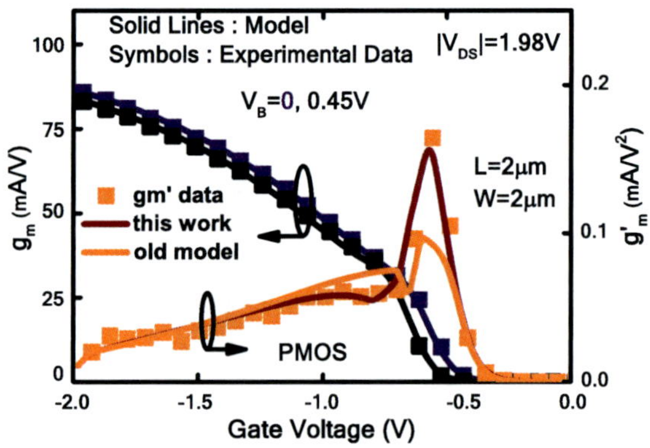


(b)

FIGURE 3.33 PMOS - Model validation with experimental data in saturation region: (a) $I_D - V_G$ (b) $g_m$ at $V_B = 0, 0.45$ V. $g_m$ and its derivative are accurately modeled by the proposed model. L=2 μm, W=2 μm.

here **K1** and **K2** are the model parameters. The model accurately captures the peak of $g_m$ as shown in Fig. 3.32(b). Fig. 3.33 reports PMOS characteristics in the saturation region. The proposed model agrees well with the measurement data over a wide range of biases.

Fig. 3.34(a) compares measured $I_D - V_G$ of the NMOS for L=2 μm and L=120 nm at $V_{DS}$=50 mV. From the figure, it can be observed that the gate length scaling has no or little impact in the weak inversion, since in weak inversion, the characteristics are determined by the halo regions, which do not scale with the gate length. In fact, such control on the sub-threshold behavior of the short channel device is the advantage of pocket implants. Fig. 3.34(b) compares $I_D - V_G$ of a L=120 nm NMOS device at two temperatures: 27°C and 150°C. At higher temperature, two competing effects come into play. First, the average thermal energy of the carrier increases,

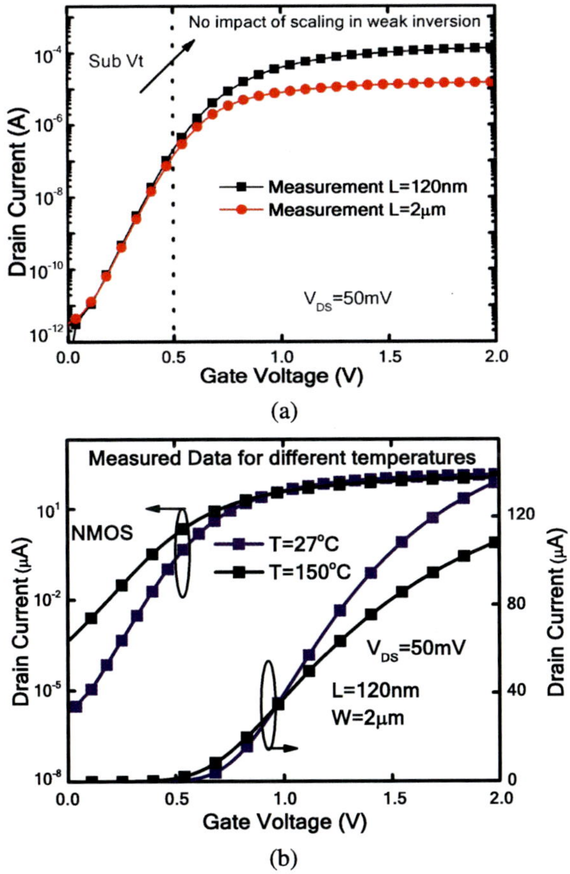


FIGURE 3.34 Impact of gate length scaling and temperature variation: Measured $I_D - V_G$ of NMOS at $V_{DS}$=50 mV (a) for channel length L = 2 μm and L = 120 nm. Subthreshold characteristics are determined by the halo regions, therefore, scaling has no impact in weak inversion. (b) Short channel $I_D - V_G$ at two different temperatures T=27°C and T=150°C.

and thus more carriers can surmount the barriers and participate in current conduction. On the other hand, the carrier mobility decreases due to the increase in lattice vibrations (phonon scattering). As a result, the subthreshold current increases, and strong inversion current reduces due to mobility reduction with temperature [2].

Fig. 3.35 shows the model-data overlay for the short-channel device. From the figure, it can be noticed that the short channel device does not show distinctive hump in the $g_m$, which is expected since as the channel length decreases, the two halo regions merge, and the device behaves like a uniformly doped device.

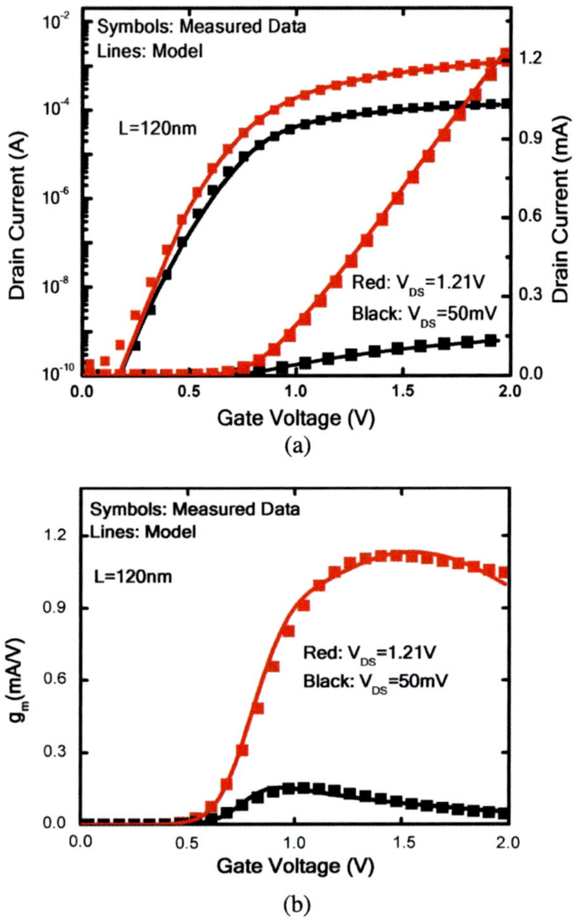


**FIGURE 3.35** NMOS - Model validation with experimental data for a short channel device in linear and saturation region: (a) $I_D - V_G$ (b) $g_m - V_G$. L=120 nm, W=1 μm.

## References


[1] C.K. Dhabhi, S.S. Parihar, H. Agrawal, N. Paydavosi, T.H. Morshed, D. Lu, et al., BSIM4 technical manual. [Online]. Available: https://bsim.berkeley.edu/models/bsim4/.

[2] Y. Tsividis, Operation and Modeling of the MOS Transistor, McGraw-Hill, Inc., 1987.

[3] A. Sharma, R. Goel, C. Dabhi, G. Pahwa, Y. Chauhan, S. Salahuddin, C. Hu, BSIM-BULK107.1.0 MOSFET compact model, 2022.

[4] Y.S. Chauhan, S. Venugopalan, M.A. Chalkiadaki, M.A.U. Karim, H. Agarwal, S. Khandelwal, N. Paydavosi, J.P. Duarte, C.C. Enz, A.M. Niknejad, C. Hu, BSIM6: analog and RF compact model for bulk MOSFET, IEEE Transactions on Electron Devices 61 (2) (Feb 2014) 234–244.

[5] Y. Cheng, M. Chan, K. Hui, M. Jeng, Z. Liu, J. Huang, K. Chen, J. Chen, R. Tu, P. Ko, et al., BSIM3 version 3.1 manual. [Online]. Available: http://wwwdevice.eecs.berkeley.edu/bsim/?page=BSIM3.

[6] Y. Cheng, M.-C. Jeng, Z. Liu, J. Huang, M. Chan, K. Chen, P.K. Ko, C. Hu, A physical and scalable I-V model in BSIM3v3 for analog/digital circuit simulation, IEEE Transactions on Electron Devices 44 (2) (Feb 1997) 277–287.
[7] C. Gupta, H. Agarwal, R. Goel, C. Hu, Y.S. Chauhan, Improved modeling of bulk charge effect for BSIM-bulk model, IEEE Transactions on Electron Devices 66 (6) (2019) 2850–2853.
[8] U. Manual, J. Version, Synopsys TCAD sentaurus, in: Synopsys, San Jose, CA, USA, 2014.
[9] R.R. Troutman, VLSI limitations from drain-induced barrier lowering, IEEE Journal of Solid-State Circuits 14 (2) (1979) 383–391.
[10] T.A. Fjeldly, M. Shur, Threshold voltage modeling and the subthreshold regime of operation of short-channel MOSTETs, IEEE Transactions on Electron Devices 40 (1) (1993) 137–145.
[11] J. Kendall, A. Boothroyd, A two-dimensional analytical threshold voltage model for MOSFET's with arbitrarily doped substrates, IEEE Electron Device Letters 7 (7) (1986) 401–403.
[12] Z.-H. Liu, C. Hu, J.-H. Huang, T.-Y. Chan, M.-C. Jeng, P.K. Ko, Y.C. Cheng, Threshold voltage model for deep-submicrometer MOSTETs, IEEE Transactions on Electron Devices 40 (1) (1993) 86–95.
[13] Y. Cheng, C. Hu, MOSFET Modeling & BSIM3 User's Guide, Springer Science & Business Media, 1999.
[14] K.M. Cao, W. Liu, X. Jin, K. Vashanth, K. Green, J. Krick, T. Vrotsos, C. Hu, Modeling of pocket implanted MOSTETs for anomalous analog behavior, in: International Electron Devices Meeting 1999. Technical Digest (Cat. No. 99CH36318), IEEE, 1999, pp. 171–174.
[15] Y. Chauhan, M. Karim, S. Venugopalan, S. Khandelwal, P. Thakur, N. Paydavosi, A. Sachid, A. Niknejad, C. Hu, BSIM6: symmetric bulk MOSFET model, in: Workshop on Compact Modeling, Santa Clara, USA, 2012.
[16] C.C. Enz, E.A. Vittoz, Charge-Based MOS Transistor Modeling: the EKV Model for Low-Power and RF IC Design, John Wiley & Sons, 2006.
[17] Huang, Liu, Jeng, Ko, Hu, A physical model for MOSFET output resistance, in: 1992 International Technical Digest on Electron Devices Meeting, 1992, pp. 569–572, https://doi.org/10.1109/IEDM.1992.307426.
[18] A. Bryant, W. Haensch, S. Geissler, J. Mandelman, D. Poindexter, M. Steger, The current-carrying corner inherent to trench isolation, IEEE Electron Device Letters 14 (8) (1993) 412–414.
[19] S. Khandelwal, H. Agarwal, J.P. Duarte, K. Chan, S. Dey, Y.S. Chauhan, C. Hu, Modeling STI edge parasitic current for accurate circuit simulations, IEEE Transactions on Computer-Aided Design of Integrated Circuits and Systems 34 (8) (2015) 1291–1294.
[20] K. Benaissa, et al., New cost-effective integration schemes enabling analog and high-voltage design in advanced CMOS SOC technologies, in: Symposium on VLSI Technology (VLSIT), June 2010, pp. 221–222.
[21] S. Ekbote, et al., 45nm low-power CMOS SoC technology with aggressive reduction of random variation for SRAM and analog transistors, in: Symposium on VLSI Technology (VLSIT), 2008, pp. 160–161.
[22] A. Cathignol, et al., Abnormally high local electrical fluctuations in heavily pocket-implanted bulk long MOSFET, Solid-State Electronics 53 (2) (2009) 127–133.
[23] R. Rios, et al., A three-transistor threshold voltage model for halo processes, in: IEEE Electron Devices Meeting, 2002, pp. 113–116.
[24] BSIM6 Technical Manual. [Online]. Available: http://www-device.eecs.berkeley.edu/bsim/?page=BSIM6_LR, 2014.
[25] G. Gildenblat, et al., PSP: an advanced surface-potential-based MOSFET model for circuit simulation, IEEE Transactions on Electron Devices 53 (9) (2006) 1979–1993, https://doi.org/10.1109/TED.2005.881006.

[26] K.M. Cao, W. Liu, X. Jin, K. Vashanth, K. Green, J. Krick, T. Vrotsos, C. Hu, Modeling of pocket implanted MOSFETs for anomalous analog behavior, in: Proc. IEEE IEDM, 1999, pp. 171–174.

[27] C. McAndrew, P. Drennan, Analysis of halo implanted MOSFETs, in: NSTI Nanotechnology Conference and Trade Show, 2007, pp. 594–598.

[28] S. Mudanai, et al., Analytical modeling of output conductance in long-channel halo-doped MOSFETs, IEEE Transactions on Electron Devices 53 (9) (April 2006) 2091–2097.

[29] H. Agarwal, C. Gupta, S. Dey, S. Khandelwal, C. Hu, Y.S. Chauhan, Anomalous transconductance in long channel halo implanted MOSFETs: analysis and modeling, IEEE Transactions on Electron Devices 64 (2) (2017) 376–383, https://doi.org/10.1109/TED.2016.2640279.

[30] V. Quenette, et al., Electrical characterization and compact modeling of MOSFET body effect, in: Ultimate Integration of Silicon, 2008, pp. 163–166.

[31] S. Khandelwal, et al., Analysis and modeling of vertical non-uniform doping in bulk MOSFETs for circuit simulation, in: International Caribbean Conference on Devices, Circuits and Systems, 2012, pp. 1–5.

CHAPTER

# 4

# Leakage Current

## 4.1 Introduction

In the past decade, the size of the MOSFET has kept shrinking, leading to the speed improvement and increase the density of devices per unit area. However, continuous scaling reduces the gate control over the channel, as a result, the device exhibits significant leakage even when it is in the OFF state. Due to the leakage current the static power dissipation is also increasing.

Fig. 4.1 shows the plot for the drain current of a MOSFET device (on a logarithmic scale) versus gate voltage for the nonzero drain voltage. The value of drain current, corresponding to which the applied gate-to-source voltage is zero, and the drain-to-source voltage is equal to the supply voltage $V_{DD}$, represents the OFF-state leakage current. Ideally, the transistor should be completely OFF for these bias conditions, but due to different mechanisms, there always present different leakage components across the device. These include the weak-inversion current between the drain terminal and source terminal, the substrate and drain junction leakage currents, the gate-induced drain leakage (GIDL) current between the drain terminal and the substrate terminal, and a portion of the gate oxide tunneling current.

## 4.2 Weak-inversion current

As already discussed above, if the applied gate-to-source voltage is below the threshold voltage of the transistor ($V_{TH}$), the real device exhibits significant leakage, which is represented by the $I_{OFF}$ current. This $I_{OFF}$ current is attributed to different mechanisms owing to different leakage components. For the n-channel MOSFET, for $V_{GS} < V_{TH}$ and $V_{DS} > 0$ at room temperature, there are some electrons, which acquire enough energy to overcome the source-body barrier and collected at the drain side, due to their Boltzmann distribution, and which leads to nonzero drain

BSIM-Bulk MOSFET Model for IC Design - Digital, Analog, RF and High-Voltage
https://doi.org/10.1016/B978-0-32-385677-5.00008-1

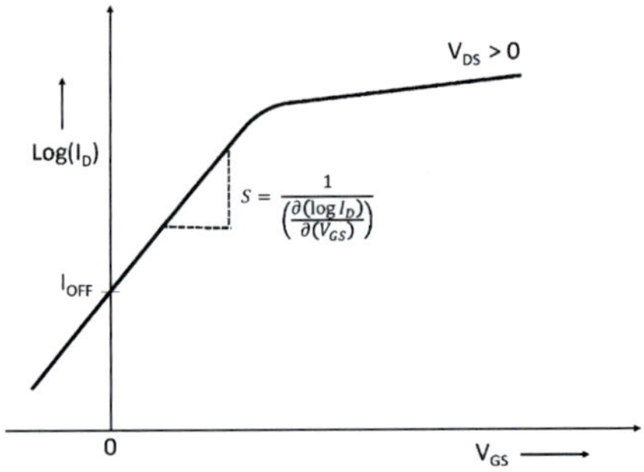


FIGURE 4.1 The transistor's drain current (on a logarithmic scale) as a function of its gate voltage, illustrating the off-state leakage current, threshold voltage, and sub-threshold swing.

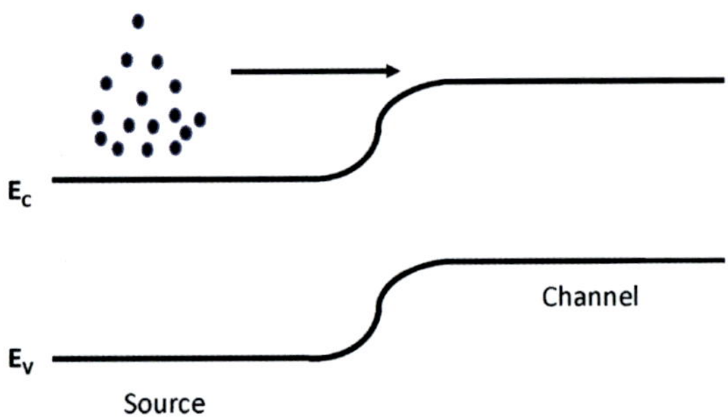


FIGURE 4.2 The generation of drain-source sub-threshold leakage caused by the Boltzmann carrier distribution.

current for $V_{DS} > 0$, as shown in Fig. 4.2. This current is known as a sub-threshold drain-to-source leakage current. In fact, this leakage current component is one of the major contributors of $I_{OFF}$. The sub-threshold drain-to-source leakage current follows the straight-line trend with the finite slope (on a logarithmic scale), as shown in Fig. 4.1. The reverse of the slope of this line ($\partial I_D/\partial V_{GS}$) is known as the sub-threshold swing S, and it has units of millivolts of the gate voltage per decade of the drain current. The theoretical value of S is approximately 60 mV per decade. However, in a real device, the S value may be higher than the 60 mV per decade attributed to the short-channel effects. As the size of the MOSFET shrinks, the control of gate on the channel gets reduced, and source/drain fields get more authority. For the short-channel MOSFETs, drain voltage reduces the source-side barrier, lowering the threshold voltage and hence degrading the subthreshold slope, owing to a higher leakage current.

## 4.3 Modeling of bulk current

In the MOSFET, a bulk current flow between the drain and substrate due to the effects of impact ionization and gate-induced drain leakage (GIDL). Both of these effects lead to the formation of electron-hole pairs. The impact ionization effect forms the electron-hole pair in the channel due to the weak avalanche breakdown phenomenon, and for the nMOSFET, the holes start to flow towards the substrate and hence produce the bulk current. The GIDL, on the other hand, generates the electron-hole pairs in the gate-drain overlap region due to band-to-band tunneling attributed to the high vertical electric field in this region, and here too, holes start to move towards the bulk and therefore contribute to the substrate current.

### 4.3.1 Gate-induced source and drain leakage model

For the zero-gate voltage and $V_{DS} > 0$, the effective gate-to-drain voltage will be significantly high, and for $V_{DS}$ much lower than the breakdown voltage, a large current flows from the drain to bulk [1–3]. This drain leakage current is named gate-induced drain leakage (GIDL) since it is due to a gate-induced high electric field present in the gate-to-drain overlap region. Fig. 4.3(a) represents a simple NMOS structure, where the drain/source overlap regions have clearly illustrated. Fig. 4.3(b), on the other hand, demonstrates the zoomed picture of the drain overlap region, where the mechanism of GIDL is happening. For high negative gate-to-drain voltage, a large depletion region will be formed underneath the drain over-

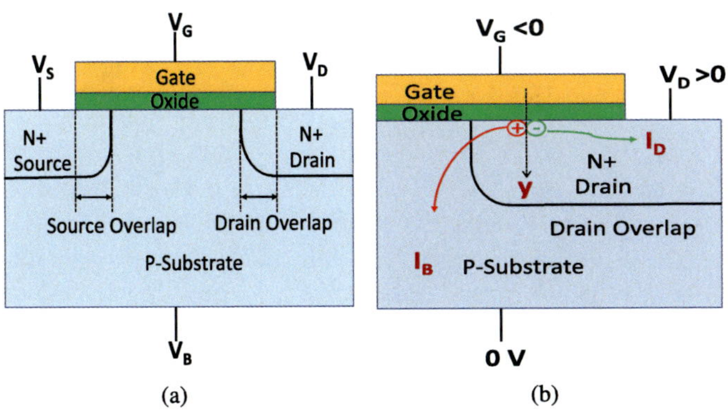


FIGURE 4.3 GIDL phenomena: (a) Conventional MOSFET structure demonstrating the source/drain overlap regions (b) Cross-section of gate-drain overlap region. Generation of leakage current between drain and bulk due to electron-hole pairs generated by tunneling of valence band electrons into the conduction band. This effect is called gate-induced drain leakage (GIDL).

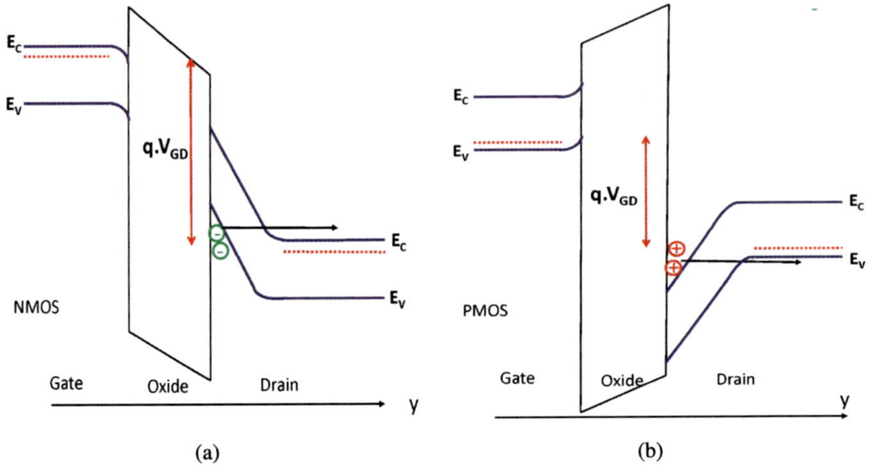


FIGURE 4.4 Energy-band diagram along the y-direction illustrating the GIDL phenomenon (a) In NMOS Transistor (b) In PMOS Transistor.

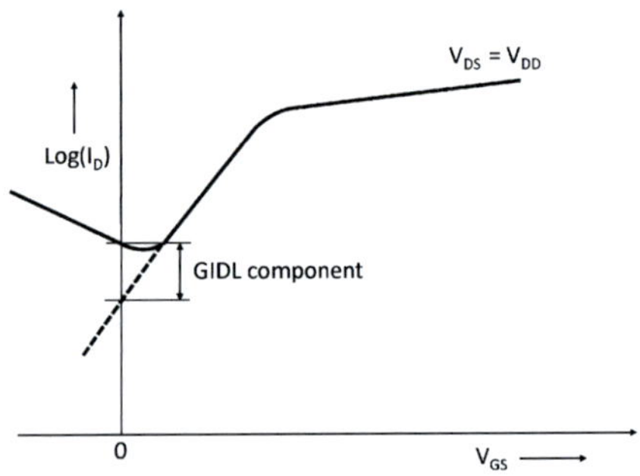


FIGURE 4.5 Contributions of GIDL to the transistor's subthreshold leakage current. The position of the dip caused by GIDL will vary around $V_{GS} = 0$ depending on $V_{DS}$, the channel material, doping, and trap density.

lap region, and this leads to a high vertical field in the depletion region. When the valence band electrons tunnel into the conduction band, it results in the formation of electron-hole pairs. The above-mentioned tunneling is mainly controlled by the band-to-band tunneling (BTBT), as shown with the help of the energy band diagrams in Fig. 4.4(a) (for NMOS) and Fig. 4.4(b) (for PMOS). Fig. 4.5 clearly shows that this GIDL component leads to a vital contributor to the OFF leakage current. The GIDL current density is modeled using tunneling physics. The tunneling current density is given by [1–4]:

$$J = A.E_S.exp\left(-\frac{B}{E_S}\right) \tag{4.1}$$

where $A$ is a pre-exponential parameter, $B$ represents the tunneling coefficient, and $E_S$ is the transversal electric field. In the BSIM-BULK model, $E_S$ is formulated by the following expression

$$E_S = \frac{V_{DS} - V_{GS} - V_{FBSD} - E_g}{\epsilon_{ratio}.EOT} \tag{4.2}$$

Band-to-band tunneling will occur when the band bending in the drain is equal to $E_g$. Here $V_{FBSD}$ is the flat-band voltage between the gate and the drain, $\epsilon_{ratio}$ is the ratio of the dielectric constant of the substrate material over that of silicon dioxide, and $EOT$ is the equivalent oxide thickness. Eqs. (4.1) and (4.2) result in the following formulation in the BSIM-BULK model [2]

$$\begin{aligned} I_{gidl0} = {} & W.NF.AGIDL.\Big(\frac{V_{DG} - V_{FBSD} - EGIDL}{\epsilon_{ratio}.EOT}\Big)^{PGIDL} \\ & .exp\Big(-\frac{BGIDL.\epsilon_{ratio}.EOT}{V_{DG} - V_{FBSD} - EGIDL}\Big) \end{aligned} \tag{4.3}$$

where $W$ and $NF$ represent the effective width and number of fingers, respectively, and parameters $A$, $B$ and $E_g$ have been replaced by $AGIDL$, $BGIDL$ and $EGIDL$, respectively. To further enhance the fitting flexibility of our model, the model parameter $PGIDL$ has also been introduced [2].

Similarly, the GISL current is calculated:

$$\begin{aligned} I_{gisl0} = {} & W.NF.AGISL.\Big(\frac{V_{SG} - V_{FBSD} - EGISL}{\epsilon_{ratio}.EOT}\Big)^{PGISL} \\ & .exp\Big(-\frac{BGISL.\epsilon_{ratio}.EOT}{V_{SG} - V_{FBSD} - EGISL}\Big) \end{aligned} \tag{4.4}$$

In addition to the $V_{dg}$ dependence present, the GIDL current is also dependent on the drain to substrate voltage. The effective GIDL current is obtained multiplying (4.3) by an empirical factor as follows [2]

$$\begin{aligned} I_{GIDL} = {} & AGIDL \cdot W \cdot NF \cdot \frac{V_{ds} - V_{gse} - EGIDL}{3 \cdot T_{oxe}} \\ & \cdot exp\left(-\frac{3 \cdot T_{oxe} \cdot BGIDL}{V_{ds} - V_{gse} - EGIDL}\right) \cdot \frac{V_{db}^3}{CGIDL + V_{db}^3} \end{aligned} \tag{4.5}$$

$$\begin{aligned} I_{GISL} = {} & AGISL \cdot W \cdot NF \cdot \frac{-V_{ds} - V_{gde} - EGISL}{3 \cdot T_{oxe}} \\ & \cdot exp\left(-\frac{3 \cdot T_{oxe} \cdot BGISL}{-V_{ds} - V_{gde} - EGISL}\right) \cdot \frac{V_{sb}^3}{CGISL + V_{sb}^3} \end{aligned} \tag{4.6}$$

where, $CGIDL$ and $CGISL$ account for the body-bias dependence of $I_{GIDL}$ and $I_{GISL}$, respectively.

### 4.3.2 Impact ionization

For the transistor biased in the saturation region, the lateral electric field near the drain end of the channel may achieve very high values. As a result, the electrons traveling from the source to drain across the channel can gain sufficient energy so that they could ionize the lattice and lead to extra electron-hole pairs by releasing electrons from the valence band into the conduction band. The phenomenon is commonly called impact ionization. The generated holes out of electron-hole pairs move towards the substrate, resulting in the substrate current. The extra high-energy electrons (hot carriers), on the other hand, become the part of the drain current and get collected by the drain terminal. Thus the drain current ($I_D$) will no longer equal to the channel current ($I_{DS}$), as shown in Fig. 4.6:

$$I_D = I_{DS} + I_B \tag{4.7}$$

For low-level avalanche multiplication, the impact ionization current can be expressed as [1–3]

$$I_{ii} = I_{DS}.\int_0^L \alpha_n dx \tag{4.8}$$

The local impact ionization current $I_{ii}$ is a function of the channel current and the electron impact ionization coefficient per unit length ($\alpha_n$). The effect of bulk current is more dominate in n-channel than in p-channel MOSFETs because impact ionization coefficient is much higher for electrons as compared to the holes. The impact ionization coefficient is a strong function of the lateral electric field $E_x$ [1–4]

$$\alpha_n = A_i.exp(-\frac{B_i}{E_x}) \tag{4.9}$$

where $A_i$ and $B_i$ are called the impact ionization constants. Since $\alpha_n$ exponentially depends on the electric field, the impact ionization exhibits its more impact at the position where the electric field is maximum, i.e., at the drain end (the stronger the electric field, the higher the kinetic energy of the carriers).

One would write the total impact ionization current as [1–4]

$$I_{ii} = I_{DS}.A_i.\int_{x=0}^{x=l'} exp(-\frac{B_i}{E_x})dx \tag{4.10}$$

where $x = 0$ is the starting point of the velocity saturation region, and $l'$ is the length of this region. The integration in Eq. (4.10) gives the following

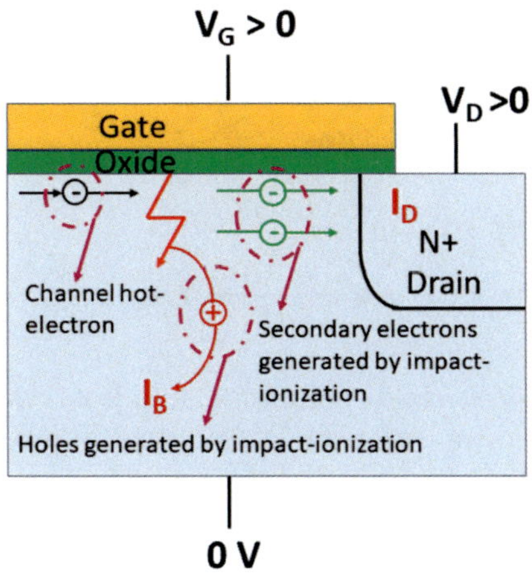


FIGURE 4.6 Schematic illustrates the impact ionization phenomenon in an NMOS transistor. Electrons traveling through the channel from source to drain are accelerated and gain high energy due to the presence of a high electric field near the drain end of the channel so that they can create extra electron-hole pairs by exciting electrons from the valence band into the conduction band. The generated electrons are collected by drain terminal, and holes are collected by the bulk terminal.

equation for the BSIM-BULK impact ionization model [2,4]:

$$I_{ii} = \frac{A_i}{B_i}.I_{DS}.(V_{DS} - V_{DSAT}).exp\left(-\frac{B_i.\lambda}{(V_{DS} - V_{DSAT})}\right) \tag{4.11}$$

where $V_{DSAT}$ is the saturation voltage, and $\lambda$ is the characteristic length.

The final expression of impact ionization current model in BSIM-BULK, which is very similar to that in BSIM4, is modeled by [2,4]

$$I_{ii} = ALPHA0.I_{DS}.(V_{DS} - V_{DSAT}).exp\left(-\frac{BETA0}{(V_{DS} - V_{DSAT})}\right) \tag{4.12}$$

where parameters $ALPHA0$ and $BETA0$ are impact ionization coefficients.

## 4.4 Gate tunneling current model

As the gate oxide thickness is scaled down to 3 nm and below, gate leakage current due to carrier direct tunneling becomes important. This tunneling happens between the gate and silicon beneath the gate oxide. To reduce the tunneling current, high-k dielectrics are being used in place of the gate oxide. To maintain a good interface with the substrate, multilayer dielectric stacks are being used. However, the gate tunneling leakage

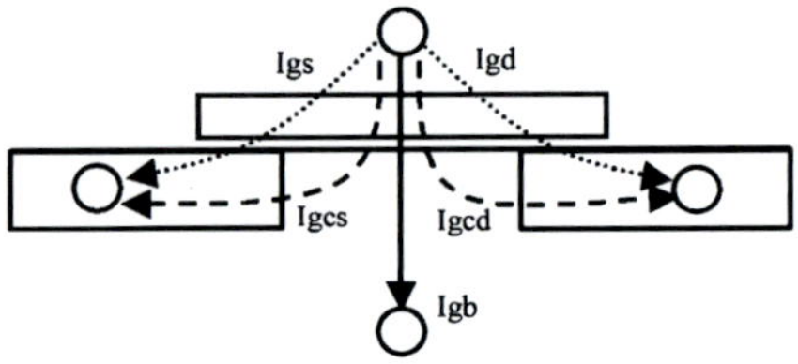


FIGURE 4.7 Schematic gate current components flowing between MOSFET terminals.

through the gate oxide remains significant and bothers us as the advanced technology nodes require a smaller $EOT$. The BSIM-BULK gate tunneling model (taken from BSIM4) has also been shown to work for multi-layer gate stacks. The tunneling carriers can be either electrons or holes, or both, either from the conduction band or valence band, depending on (the type of the gate and) the bias regime.

### 4.4.1 Modeling of gate oxide tunneling in BSIM-BULK

BSIM4 adopts the gate oxide tunneling model developed by Lee and Hu [5] that can predict all the significant tunneling components, such as electron tunneling from the conduction band (ECB), hole tunneling from the valence band (HVB), and electron tunneling from the valence band (EVB). The gate oxide tunneling in the BSIM-BULK model inherits a similar formulation to that of BSIM4. In BSIM-BULK, the gate tunneling current components include the tunneling current between gate and substrate ($I_{gb}$), and the current between gate and channel ($I_{gc}$), which is partitioned between the source and drain terminals by $I_{gc} = I_{gcs} + I_{gcd}$. The third component happens between the gate and source/drain diffusion regions ($I_{gs}$ and $I_{gd}$). Fig. 4.7 shows the schematic of gate tunneling current flows.

The unified BSIM-BULK model equation for the tunneling current density is given by:

$$
\begin{aligned}
J_T = A \cdot \left(\frac{T_{oxref}}{T_{ox}.P}\right)^{NTOX} \cdot \frac{V_{aux}.V_{appl}}{(T_{ox}.P)^2} \\
\cdot exp[-B \cdot (\alpha - \beta.|V_{ox}|).(1+\gamma.|V_{ox}|).T_{ox}.P]
\end{aligned}
\tag{4.13}
$$

where $A = q^2/(8.\pi.h.\phi_b)$, $B = 8\pi\sqrt{2qm_{ox}}(q\phi_b)^{3/2}/(3h)$, $h$ is the Planck constant, $T_{oxref}$ is the reference oxide thickness for parameter extraction, $NTOX$ is a fitting parameter, and $V_{aux}$ is an auxiliary function for modeling the density of tunneling carriers and available states in various regions of operation. The model parameters $\alpha$, $\beta$, $\gamma$, and P have different values depending on the tunneling mechanism, regions of operation, and current components described in Fig. 4.7. They are $AIGACC$, $AIGBINV$, $AIGC$

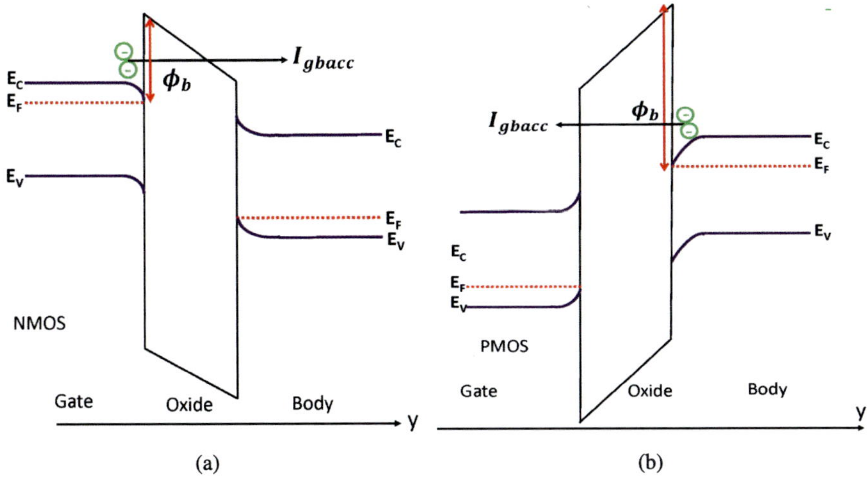


FIGURE 4.8 The gate-to-body tunneling current in accumulation region leads to $I_{gbacc}$ current. (a) In NMOS transistor conduction-band, electrons tunnel from the gate into the body (b) In PMOS transistor conduction-band, electrons tunnel from the body into the gate.

and $AIGSD$ for $\alpha$; $BIGACC$, $BIGBINV$, $BIGC$, and $BIGSD$ for $\beta$; and $CIGACC$, $CIGBINV$, $CIGC$, and $CIGSD$ for $\gamma$, respectively, in the current component $J_{GB}$ in accumulation, $J_{GB}$ in inversion/depletion, $J_{GC}$, and $J_{(GS,D)}$.

#### 4.4.1.1 Gate-to-Substrate Current ($I_{gb} = I_{gbacc} + I_{gbinv}$)

Gate-to-substrate tunneling current in the accumulation region ($I_{gbacc}$) is determined by ECB. As it is evident from Fig. 4.8 (a), in NMOS, the electrons tunnel from the conduction band of the gate material into the conduction band of the body, and, in PMOS, they tunnel in the reverse direction, as shown in Fig. 4.8 (b). In BSIM-BULK for the modeling of $I_{gbacc}$, (4.13) is modified as:

$$\begin{aligned} I_{gbacc} = {} & NF \cdot W_{eff} L_{eff} \cdot A \cdot T_{oxRatio} \cdot V_{gb} \cdot V_{aux} \cdot i_{gtemp} \\ & \cdot exp[-B \cdot TOXE(AIGBACC - BIGBACC \cdot V_{oxacc}) \\ & \cdot (1 + CIGBAC \cdot V_{oxacc})] \end{aligned} \tag{4.14}$$

where the physical constants $A = 4.97232 \times 10^{-7}\ A/V^2$, $B = 7.45669 \times 10^{11}(g/F - s^2)^{0.5}$, $P = 1$, $\alpha = AIGBACC$, $\beta = BIGBACC$, $\gamma = CIGACC$ and

$$T_{oxRatio} = \left(\frac{TOXREF}{TOXE}\right)^{NTOX} \cdot \frac{1}{TOXE^2} \tag{4.15}$$

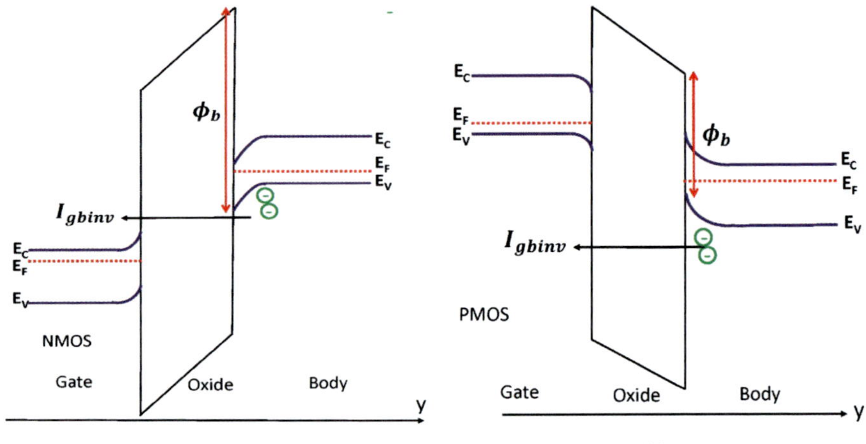


FIGURE 4.9 The gate-to-body tunneling current in inversion due to tunneling of valence-band electrons from the body into the gate leads to $I_{gbinv}$ current. (a) In NMOS transistor (b) In PMOS transistor.

$$V_{aux} = NIGBACC \cdot V_t \cdot log\left(1 + exp\left(-\frac{V_{oxacc}}{NIGBACC \cdot V_t}\right)\right) \tag{4.16}$$

Gate-to-substrate tunneling current in the depletion/inversion region ($I_{gbinv}$) is determined by EVB and is significant in the depletion/inversion region. Fig. 4.9 illustrates the dominant leakage mechanism between the gate and the body in depletion/inversion region, given by $I_{gbinv}$. The electrons tunnel from the valence band of the body into the gate material (as the gate is biased at positive bias for the NMOS) for both p-type MOS (PMOS) and n-type MOS (NMOS) transistors. In BSIM-BULK for the modeling of $I_{gbinv}$, (4.13) is modified as:

$$\begin{aligned} I_{gbinv} = {} & NF \cdot W.L \cdot A \cdot T_{oxRatio} \cdot V_{gb} \cdot V_{aux} \cdot i_{gtemp} \\ & \cdot exp[-B \cdot TOXE(AIGBINV - BIGBINV \cdot V_{oxdepinv}) \\ & \cdot (1 + CIGBINV\, V_{oxdepinv})] \end{aligned} \tag{4.17}$$

where $A = 3.75956 \times 10^{-7}\ A/V^2$, $B = 9.82222 \times 10^{11}\ (g/F - s^2)^{0.5}$, $P = 1$, $\alpha = AIGBINV, \beta = BIGBINV, \gamma = CIGBINV$ and

$$V_{aux} = NIGBINV \cdot V_t \cdot log\left(1 + exp\left(\frac{V_{oxdepinv} - EIGBINV}{NIGBINV \cdot V_t}\right)\right) \tag{4.18}$$

$$I_{gb} = I_{gbacc} + I_{gbinv} \tag{4.19}$$

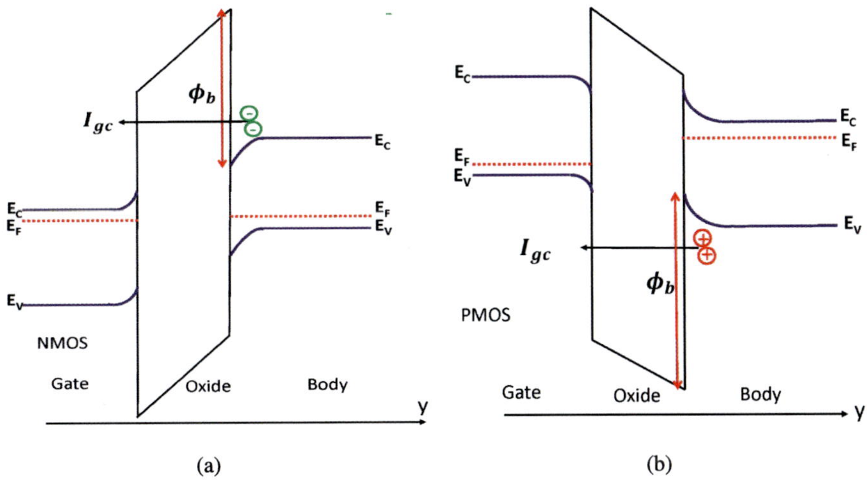


FIGURE 4.10 The gate-to-channel tunneling current in inversion leads to $I_{gc}$ current. (a) In NMOS transistor conduction band, electrons tunnel from the channel into the gate (b) In PMOS transistor the valence-band, holes tunnel from the channel into the gate.

#### 4.4.1.2 Gate-to-Channel Current ($I_{gc0}$) and Gate-to-S/D ($I_{gs}$ and $I_{gd}$)

$I_{gc0}$ is determined by ECB for NMOS and HVB for PMOS at $V_{ds} = 0$. As shown in Fig. 4.10 (a), in inversion, the electrons (holes in PMOS) tunnel from the inversion channel into the conduction band of the gate (and valence band for PMOS). In BSIM-BULK for the modeling of $I_{gc0}$, (4.13) is modified as:

$$I_{gc0} = NF \cdot W.L. \cdot A \cdot T_{oxRatio} \cdot V_{gse} \cdot V_{aux} \cdot i_{gtemp} \\ \cdot exp[-B \cdot TOXE(AIGC - BIGC \cdot V_{oxdepinv}) \cdot (1 + CIGC V_{oxdepinv})] \quad (4.20)$$

where A = 4.97232 $A/V^2$ for NMOS and 3.42537 $A/V^2$ for PMOS, B = 7.45669e11, $(g/F - s^2)^0.5$ for NMOS and 1.16645e12 $(g/F - s^2)^{0.5}$ for PMOS, $\alpha = AIGC, \beta = BIGC, \gamma = CIGC$.

$$V_{aux} = n_q \cdot nVt \cdot (q_s + q_{deff}) \quad (4.21)$$

**Partition of $I_{gc}$:** To consider the drain-bias effect, $I_{gc}$ is split into two components, $I_{gcs}$ and $I_{gcd}$, that is $I_{gc} = I_{gcs} + I_{gcd}$. Here, we use the perturbation theory to dig out the expressions of these components. To obtain a gate current partition, we first extract the perturbation on the electron quasi-Fermi $V_{CS(y)}$ due to the presence of gate current. The $V_{CS(y)}$ must satisfy the following current continuity equation:

$$\frac{1}{W}.\frac{dI(x)}{dx}.J_G(x)=0 \tag{4.22}$$

After incorporating the current expression in terms of the channel potential, the above equation can be written as:

$$\frac{d[(V_{GST}-n.V_{CS})(dV_{CS}/dx)]}{dx}+\frac{J_G(x)}{\mu_{eff}.C'_{ox}}=0 \tag{4.23}$$

One would consider that in the presence of gate current there will be perturbation in the channel potential, which is given by:

$$V_{CS}(x)=V_{CS0}(x)+V_{CS1}(x) \tag{4.24}$$

where $V_{CS1}(x)$ is the perturbation introduced. For the assumption $V_{CS0}(x)>>V_{CS1}(x)$, (4.23) takes the following form:

$$(V_{GST}-V_{CS0})\frac{dV_{CS1}^2}{dx^2}-V_{CS1}\frac{d^2V_{CS0}}{dx^2}-2.\frac{dV_{CS0}}{dx}.\frac{dV_{CS1}}{dx}+\frac{J_G(x)}{\mu_{eff}.C'_{ox}}=0 \tag{4.25}$$

The differential equation (4.25) is solved using the boundary conditions: $V_{CS1}(0)=0$ and $V_{CS1}(L)=0$. The solution is given by:

$$V_{CS1}=-\frac{J_{G0}\left[(e^{-B^*K.x}-1)+\frac{y}{L}(1-e^{-B^*K.L})\right]}{\mu_{eff}C'_{ox}B^{*2}K^2(V_{GST}-K.x)} \tag{4.26}$$

$$\text{where, } K=(V_{GST}-V_{DS}/2)V_{DS}/LV_{GST}$$

where $J_{G0}$ is the gate current density for $V_{DS}=0$, $B^*=PIGCD.B.T_{ox}/V_{oxs}^2$, $V_{oxs}\approx V_{GS}$ is the potential drop across the oxide at $x=0$, and $PIGCD$ is the model fitting parameter.

The partitioned current components $I_{gcs}$ and $I_{gcd}$ can be derived by taking the derivative of the quasi-Fermi level of electrons with respect to the lateral distance $x$

$$I_{gcs}=\mu_{eff}.C'_{ox}V_{GST}W.\left.\frac{dV_{CS1}}{dx}\right|_{x=0}=\frac{J_{G0}WL(B^*V_{dseff}+e^{-B^*V_{dseff}}-1)}{B^{*2}V_{dseff}^2} \tag{4.27}$$

$$I_{gcd}=-\mu_{eff}.C'_{ox}(V_{GST}-V_{dseff})W.\left.\frac{dV_{CS1}}{dx}\right|_{x=L}=-\frac{J_{G0}WL(B^*V_{dseff}e^{-B^*V_{dseff}}+e^{-B^*V_{dseff}}-1)}{B^{*2}V_{dseff}^2} \tag{4.28}$$

Further, in BSIM-BULK model equations, (4.27) and (4.28) are slightly manipulated to get the following form:

$$I_{gcs} = I_{gc0} \cdot \frac{PIGCD \cdot V_{dseffx} + exp(-PIGCD \cdot V_{dseffx}) - 1 + 10^{-4}}{PIGCD \cdot V_{dseffx}^{2} + 2 \cdot 10^{-4}} \tag{4.29}$$

and

$$I_{gcd} = I_{gc0} \cdot \frac{1 - (PIGCD \cdot V_{dseffx} + 1) \cdot exp(-PIGCD \cdot V_{dseffx}) + 10^{-4}}{PIGCD \cdot V_{dseffx}^{2} + 2 \cdot 10^{-4}} \tag{4.30}$$

where

$$V_{dseffx} = \sqrt{V_{dseff} + 0.01} - 0.1 \tag{4.31}$$

At $V_{ds} = 0$, $I_{gcs} = I_{gcd} = \frac{1}{2} I_{gc0}$. Thus $I_{gc0}$ is the gate to channel current $I_{gc}$ at $V_{ds} = 0$.

#### 4.4.1.3 $I_{gs}$ and $I_{gd}$

$I_{gs}$ represents the gate tunneling current between the gate and the source diffusion region, while $I_{gd}$ represents the gate tunneling current between the gate and the drain diffusion region. $I_{gs}$ and $I_{gd}$ are determined by ECB for NMOS and HVB for PMOS, respectively. The n+ (p+) gate to n+ (p+) source and drain currents $I_{gs}$ and $I_{gd}$ are principally caused by the tunneling of the conduction-band electrons in NMOS and valence-band holes in PMOS, as shown in Fig. 4.11. In BSIM-BULK, for the modeling of $I_{gs}$ and $I_{gd}$, (4.13) is modified as:

$$\begin{aligned} I_{gs} = NF \cdot W_{eff} DLCIG \cdot A \cdot T_{oxRatioEdge} \cdot V_{gs} \cdot V'_{gs} \cdot i_{gtemp} \\ \cdot exp[-B \cdot TOXE \cdot POXEDGE \cdot (AIGS - BIGS \cdot V'_{gs}) \\ \cdot (1 + CIGSV'_{gs})] \end{aligned} \tag{4.32}$$

and

$$\begin{aligned} I_{gd} = NF \cdot W_{eff} DLCIGD \cdot A \cdot T_{oxRatioEdge} \cdot V_{gd} \cdot V'_{gd} \cdot i_{gtemp} \\ \cdot exp[-B \cdot TOXE \cdot POXEDGE \cdot (AIGD - BIGD \cdot V'_{gd}) \\ \cdot (1 + CIGDV'_{gd})] \end{aligned} \tag{4.33}$$

where A = 4.97232 $A/V^2$ for NMOS and 3.42537 $A/V^2$ for PMOS, B = 7.45669e11 $(g/F - s^2)^{0.5}$ for NMOS and 1.16645e12 $(g/F - s^2)^{0.5}$ for PMOS, $P = 1, \alpha = AIGS(AIGD), \beta = BIGS(BIGD), \gamma = CIGS(CIGD)$ and

$$T_{oxRatioEdge} = \left(\frac{TOXREF}{TOXE \cdot POXEDGE}\right)^{NTOX} \cdot \frac{1}{(TOXE \cdot POXEDGE)^2} \tag{4.34}$$

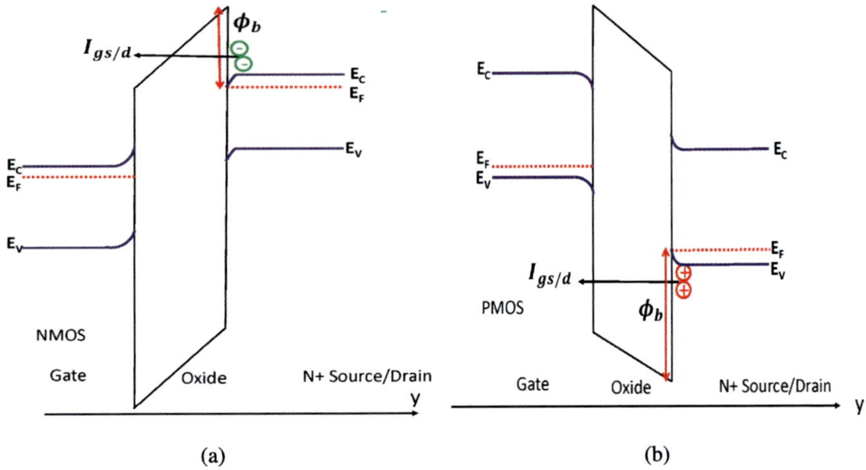


FIGURE 4.11 The source/drain-gate tunneling current in inversion leads to $I_{gs/d}$ current. Band diagrams for an inverted channel. In accumulation, the direction of the tunneling is reversed. (a) In NMOS transistor the tunneling of the conduction-band electrons (b) In PMOS transistor the tunneling of the valence-band holes make the source/drain-gate overlap leakage.

$$V'_{gs} = \sqrt{(V_{gs} - V_{fbsd})^2 + 10^{-4}} \tag{4.35}$$

$$V'_{gd} = \sqrt{(V_{gd} - V_{fbsd})^2 + 10^{-4}} \tag{4.36}$$

$V_{fbsd}$ is the flat-band voltage between gate and S/D diffusions calculated as

**If** $NGATE > 0.0$

$$V_{fbsd} = -devsign \cdot \frac{k_B T}{q} log\left(\frac{NGATE}{NSD}\right) + VFBSDOFF \tag{4.37}$$

**Else** $V_{fbsd} = 0.0$

### 4.4.2 Subsurface leakage

Growing demand for high-performance devices with low power consumption leads to the evolution of zero-threshold voltage MOS structures. These types of devices exhibit a higher $g_m/C_{OX}$ peak compared to the conventional bulk MOSFET structures. They have been used for a lot of applications like RF and analog applications, high-speed circuits, voltage regulators, multiplexers, input/output circuits, input buffer circuits, etc. [6], [7]. To design a circuit using these zero $V_{TH}$ devices, we need accurate models which accounts for the accurate simulation results. Therefore the modeling of the subsurface current component is essential. A computationally efficient compact model for circuit design is implemented in BSIM-BULK, which can account correct behavior of subsurface leakage current, caused due to low channel doping in zero-$V_{TH}$ MOSFET devices.

#### 4.4.2.1 Drain-to-source voltage and gate length dependence

It can be observed that there are electron current paths at a distance away from the $Si/SiO_2$ interface defined as subsurface leakage current. Moreover, the subsurface leakage current is gate length-dependent. This can be directly attributed to the barrier lowering induced by $V_{DS}$, which is similar to drain-induced barrier lowering effect [8] but at different depths. For a shorter channel device, even a small $V_{DS}$ can sufficiently reduce the barrier height between the source and the drain so that the electrons are able to surmount the barrier from the source side, and the corresponding current will be similar to the diode current form. Thus the subsurface leakage current can be expressed as [2,9]

$$I_{DS}(Leakage) = SSL0.e^{-SSL1.L_G}.\left(e^{\frac{SSL2.V_{DS}}{V_t}} - 1\right).W \tag{4.38}$$

where $W$ is the device width in units of $m$, $V_t$ is the thermal voltage ($= kT/q$), and $SSL0$ and $SSL1$ in units of $A/m$ and $m^{-1}$ are the variables called inverse characteristic length and intrinsic leakage, respectively. Further, $SSL2$ (dimensionless) is a fitting parameter defined as a barrier-change coefficient. It should be noted that (4.38) guarantees that the leakage current must be zero when $V_{DS} = 0$.

#### 4.4.2.2 Impact of gate-to-source voltage

As the gate voltage sweeps from the strong accumulation bias to around the flat-band voltage ($V_{FB}$), the local minimum of the conduction band decreases further and moves toward the $Si/SiO_2$ interface, which means that the leakage current is gate voltage-dependent at the transition region from accumulation to weak inversion. Before the accumulation layer is completely formed, the gate field can penetrate into the substrate, leading to $V_{GS}$-dependent barrier height reduction. If the accumulation layer is built, the gate electric field will be screened out by the accumulated holes, resulting in $V_{GS}$-independent behavior. Nevertheless, the $V_{GS}$-dependent effect will no longer be prominent when $V_{GS}$ is close to or higher than the threshold voltage ($V_{TH}$). In other words, if the channel is inverted, the subsurface leakage would mix with the normal transistor current and is negligible. The reduction should gradually saturate when $V_{GS}$ is close to or higher than $V_{TH}$, where the band bending is for the formation of electron inversion layer. We model $V_{GS}$ effect on barrier height change as an exponential function inside the hyperbolic tangent function [2,9]

$$\Delta\phi = SSL3.tanh(exp(SSL4.(V_{GS} - V_{TH}))) \quad (4.39)$$

where $SSL3$ and $SSL4$ in units of $V$ and $V^{-1}$ are the $V_{GS}$-associated fitting parameters that can be determined from the TCAD data. The threshold voltage ($V_{TH}$) will be automatically calculated by the BSIM-BULK model [10]. Therefore, from (4.38), the subsurface leakage current can be rewritten as [2,9]

$$I_{DS}(Leakage) = SSL0.e^{-SSL1.L_G}.e^{\frac{\Delta\phi}{V_t}}.\left(e^{\frac{SSL2.V_{DS}}{V_t}} - 1\right).W \quad (4.40)$$

#### 4.4.2.3 Substrate doping and temperature dependence

The subsurface leakage current is inversely proportional to substrate doping $N_{sub}$ because a heavily doped substrate can reduce the depletion width of source-to-body and drain-to-body and, hence, coupling. However, if $N_{sub}$ is high enough (for example, $10^{18}$ $cm^{-3}$), the GIDL current can overwhelm the subsurface leakage current in the drain current due to high electric field [6], [8]. If $N_{sub}$ is not fairly high, the leakage currents in the accumulation region are still linear but with different slopes in a log scale for different $N_{sub}$ values, which implies that the doping effect can be simply modeled by letting the variables $SSL0(N_{sub})$ and $SSL1(N_{sub})$ be doping concentration-dependent with respect to $10^{23}(m^{-3})$

$$SSL0(N_{sub}) = SSL0.exp\left(-\left(\frac{N_{sub}}{10^{23}}\right)^{SSLEXP1}\right) \quad (4.41)$$

$$SSL1(N_{sub}) = SSL1.\left(\frac{N_{sub}}{10^{23}}\right)^{SSLEXP1} \quad (4.42)$$

where $N_{sub}$ is in units of $m^{-3}$, and $SSL0$, $SSL1$, and $SSLEXP1$ are the fitting parameters. If the doping concentration between the source and the drain is high enough to avoid coupling of the drain-to-body and source-to-body junction depletion regions, the subsurface leakage current can be blocked by the barrier.

In addition to $N_{sub}$, the temperature dependence should also be included in the model. Due to more energetic electrons at higher temperature passing the barrier, the subsurface leakage would increase as the temperature rises. It is observed that the leakage currents in the accumulation region across different temperatures are linear but with different slopes in a log scale, just like $N_{sub}$ effect, so a fitting parameter $SSLEXP2$ is introduced to capture the temperature effect. As a result, variables $SSL0(N_{sub})$ and $SSL1(N_{sub})$, by incorporating the effect of temperature can finally be

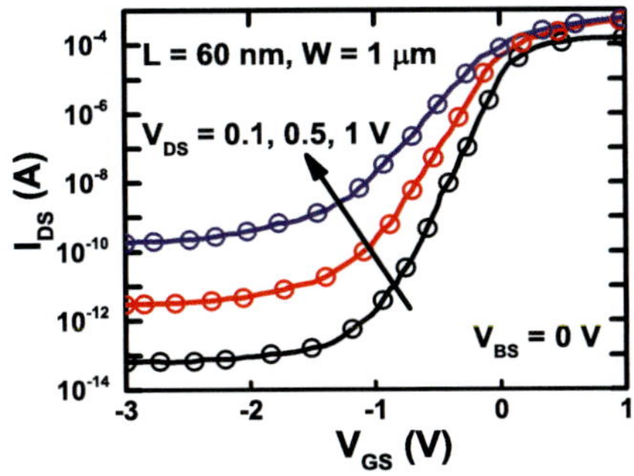


FIGURE 4.12 Verification of the proposed model for subsurface leakage in the accumulation region: $I_{DS} - V_{GS}$ characteristics for zero $V_{TH}$ nMOSFET for different $V_{DS}$ (0.1, 0.5, 1 V) and $V_{SB} = 0\,\text{V}$ [9].

expressed as

$$SSL0(N_{sub}, T) = SSL0.exp\left(-\left(\frac{N_{sub}}{10^{23}}\right)^{SSLEXP1} \cdot \left(\frac{300}{T}\right)^{SSLEXP2}\right) \tag{4.43}$$

$$SSL1(N_{sub}, T) = SSL1.\left(\frac{N_{sub}}{10^{23}}\right)^{SSLEXP1} \cdot \left(\frac{300}{T}\right)^{SSLEXP2} \tag{4.44}$$

where $T$ is in units of $K$. From (4.40), the complete form of subsurface leakage current is

$$I_{DS}(Leakage) = SSL0(N_{sub}, T).e^{-SSL1(N_{sub},T).L_G}.e^{\frac{\Delta\phi}{V_t}}.\left(e^{\frac{SSL2.V_{DS}}{V_t}} - 1\right).W \tag{4.45}$$

Fig. 4.12 and Fig. 4.13 show the comparison of the TCAD simulation data with the SPICE simulation results for different $V_{DS}$ values (from linear to saturation bias regimes) and temperatures for an nMOSFET with the gate length of 60 nm, the device width of 1 μm, the substrate doping of $10^{23}\,\text{m}^3$, and the gate oxide thickness of 2 nm, respectively. Fig. 4.12 and Fig. 4.13 are with the same set of fitting parameters. The developed model exhibits good match with the TCAD simulation data. The $V_{GS}$ effect appears at around flat-band voltage $V_{FB}$ ($\approx -1$ V) and is successfully included in the model. Eq. (4.40) captures not only the barrier lowering due to $V_{DS}$ but also the current increment caused by rising temperature. It is worth noting that the influence of the junction depths $X_j$ of source and drain is not explicitly modeled in this paper. As $X_j$ is shallower, the leakage path is much closer to the surface, and the leakage is suppressed because of better gate control.

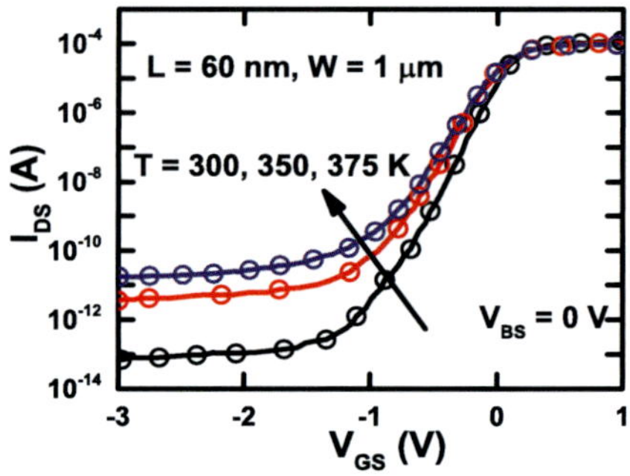


FIGURE 4.13 Verification of proposed model for subsurface leakage in the accumulation region: $I_{DS} - V_{GS}$ characteristics for different temperatures [9]. Solid Line: Model and Symbol: TCAD data.

## References

[1] R. Van Langevelde, A. Scholten, D. Klaassen, Physical background of MOS model 11, Measurements 6 (2003) 9.

[2] A. Sharma, R. Goel, C. Dabhi, G. Pahwa, Y. Chauhan, S. Salahuddin, C. Hu, BSIM-BULK107.1.0 MOSFET compact model, 2022.

[3] S.M. Sze, Y. Li, K.K. Ng, Physics of Semiconductor Devices, John Wiley & Sons, 2021.

[4] C.C. Enz, E.A. Vittoz, Charge-Based MOS Transistor Modeling: the EKV Model for Low-Power and RF IC Design, John Wiley & Sons, 2006.

[5] W.-C. Lee, C. Hu, Modeling CMOS tunneling currents through ultrathin gate oxide due to conduction- and valence-band electron and hole tunneling, IEEE Transactions on Electron Devices 48 (7) (2001) 1366–1373.

[6] V.P.-H. Hu, M.-L. Fan, P. Su, C.-T. Chuang, Band-to-band-tunneling leakage suppression for ultra-thin-body GeOI MOSFETs using transistor stacking, IEEE Electron Device Letters 33 (2) (2012) 197–199.

[7] G. Eneman, G. Hellings, A. De Keersgieter, N. Collaert, A. Thean, Quantum-barriers and ground-plane isolation: a path for scaling bulk-FinFET technologies to the 7 nm-node and beyond, in: 2013 IEEE International Electron Devices Meeting, IEEE, 2013, pp. 12.3.1–12.3.4.

[8] Y. Tsividis, Operation and Modeling of the MOS Transistor, McGraw-Hill, Inc., 1987.

[9] Y.-K. Lin, S. Khandelwal, A.S. Medury, H. Agarwal, H.-L. Chang, Y.S. Chauhan, C. Hu, Modeling of subsurface leakage current in low VTH short channel MOSFET at accumulation bias, IEEE Transactions on Electron Devices 63 (5) (2016) 1840–1845.

[10] H. Agarwal, C. Gupta, P. Kushwaha, C. Yadav, J.P. Duarte, S. Khandelwal, C. Hu, Y.S. Chauhan, Analytical modeling and experimental validation of threshold voltage in BSIM6 MOSFET model, IEEE Journal of the Electron Devices Society 3 (3) (2015) 240–243.

CHAPTER

# 5

# BSIM-BULK Charge and Capacitance Model

## 5.1 Charge model

**Inversion Charge:** Total inversion charge (excluding velocity saturation, CLM and poly-depletion) can be expressed explicitly in terms of normalized charge densities at source and drain sides as follows:

$$Q_I = W \cdot \int_0^L Q_i \, dx = -WL \cdot C_{ox} \cdot V_t \int_0^1 2.n_q.q \, d\xi \tag{5.1}$$

$$-\frac{Q_I}{WL \cdot C_{ox} V_t} = q_I = 2.n_q \cdot \int_0^1 q \, d\xi \tag{5.2}$$

Here $\xi = \frac{x}{L}$. The inversion charge density is normalized to $-2.n_q.C_{ox}.V_t$ and voltages to $V_t$. From Chapter 2, we know that

$$I_{ds} = -2 \cdot n_q \cdot \mu_{eff} \cdot \frac{W_{eff}}{L_{eff}} \cdot C_{ox} \cdot n V_t^2 \cdot (2q+1)\frac{dq}{d\xi} \tag{5.3}$$

Normalizing current with $2 \cdot n_q \cdot \mu_{eff} \cdot \frac{W_{eff}}{L_{eff}} \cdot C_{ox} \cdot n V_t^2$,

$$i = -(2q+1)\frac{dq}{d\xi} \tag{5.4}$$

which gives $d\xi = -\frac{(2q+1)}{i} dq = -\frac{(2q+1)}{(q_s - q_d)(1+q_s+q_d)}$. Substituting in (5.2)

$$q_I = -\frac{2n_q}{(q_s - q_d)(1+q_s+q_d)} \int_{q_s}^{q_d} q(2q+1)\, dq \tag{5.5}$$

$$= -\frac{2n_q}{(q_s - q_d)(1+q_s+q_d)} \left[\frac{2}{3}({q_d}^3 - {q_s}^3) + \frac{1}{2}({q_d}^2 - {q_s}^2)\right] \tag{5.6}$$


https://doi.org/10.1016/B978-0-32-385677-5.00009-3

on rearranging,

$$q_I = n_q \cdot \left[ q_s + q_d + \frac{1}{3} \cdot \frac{(q_s - q_d)^2}{1 + q_s + q_d} \right] \tag{5.7}$$

**Source and Drain Charges:** To obtain terminal source ($Q_s$) and drain ($Q_d$) charges, the Ward-Dutton charge partition scheme [1] is used.

$$Q_d = W \int_0^L \frac{x}{L} Q_I dx = WL \int_0^1 \xi Q_I d\xi \tag{5.8}$$

$$Q_s = W \int_0^L \left(1 - \frac{x}{L}\right) Q_I dx = WL \int_0^1 (1 - \xi) Q_I d\xi \tag{5.9}$$

which leads to

$$Q_s = \frac{n_q}{3} \left[ 2 \cdot q_s + q_{deff} + \frac{1}{2} \cdot \left(1 + \frac{4}{5} \cdot q_s + \frac{6}{5} \cdot q_{deff}\right) \left(\frac{q_s - q_{deff}}{1 + q_s + q_{deff}}\right)^2 \right] \tag{5.10}$$

$$Q_d = \frac{n_q}{3} \left[ q_s + 2 \cdot q_{deff} + \frac{1}{2} \cdot \left(1 + \frac{6}{5} \cdot q_s + \frac{4}{5} \cdot q_{deff}\right) \left(\frac{q_s - q_{deff}}{1 + q_s + q_{deff}}\right)^2 \right] \tag{5.11}$$

**Bulk Charge:** Bulk charge density is given as

$$Q_b = -C_{ox} \cdot (V_G - V_{FB} - \psi_S) - Q_i \tag{5.12}$$

using charge linearization

$$Q_b = -C_{ox} \cdot (V_G - V_{FB} - \psi_P) - Q_i \left(1 - \frac{1}{n_q}\right) \tag{5.13}$$

Total bulk charge,

$$Q_B = W \cdot \int_0^L Q_b \, dx \tag{5.14}$$

$$= -WL \cdot C_{ox} \cdot \left[ V_G - V_{FB} - \psi_P + \left(1 - \frac{1}{n_q}\right) \cdot \int_0^1 \frac{Q_i}{C_{ox}} d\xi \right] \tag{5.15}$$

Normalizing the bulk charge to $-W.L.C_{ox}.V_t$,

$$q_B = v_g - v_{fb} - \psi_p - 2(n_q - 1) \cdot \int_0^1 q \, d\xi \tag{5.16}$$

We know that $i_{ds} = -(2q+1)\frac{dq}{d\xi}$, thus $d\xi = -\frac{-(2q+1)}{i_{ds}}dq$,

$$q_B = v_g - v_{fb} - \psi_p + \frac{2(n_q-1)}{i_{ds}} \cdot \int_{q_s}^{q_d} q(2q+1)\,dq \tag{5.17}$$

$$= v_g - v_{fb} - \psi_p + \frac{2(n_q-1)}{(q_s-q_d)(1+q_s+q_d)} \int_{q_s}^{q_d} \left[\frac{2q^3}{3} + \frac{q^2}{2}\right] \tag{5.18}$$

$$= v_g - v_{fb} - \psi_p + \frac{2(n_q-1)}{(q_s-q_d)(1+q_s+q_d)} \times \left[\frac{2}{3} \cdot (q_d-q_s)(q_d^2+q_s^2+q_s.q_d) + \frac{1}{2} \cdot (q_d-q_s)(q_d+q_s)\right] \tag{5.19}$$

which on rearrangement becomes,

$$q_B = v_g - v_{fb} - \psi_p - (n_q-1)\left[q_s + q_d + \frac{1}{3} \cdot \frac{(q_s-q_d)^2}{1+q_s+q_d}\right] \tag{5.20}$$

**Bulk charge with poly-depletion effect:**

$$q_B = A + B + \frac{1}{3}.\frac{\Delta q^2}{C^3} \cdot \left[\frac{4}{5} \cdot \left(C^2 + P.Q\right) \cdot \frac{1}{1+q_s+q_d} + \frac{2}{\gamma_g^2}\right] - n_q.\left[q_s + q_d + \frac{1}{3} \cdot \frac{(q_s-q_d)^2}{1+q_s+q_d}\right] \tag{5.21}$$

where

$$P = \sqrt{\frac{1}{4} + \frac{v_g - v_{fb} - \psi_p + 2.q_s}{\gamma_g^2}} \tag{5.22}$$

$$Q = \sqrt{\frac{1}{4} + \frac{v_g - v_{fb} - \psi_p + 2.q_d}{\gamma_g^2}} \tag{5.23}$$

$$A = \frac{v_g - v_{fb} - \psi_p + 2.q_s}{1 + 2.\sqrt{\frac{1}{4} + \frac{v_g - v_{fb} - \psi_p + 2.q_s}{\gamma_g^2}}} \tag{5.24}$$

$$B = \frac{v_g - v_{fb} - \psi_p + 2.q_d}{1 + 2.\sqrt{\frac{1}{4} + \frac{v_g - v_{fb} - \psi_p + 2.q_d}{\gamma_g^2}}} \tag{5.25}$$

$$C = \sqrt{\frac{1}{4} + \frac{v_g - v_{fb} - \psi_p + 2.q_s}{\gamma_g^2}} + \sqrt{\frac{1}{4} + \frac{v_g - v_{fb} - \psi_p + 2.q_d}{\gamma_g^2}} \tag{5.26}$$

**Bulk charge with poly-depletion effect, CLM and Velocity Saturation effects:**

$$d_{qgeff} = (v_g - v_{fb} - \psi_p + 2.q_d)\left[\frac{2\sqrt{\frac{1}{4}+\frac{v_g-v_{fb}-\psi_p+2.q_d}{\gamma_g^2}}-1}{2\sqrt{\frac{1}{4}+\frac{v_g-v_{fb}-\psi_p+2.q_d}{\gamma_g^2}}+1}\right] \tag{5.27}$$

$$q_{beff} = \left[(v_g - v_{fb} - \psi_p) - 2(n_q - 1)q_d + d_{qgeff}\right] \tag{5.28}$$

$$q_B = \frac{1}{MDL}\left[A + B + \frac{1}{3}.\frac{\Delta q^2}{C^3}\cdot\left(\frac{4}{5}\cdot(C^2+P.Q)\cdot\frac{1}{1+q_s+q_d}+\frac{2}{\gamma_g^2}\right) - n_q.\left(q_s + q_d + \frac{1}{3}\cdot\frac{(q_s-q_d)^2}{1+q_s+q_d}\right)\right] + (MDL-1)q_{beff} \tag{5.29}$$

where

$$P = \sqrt{\frac{1}{4}+\frac{v_g - v_{fb} - \psi_p + 2.q_s}{\gamma_g^2}} \tag{5.30}$$

$$Q = \sqrt{\frac{1}{4}+\frac{v_g - v_{fb} - \psi_p + 2.q_d}{\gamma_g^2}} \tag{5.31}$$

$$A = \frac{v_g - v_{fb} - \psi_p + 2.q_s}{1+2.\sqrt{\frac{1}{4}+\frac{v_g-v_{fb}-\psi_p+2.q_s}{\gamma_g^2}}} \tag{5.32}$$

$$B = \frac{v_g - v_{fb} - \psi_p + 2.q_d}{1+2.\sqrt{\frac{1}{4}+\frac{v_g-v_{fb}-\psi_p+2.q_d}{\gamma_g^2}}} \tag{5.33}$$

$$C = \sqrt{\frac{1}{4}+\frac{v_g - v_{fb} - \psi_p + 2.q_s}{\gamma_g^2}} + \sqrt{\frac{1}{4}+\frac{v_g - v_{fb} - \psi_p + 2.q_d}{\gamma_g^2}} \tag{5.34}$$

Detailed derivation of bulk charge with including velocity saturation effect is given in Appendix 5.A.

**Source and Drain Charges with CLM and Velocity Saturation Model**

$$Q_i = \frac{n_q}{MDL}\left[(q_s + q_{deff}) + \frac{1}{3}(q_s - q_{deff})^2\frac{DVSAT}{MDL\cdot(1+q_s+q_{deff})} + 2n_q(MDL-1)q_{deff}\right] \tag{5.35}$$

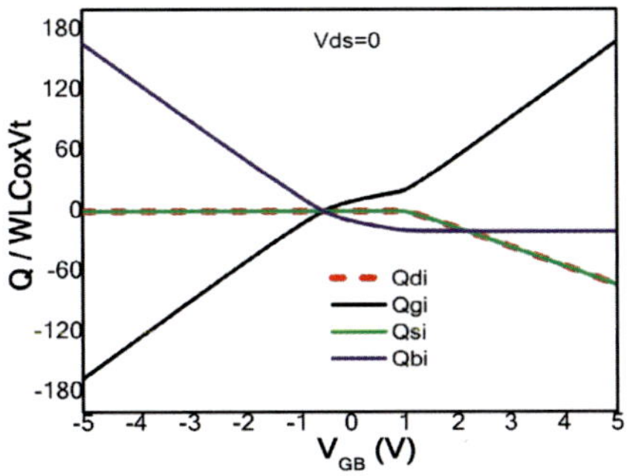


FIGURE 5.1 Plot of intrinsic charge vs. gate bias. Default parameters used along with L=W=1 µm, **CVMOD**=1 and **RDSMOD**=2. L=W=1 µm, **CVMOD**=1, **RDSMOD**=2, **ADOS**=0, **NGATE**=0.

$$Q_d = \frac{n_q}{3 \cdot MDL^2}\left[q_s + 2 \cdot q_{deff} + \frac{1}{2} \cdot \left(\frac{1}{5} \cdot (q_s - q_{deff}) + \frac{MDL}{DVSAT} \cdot (1 + q_s + q_{deff})\right)\left(\frac{q_s - q_{deff}}{1 + q_s + q_{deff}}\right)^2 \left(\frac{DVSAT}{MDL}\right)^2\right] + n_q\left[MDL - \frac{1}{MDL}\right]$$

$$Q_s = Q_i - Q_d \tag{5.36}$$

Since the above expressions do not include contribution of the parasitic component, these charges are referred as intrinsic charges. Fig. 5.1 shows a plot of all four intrinsic charges as a function of applied gate voltage at zero drain bias. Fig. 5.2 shows all 16 capacitances and it can be seen that the model correctly predicts transcapacitance reciprocity at zero drain bias. Fig. 5.3 shows capacitance behavior as a function of drain-bias.

## 5.2 Quantum mechanical effect (QME)

In deriving the inversion charge and pinch-off potential so far, the inversion layer was assumed to be located right at the interface between oxide and silicon. From quantum-mechanical physics, we know that it is an incorrect representation since electron/hole concentration lies away from the interface. The average thickness of inversion layer is approximated by [2,3].

$$X_{DC}^{inv} = \frac{\int \rho(x)xdx}{\int \rho(x)x} \tag{5.37}$$

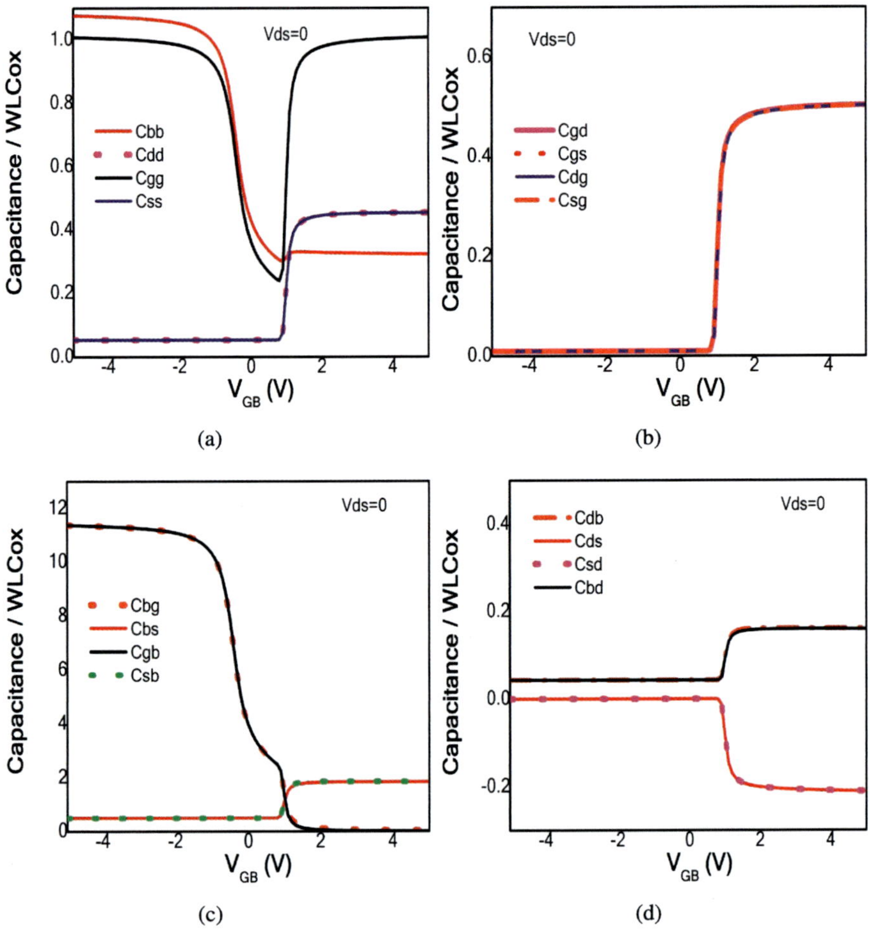


FIGURE 5.2 Core model results: plot of all 16 capacitances vs. gate voltage at Vds=0. Capacitance behavior is smooth across biases. Default parameters used along with L=W=1 μm, **CVMOD**=1, **RDSMOD**=2, **ADOS**=0, **NGATE**=0.

where $\rho$ represents inversion charge density. The universal model that fits the inversion layer thickness for variety of doping and oxide-thickness range is given as [2]

$$X_{DC}^{inv} = \frac{1.9 \cdot 10^{-9} \cdot \mathbf{ADOS}}{1 + \left[V_{gsteff} + 4(V_{th0} - VFB - \psi_s)\right]^{0.7 \cdot \mathbf{BDOS}}} \tag{5.38}$$

Two parameters, **ADOS** and **BDOS** are introduced for modeling accuracy and flexibility. The final equation in the BSIM-BULK model is given as

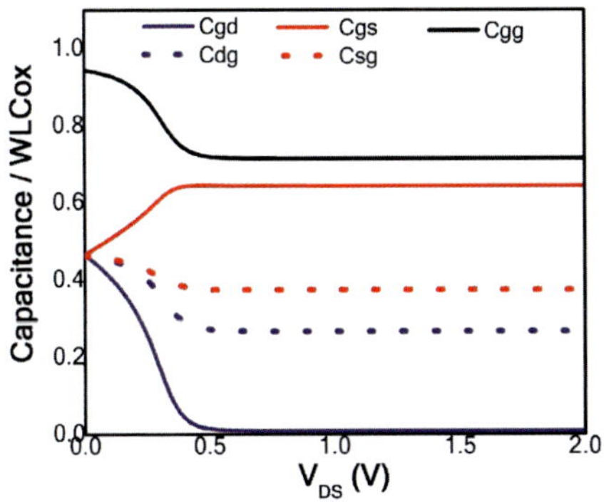


FIGURE 5.3 Core model results: plot of trans capacitances Cgs, Cgd, Cgb, Csg, Cgg vs. Vds at Vg=1 V. Capacitance behavior is smooth across biases. Default parameters used along with L=W=1 µm, **CVMOD**=1, **RDSMOD**=2, **ADOS**=0, **NGATE**=0.

follows:

$$X_{DC}^{inv} = \frac{\mathbf{ADOS} \cdot (1.9 \cdot 10^{-9})}{1 + \left[\frac{Q_i + \mathbf{ETAQM} \cdot Q_B}{\mathbf{QM0}}\right]^{0.7 * \mathbf{BDOS}}} \tag{5.39}$$

Here ETAQM is a parameter for bulk-charge coefficient, and QM0 is a charge centroid parameter, which governs the starting point for quantum-mechanical effects in the inversion region. $Q_i$ is the sum of $Q_s$ and $Q_d$. The effect of finite inversion layer thickness is to add capacitance in series ($C_{charge}$) with the oxide capacitance. The effective oxide capacitance is given by:

$$C_{ox}^{inv} = \frac{3.9 \cdot \epsilon_0}{BSIMBULKTOXP \cdot \frac{3.9}{\mathbf{EPSROX}} + \frac{X_{DC}^{inv}}{\epsilon_{ratio}}} \tag{5.40}$$

In the above equation, if physical oxide thickness parameter (**TOXP**) is supplied by the user, then BSIMBULKTOXP takes that value. Otherwise, it is calculated from electrical oxide thickness as BSIMBULKTOXP = (**TOXE** * **EPSROX** / 3.9). Fig. 5.4 shows the impact of finite inversion charge layer thickness and poly-gate depletion on pinch-off potential and normalized inversion charge. Fig. 5.5 shows the impact of poly-depletion and QME on the gate capacitance $C_{gg}$.

## 5.3 Parasitic capacitance model

An accurate overlap capacitance model is essential to capture the CV characteristics of the transistor. This is especially true for the drain side,

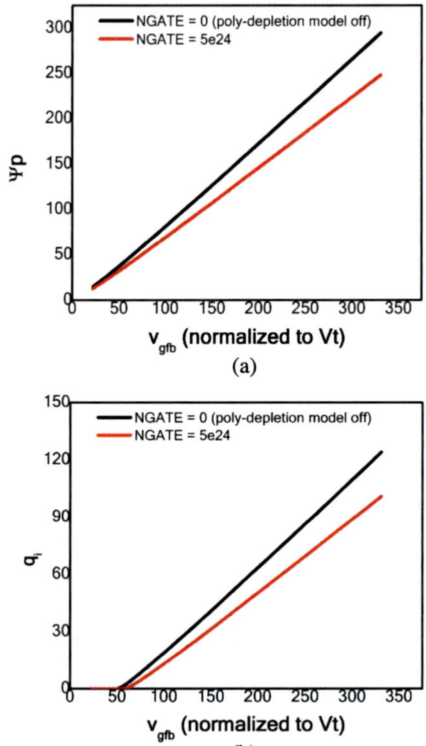


FIGURE 5.4 Impact of poly-depletion effect: (a) pinch-off potential (b) inversion charge density for **NGATE**=0, $5 \cdot 10^{24}$. **NGATE** also acts as a model selector and setting it to zero disables the poly-depletion model.

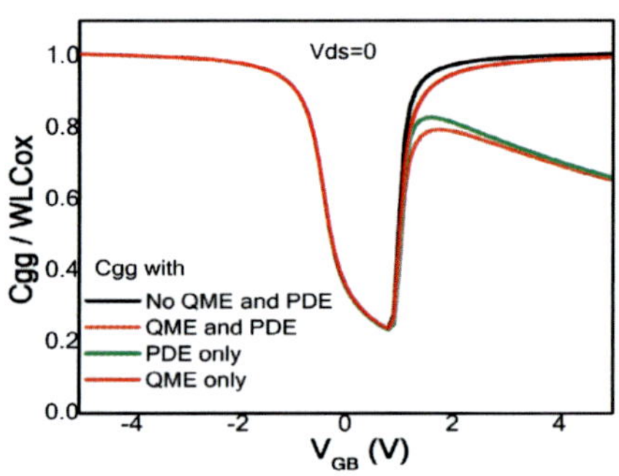


FIGURE 5.5 Total gate capacitance: Impact of poly-gate depletion and finite inversion charge thickness on Cgg. Default parameters used along with L=W=1 µm, **CVMOD**=1 and **RDSMOD**=2. For enabling QME/PDE, **ADOS**=40 and **NGATE**=5e25 are used.

where the effect of the capacitance is amplified by the transistor gain. In LDD MOSFETs, a substantial portion of the LDD region can be depleted, both in the vertical and lateral directions. This can lead to a large reduction of the overlap capacitance. Further, the LDD region can be in accumulation or depletion. Therefore the overlap capacitance is a strong function of gate to source and gate to drain biases.

The BSIM-BULK model offers two methods to model the overlap capacitances, which can be selected using model selector parameter COVMOD. One is the bias-dependent model (COVMOD=1), and the other is the simplified bias-independent model (COVMOD=0, default). Let us first define the following quantities for the ease of understanding the model:

$$q_{ov,s}^{bias-indep} = C_{gsof} \cdot V_{gs,ov}^{noswap}; \quad C_{gsof} = CGSO + CF \tag{5.41}$$

$$q_{ov,d}^{bias-indep} = C_{gdof} \cdot V_{gd,ov}^{noswap}; \quad C_{gdof} = CGDO + CF \tag{5.42}$$

here CGSO, CGDO, and CF are the model parameters representing gate-to-source, gate-to-drain overlap capacitance, and outer fringe capacitance, respectively. The fringing capacitance consists of a bias-independent outer fringing capacitance and a bias-dependent inner fringing capacitance. Only the bias-independent outer fringing capacitance is modeled in BSIM-BULK. If CF is not given, then outer fringe capacitance is calculated as:

$$CF = \frac{2 \cdot EPSROX \cdot \epsilon_0}{\pi} \cdot ln\left[CFRCOEFF \cdot (1 + \frac{4 \cdot 10^{-7}}{TOXE})\right] \tag{5.43}$$

$$q_{ov,s}^{bias-dep} = CGSL \cdot \left[ V_{gs,ov}^{noswap} - V_{fbsd} - V_{gs,ov} - \frac{CKAPPAS}{2}\left(\sqrt{1 - \frac{4V_{gs,ov}}{CKAPPAS}} - 1\right)\right] \tag{5.44}$$

$$q_{ov,d}^{bias-dep} = CGDL \cdot \left[ V_{gd,ov}^{noswap} - V_{fbsd} - V_{gd,ov} - \frac{CKAPPAD}{2}\left(\sqrt{1 - \frac{4V_{gd,ov}}{CKAPPAD}} - 1\right)\right] \tag{5.45}$$

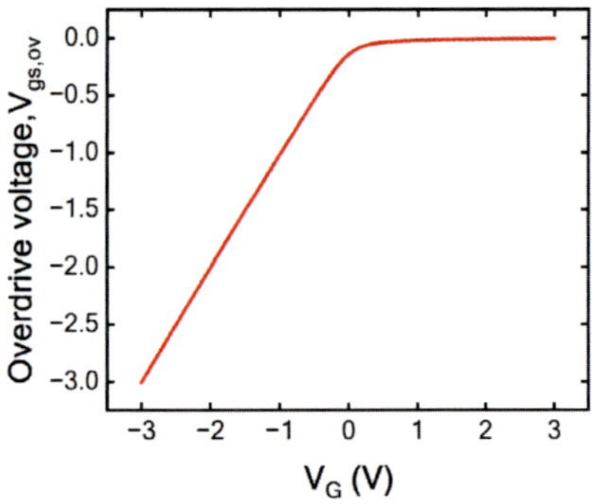


FIGURE 5.6 Overdrive voltage used in the overlap-capacitance model. $V_{gs,ov}$ ($V_{gd,ov}$) smoothly limits $V_{gs}$ ($V_{gd}$) to zero at large voltages.

where

$$V_{gs,ov} = \frac{1}{2}\left[V_{gs,ov}^{noswap} - V_{fbsd} + \delta_1 - \sqrt{(V_{gs,ov}^{noswap} - V_{fbsd} + \delta_1)^2 + 4\delta_1}\right] \tag{5.46}$$

$$V_{gd,ov} = \frac{1}{2}\left[V_{gd,ov}^{noswap} - V_{fbsd} + \delta_1 - \sqrt{(V_{gd,ov}^{noswap} - V_{fbsd} + \delta_1)^2 + 4\delta_1}\right] \tag{5.47}$$

$$\delta_1 = 0.02\,\mathrm{V} \tag{5.48}$$

The above bias-dependent overlap charge model in BSIM-BULK is adopted from BSIM4 [4–6]. $V_{gs,ov}^{noswap}$ and $V_{gd,ov}^{noswap}$ in the above equations represent gate-source and gate-drain voltages without swapping terminals for the negative $V_{ds}$. $V_{gs,ov}$ ($V_{gd,ov}$) is a smoothing function, which tracks $V_{gs,ov}^{noswap}$ ($V_{gd,ov}^{noswap}$) for negative voltage range and smoothly clamps to a small positive number for positive voltage range as shown in Fig. 5.6. Fig. 5.7 shows the bias-dependent overlap capacitance behavior for two values of parameter CKAPPAS. Overlap capacitance at high gate voltage is determined by CGSL (CGDL) and CKAPPAS (CKAPPAD) governs the bias dependency.

Depending on the choice of COVMOD, the normalized gate-to-source ($q_{ov,s}$) overlap and gate-to-drain ($q_{ov,d}$) overlap charges are defined as:

**COVMOD=0:** This is a simplified model where the overlap capacitances are bias-independent. The normalized overlap charges are given as:

$$q_{ov,s} = q_{ov,s}^{bias-indep} \tag{5.49}$$

$$q_{ov,d} = q_{ov,d}^{bias-indep} \tag{5.50}$$

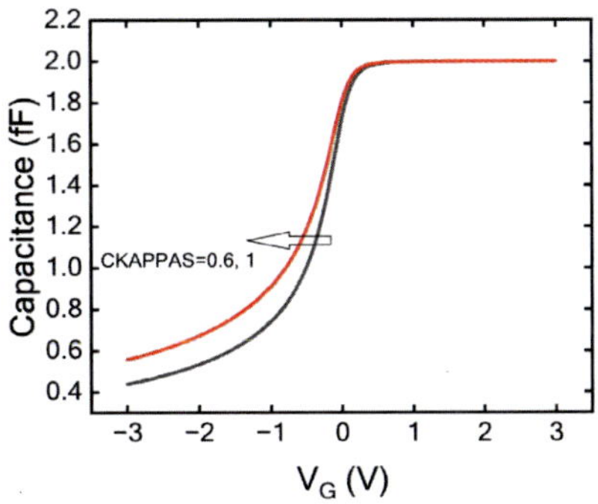


FIGURE 5.7 Overlap capacitance model consisting of bias-dependent (constant capacitance) and bias-dependent parts. This figure shows the bias-dependent model behavior for two values of the parameter CKAPPAS.

**COVMOD=1:** The normalized overlap charges are given as:

$$q_{ov,s} = q_{ov,s}^{bias-dep} + q_{ov,s}^{bias-indep} \tag{5.51}$$

$$q_{ov,d} = q_{ov,d}^{bias-dep} + q_{ov,d}^{bias-indep} \tag{5.52}$$

Total overlap charges are defined as:

$$Q_{ov,s} = q_{ov,s} \cdot NF \cdot W_{effCV} \tag{5.53}$$

$$Q_{ov,d} = q_{ov,d} \cdot NF \cdot W_{effCV} \tag{5.54}$$

Finally, the body overlap charge is defined as:

$$Q_{ov,b} = -NF \cdot L \cdot CGBO \cdot V(gm, bi) \tag{5.55}$$

here CGBO is a gate-to-body overlap capacitance model parameter.

These charges are then time-differentiated in the verilog-A to instantiate capacitance:

$$I(gm, di) < +ddt(-Q_{ov,d}) \tag{5.56}$$

$$I(gm, si) < +ddt(-Q_{ov,s}) \tag{5.57}$$

$$I(gm, bi) < +ddt(-Q_{ov,b}) \tag{5.58}$$

Fig. 5.8 illustrates the impact of the overlap capacitance on the total gate capacitance of the device. Table 5.1 lists overlap and fringe capacitance model parameters.

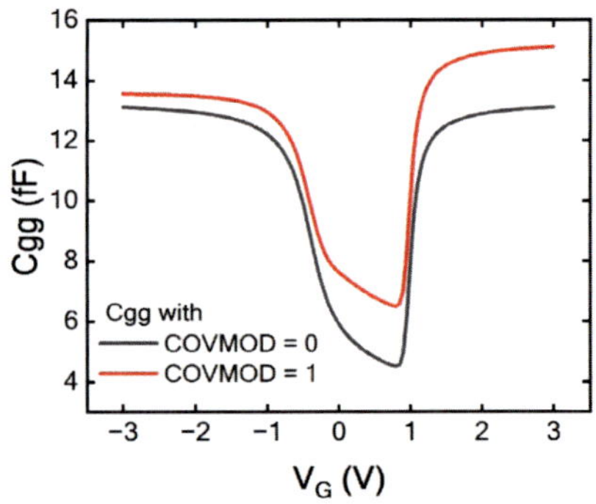


FIGURE 5.8 Total gate capacitance vs. gate voltage for COVMOD=0 and 1.

TABLE 5.1 Overlap and Fringe capacitance model parameters.

| Name | Default | Description |
|---|---|---|
| COVMOD | 0 | 0: Use bias-independent overlap capacitances, 1: Use bias-dependent overlap capacitances |
| CGSO | 0 | Gate-to-source overlap capacitance |
| CGDO | 0 | Gate-to-drain overlap capacitance |
| CGBO | 0 | Gate-to-body overlap capacitance |
| CGSL | 0 | Overlap capacitance between gate and lightly-doped source region |
| CGDL | 0 | Overlap capacitance between gate and lightly-doped drain region |
| CKAPPAS | 0.6 | Coefficient of bias-dependent overlap capacitance for the source side |
| CKAPPAD | 0.6 | Coefficient of bias-dependent overlap capacitance for the drain side |
| CFRCOEFF | 1 | Coefficient for outer fringe capacitance |
| CF | 0 | Outer fringe capacitance. If CF is not provided, Eq. (5.43) is used to estimate it. |

## Appendix 5.A Bulk charge with poly-depletion effect

Total Bulk Charge,

$$Q_B = W \, . \int_0^L Q_b \, dx \tag{1}$$

Using charge balance equation, bulk charge density considering poly depletion effect is given by:

$$Q_b = -C_{ox}.(V_G - V_{FB} - \Psi_S) - Q_i - \Delta Q_g \tag{2}$$

where $\Delta Q_g$ is the charge density due to polysilicon gate. Then using inversion charge linearization:

$$Q_i = n_q.C_{ox}.(\Psi_S - \Psi_P)$$
$$or$$
$$\Psi_S = \Psi_P + \frac{Q_i}{n_q C_{ox}}$$

Substituting this $\Psi_S$ into (2):

$$Q_b = -C_{ox}.(V_G - V_{FB} - \Psi_P) - Q_i \cdot \left(1 - \frac{1}{n_q}\right) - \Delta Q_g \tag{3}$$

Now, (1) reads:

$$Q_B = -W.L.C_{ox}.\left((V_G - V_{FB} - \Psi_P) + (1 - \frac{1}{n_q}) \int_0^1 \frac{Q_i}{C_{ox}} d\xi + \int_0^1 \frac{\Delta Q_g}{C_{ox}} d\xi\right)$$

$$Q_B = -W.L.C_{ox}.\left((V_G - V_{FB} - \Psi_P) + (1 - \frac{1}{n_q}) \int_0^1 \frac{Q_i}{C_{ox}} d\xi\right)$$
$$- W.L. \int_0^1 \Delta Q_g \, d\xi$$

$$Q_B = -W.L.C_{ox}.\left((V_G - V_{FB} - \Psi_P) + (1 - \frac{1}{n_q}) \int_0^1 \frac{Q_i}{C_{ox}} d\xi\right) - \Delta Q_G \tag{4}$$

where $\Delta Q_G = W \,.\, \int_0^L \Delta \, Q_g \, dx$

$$Q_B = -W.L.C_{ox}.(V_G - V_{FB} - \Psi_P)$$
$$- W.L.2.n_q.C_{ox}.(1 - \frac{1}{n_q}) \int_0^1 \frac{Q_i}{2.n_q.C_{ox}} d\xi - \Delta Q_G$$

Normalizing the bulk charge to $-W.L.C_{ox}.V_t$,

$$q_B = \frac{Q_B}{-W.L.C_{ox}.V_t}$$
$$= (v_g - v_{fb} - \psi_p) + 2.n_q.(1 - \frac{1}{n_q}) \int_0^1 \frac{Q_i}{2.n_q.C_{ox} V_t} d\xi - \frac{\Delta Q_G}{-W.L.C_{ox}.V_t}$$

Now,

$$\Delta q_G = \frac{-\Delta Q_G}{W.L.C_{ox}.V_t} \qquad \& \qquad q = \frac{-Q_i}{2.n_q C_{ox}.V_t}$$

The above equation reads following:

$$q_B = (v_g - v_{fb} - \psi_p) - 2.(n_q - 1).\int_0^1 q\, d\xi - \Delta q_G$$

$$q_B = q_{B1} - \Delta q_G \tag{5}$$

where, $q_{B1} = (v_g - v_{fb} - \psi_p) - 2.(n_q - 1).\int_0^1 q\, d\xi$

Now, gate charge in the poly Si is given by [7]:

$$Q_{poly} = \Delta Q_g = \sqrt{2q\epsilon_s N_{poly}}\sqrt{\Psi_{poly}}$$

$$\frac{\Delta Q_g}{C_{ox}} = \gamma_g\sqrt{\Psi_{poly}}$$

Using potential balance $\frac{\Delta Q_g}{C_{ox}} = \Psi_{ox} = (V_G - V_{FB} - \Psi_s - \Psi_{poly})$

$$\{(V_G - V_{FB} - \Psi_s) - \Psi_{poly}\}^2 = \gamma_g^2.\Psi_{poly}$$

$$\frac{\Psi_{poly}}{\{(V_G - V_{FB} - \Psi_s) - \Psi_{poly}\}^2} = \frac{1}{\gamma_g^2}$$

$$\frac{\{(V_G - V_{FB} - \Psi_s) + \Psi_{poly}\}^2}{\{(V_G - V_{FB} - \Psi_s) - \Psi_{poly}\}^2} - 1 = \frac{4.(V_G - V_{FB} - \Psi_s)}{\gamma_g^2}$$

$$\frac{(V_G - V_{FB} - \Psi_s) + \Psi_{poly}}{(V_G - V_{FB} - \Psi_s) - \Psi_{poly}} = \sqrt{1 + \frac{4.(V_G - V_{FB} - \Psi_s)}{\gamma_g^2}}$$

$$\Psi_{poly} = (V_G - V_{FB} - \Psi_s).\frac{\sqrt{1 + \frac{4.(V_G - V_{FB} - \Psi_s)}{\gamma_g^2}} - 1}{\sqrt{1 + \frac{4.(V_G - V_{FB} - \Psi_s)}{\gamma_g^2}} + 1}$$

$$\Delta Q_g = -C_{ox}.(V_G - V_{FB} - \Psi_s).\frac{\sqrt{1 + \frac{4.(V_G - V_{FB} - \Psi_s)}{\gamma_g^2}} - 1}{\sqrt{1 + \frac{4.(V_G - V_{FB} - \Psi_s)}{\gamma_g^2}} + 1}$$

$$\Delta Q_g = -C_{ox}.\left(V_G - V_{FB} - \Psi_P - \frac{Q_i}{n_q.C_{ox}}\right) \times \frac{\sqrt{1+(\frac{2}{\gamma_g})^2.(V_G - V_{FB} - \Psi_P - \frac{Q_i}{n_q.C_{ox}})} - 1}{\sqrt{1+(\frac{2}{\gamma_g})^2.(V_G - V_{FB} - \Psi_P - \frac{Q_i}{n_q.C_{ox}})} + 1}$$

$\Delta Q_G = W \, . \int_0^L \Delta \, Q_g \, dx = W.L \, . \int_0^1 \Delta \, Q_g \, d\xi$

$$\Delta Q_G = -W.L.C_{ox}. \int_0^1 \left(V_G - V_{FB} - \Psi_P - \frac{Q_i}{n_q.C_{ox}}\right) \times \frac{\sqrt{1+(\frac{2}{\gamma_g})^2.(V_G - V_{FB} - \Psi_P - \frac{Q_i}{n_q.C_{ox}})} - 1}{\sqrt{1+(\frac{2}{\gamma_g})^2.(V_G - V_{FB} - \Psi_P - \frac{Q_i}{n_q.C_{ox}})} + 1} \, d\xi$$

$$\Delta q_G = \frac{-\Delta Q_G}{W.L.C_{ox}.V_t}$$

$$\Delta q_G = \int_0^1 \left(v_g - v_{fb} - \psi_p + 2.q\right) \frac{\sqrt{1+(\frac{2}{\gamma'_g})^2.(v_g - v_{fb} - \psi_p + 2.q)} - 1}{\sqrt{1+(\frac{2}{\gamma'_g})^2.(v_g - v_{fb} - \psi_p + 2.q)} + 1} \, d\xi \tag{6}$$

where $(\gamma'_g)^2 = (\gamma_g)^2/V_t$

Now, (17) of [8] reads:

$$\xi_m - \xi = s.\frac{(s+H_q)}{\Delta q.H_q} \quad where \quad H_q = (1 + 2q_{ia}/D_{vsat})$$

where $\xi_m$ is obtained by noting that at source end, $\xi = 0$ and $s = s_s = \Delta q/2$. Substituting back $\xi_m$ above gives:

$$\xi = \frac{1}{2.H_q}\left(\frac{\Delta q}{2} + H_q\right) - \frac{s.(s+H_q)}{\Delta q.H_q}$$

$$d\xi = -\frac{(2.s + H_q)}{\Delta q.H_q} \, ds$$

After substituting these, (6) reads:

$$\Delta q_G = -\int_{\frac{\Delta q}{2}}^{-\frac{\Delta q}{2}} \left(v_g - v_{fb} - \psi_p + 2.q\right)$$

$$\times \frac{\sqrt{1+(\frac{2}{\gamma_g'})^2.(v_g - v_{fb} - \psi_p + 2.q)} - 1}{\sqrt{1+(\frac{2}{\gamma_g'})^2.(v_g - v_{fb} - \psi_p + 2.q)} + 1} \frac{(2.s + H_q)}{\Delta q.H_q} ds$$

$$\Delta q_G = \int_{\frac{-\Delta q}{2}}^{\frac{\Delta q}{2}} \left(F_g + 2.q\right) \frac{\sqrt{1 + A_g.(F_g + 2.q)} - 1}{\sqrt{1 + A_g.(F_g + 2.q)} + 1} \frac{(2.s + H_q)}{\Delta q.H_q} ds \quad (7)$$

where $F_g = (v_g - v_{fb} - \psi_p)$ and $A_g = (\frac{2}{\gamma_g'})^2$

Also, $q = s + q_{ia}$

$$\Delta q_G = \int_{\frac{-\Delta q}{2}}^{\frac{\Delta q}{2}} \left(F_g + 2.(s + q_{ia})\right) \frac{\sqrt{1 + A_g.(F_g + 2.(s + q_{ia}))} - 1}{\sqrt{1 + A_g.(F_g + 2.(s + q_{ia}))} + 1} \frac{(2.s + H_q)}{\Delta q.H_q} ds \quad (8)$$

$$\Delta q_G = \int_{\frac{-\Delta q}{2}}^{\frac{\Delta q}{2}} \left(F_g + 2.(s + q_{ia})\right) \frac{\left[\sqrt{1 + A_g.(F_g + 2.(s + q_{ia}))} - 1\right]^2}{A_g.(F_g + 2.(s + q_{ia}))} \times \frac{(2.s + H_q)}{\Delta q.H_q} ds$$

$$\Delta q_G = \frac{1}{A_g.\Delta q.H_q} \int_{-\frac{\Delta q}{2}}^{\frac{\Delta q}{2}} \left[\sqrt{1 + A_g.(F_g + 2.(s + q_{ia}))} - 1\right]^2 (2.s + H_q)\, ds$$

$$\Delta q_G = \frac{1}{A_g.\Delta q.H_q} \int_{\frac{-\Delta q}{2}}^{\frac{\Delta q}{2}} \Big[2 + A_g.(F_g + 2.(s + q_{ia})) - 2.\sqrt{1 + A_g.(F_g + 2.(s + q_{ia}))}\Big] (2.s + H_q)\, ds \quad (9)$$

Let $t^2 = 1 + A_g.[F_g + 2.(s + q_{ia})]$ so that $ds = \frac{t}{A_g}.dt$ & $2.s = \frac{t^2 - 1 - A_g.(F_g + 2.q_{ia})}{A_g}$

So, $s = \frac{-\Delta q}{2}$ corresponds to $t = t_1 = 1 + A_g.[F_g + 2.(q_{ia} - \frac{\Delta q}{2})]$

And, $s = \frac{\Delta q}{2}$ corresponds to $t = t_2 = 1 + A_g.[F_g + 2.(q_{ia} + \frac{\Delta q}{2})]$

Now, using this (9) reads:

$$\Delta q_G = \frac{1}{A_g.\Delta q.H_q} \times \int_{t_1}^{t_2} \left[1 + t^2 - 2.t\right] \left(\frac{t^2 - 1 - A_g.(F_g + 2.q_{ia})}{A_g} + H_q\right) \cdot \frac{t}{A_g}.dt$$

$$\Delta q_G = \frac{1}{A_g^3.\Delta q.H_q} \int_{t_1}^{t_2} t.\left[t^2 + 1 - 2.t\right]\left(t^2 - 1 + A_g.\left[H_q - (F_g + 2.q_{ia})\right]\right).dt \tag{10}$$

$$\Delta q_G = \frac{1}{A_g^3.\Delta q.H_q} \int_{t_1}^{t_2} t.\left[t^2 - 2.t + 1\right]\left(t^2 - 1 + A_g.\left[H_q - (F_g + 2.q_{ia})\right]\right).dt$$

$$\Delta q_G = \frac{1}{A_g^3.\Delta q.H_q} \int_{t_1}^{t_2} \left[t^3 - 2.t^2 + t\right]\left(t^2 - 1 + B_g\right).dt \tag{11}$$

where $B_g = A_g.\left[H_q - (F_g + 2.q_{ia})\right]$

$$\Delta q_G = \frac{1}{A_g^3.\Delta q.H_q} \int_{t_1}^{t_2} \left[t^5 - 2.t^4 + B_g.t^3 + 2.t^2.(1 - B_g) - t.(1 - B_g)\right].dt \tag{12}$$

On integration, we have

$$\Delta q_G = \frac{1}{A_g^3.\Delta q.H_q} \left[\frac{t^6}{6} - \frac{2.t^5}{5} + B_g.\frac{t^4}{4} + 2.\frac{t^3}{3}.(1 - B_g) - \frac{t^2}{2}.(1 - B_g)\right]_{t1}^{t2} \tag{13}$$

$$\boxed{\Delta q_G = \frac{1}{A_g^3.\Delta q.H_q} \left[\frac{R_1}{6} - \frac{2.R_2}{5} + B_g.\frac{R_3}{4} + 2.\frac{R_4}{3}.(1 - B_g) - \frac{R_5}{2}.(1 - B_g)\right]} \tag{14}$$

where,

$$\boxed{R_1 = \left[1 + A_g.[F_g + 2.(q_{ia} + \frac{\Delta q}{2})]\right]^3 - \left[1 + A_g.[F_g + 2.(q_{ia} - \frac{\Delta q}{2})]\right]^3} \tag{A.1}$$

Simplification:

$$R_1 = \left[1 + A_g.(F_g + 2.q_{ia}) + A_g\Delta q)\right]^3 - \left[1 + A_g.(F_g + 2.q_{ia}) - A_g\Delta q)\right]^3$$

$$R_1 = \left[1 + A_g.\left[F_g + 2.q_{ia}\right]\right]^3.\left\{\left[1 + \frac{A_g\Delta q}{1 + A_g.\left[F_g + 2.q_{ia}\right]}\right]^3 - \left[1 - \frac{A_g\Delta q}{1 + A_g.\left[F_g + 2.q_{ia}\right]}\right]^3\right\}$$

$$R_1 = \left[1 + A_g.\left[F_g + 2.q_{ia}\right]\right]^3.\left\{\frac{2.A_g^3.\Delta q^3}{(1 + A_g.\left[F_g + 2.q_{ia}\right])^3} + \frac{6.A_g.\Delta q}{1 + A_g.\left[F_g + 2.q_{ia}\right]}\right\}$$

$$R_1 = 6.A_g\Delta q\left[1 + A_g.\left[F_g + 2.q_{ia}\right]\right]^2 + 2.(A_g\Delta q)^3$$

$$\boxed{\frac{R_1}{6} = A_g\Delta q\left[1 + A_g.\left[F_g + 2.q_{ia}\right]\right]^2 + \frac{(A_g\Delta q)^3}{3}} \tag{A.2}$$

$$\boxed{R_2 = \left[1 + A_g.\left[F_g + 2.(q_{ia} + \frac{\Delta q}{2})\right]\right]^{5/2} - \left[1 + A_g.\left[F_g + 2.(q_{ia} - \frac{\Delta q}{2})\right]\right]^{5/2}} \tag{B.1}$$

Simplification:

$$R_2 = \left[1 + A_g.(F_g + 2.q_{ia}) + A_g\Delta q)\right]^{5/2} - \left[1 + A_g.(F_g + 2.q_{ia}) - A_g\Delta q)\right]^{5/2}$$

$$R_2 = \left[1 + A_g.\left[F_g + 2.q_{ia}\right]\right]^{5/2}.\left\{\left[1 + \frac{A_g\Delta q}{1 + A_g.\left[F_g + 2.q_{ia}\right]}\right]^{5/2} - \left[1 - \frac{A_g\Delta q}{1 + A_g.\left[F_g + 2.q_{ia}\right]}\right]^{5/2}\right\}$$

Since, $(1 + x)^{5/2} - (1 - x)^{5/2} = (1 + x)^2.\sqrt{1 + x} - (1 - x)^2.\sqrt{1 - x} = (1 + x^2).(\sqrt{1 + x} - \sqrt{1 - x}) + 2.x.(\sqrt{1 + x} + \sqrt{1 - x})$

$$R_2 = \left[1 + A_g.\left[F_g + 2.q_{ia}\right]\right]^{5/2}$$
$$.\left\{\left[1 + \frac{(A_g \Delta q)^2}{(1 + A_g.[F_g + 2.q_{ia}])^2}\right].\left(\sqrt{1 + \frac{A_g \Delta q}{1 + A_g.[F_g + 2.q_{ia}]}} - \sqrt{1 - \frac{A_g \Delta q}{1 + A_g.[F_g + 2.q_{ia}]}}\right)\right.$$
$$\left. + 2.\left(\frac{A_g \Delta q}{1 + A_g.[F_g + 2.q_{ia}]}\right).\left(\sqrt{1 + \frac{A_g \Delta q}{1 + A_g.[F_g + 2.q_{ia}]}} + \sqrt{1 - \frac{A_g \Delta q}{1 + A_g.[F_g + 2.q_{ia}]}}\right)\right\}$$

$$\frac{2.R_2}{5} = \frac{2}{5}.\left[1 + A_g.\left[F_g + 2.q_{ia}\right]\right]^{5/2}$$
$$.\left\{\left[1 + \frac{(A_g \Delta q)^2}{(1 + A_g.[F_g + 2.q_{ia}])^2}\right].\left(\sqrt{1 + \frac{A_g \Delta q}{1 + A_g.[F_g + 2.q_{ia}]}} - \sqrt{1 - \frac{A_g \Delta q}{1 + A_g.[F_g + 2.q_{ia}]}}\right)\right.$$
$$\left. + 2.\left(\frac{A_g \Delta q}{1 + A_g.[F_g + 2.q_{ia}]}\right).\left(\sqrt{1 + \frac{A_g \Delta q}{1 + A_g.[F_g + 2.q_{ia}]}} + \sqrt{1 - \frac{A_g \Delta q}{1 + A_g.[F_g + 2.q_{ia}]}}\right)\right\} \tag{B.2}$$

$$R_3 = \left[1 + A_g.\left[F_g + 2.(q_{ia} + \frac{\Delta q}{2})\right]\right]^2 - \left[1 + A_g.\left[F_g + 2.(q_{ia} - \frac{\Delta q}{2})\right]\right]^2 \tag{C.1}$$

Simplification:

$$R_3 = \left[1 + A_g.(F_g + 2.q_{ia}) + A_g \Delta q)\right]^2 - \left[1 + A_g.(F_g + 2.q_{ia}) - A_g \Delta q)\right]^2$$

$$R_3 = \left[1 + A_g.\left[F_g + 2.q_{ia}\right]\right]^2.\left\{\left[1 + \frac{A_g \Delta q}{1 + A_g.\left[F_g + 2.q_{ia}\right]}\right]^2 - \left[1 - \frac{A_g \Delta q}{1 + A_g.\left[F_g + 2.q_{ia}\right]}\right]^2\right\}$$

$$R_3 = 4.\Delta q.A_g.\left[1 + A_g.\left[F_g + 2.q_{ia}\right]\right]$$

$$B_g.\frac{R_3}{4} = B_g.\Delta q.A_g.\left[1 + A_g.\left[F_g + 2.q_{ia}\right]\right]$$

Since, $B_g = A_g.\left[H_q - (F_g + 2.q_{ia})\right]$

$$B_g.\frac{R_3}{4} = A_g^2.\Delta q.\left[H_q - (F_g + 2.q_{ia})\right]\left[1 + A_g.\left[F_g + 2.q_{ia}\right]\right]$$

$$\boxed{B_g.\frac{R_3}{4} = A_g^2.\Delta q.\left[H_q - (F_g + 2.q_{ia})\right]\left[1 + A_g.\left[F_g + 2.q_{ia}\right]\right]} \tag{C.2}$$

$$\boxed{R_4 = \left[1 + A_g.\left[F_g + 2.(q_{ia} + \frac{\Delta q}{2})\right]\right]^{3/2} - \left[1 + A_g.\left[F_g + 2.(q_{ia} - \frac{\Delta q}{2})\right]\right]^{3/2}} \tag{D.1}$$

Simplification:

$$R_4 = \left[1 + A_g.(F_g + 2.q_{ia}) + A_g\Delta q)\right]^{3/2} - \left[1 + A_g.(F_g + 2.q_{ia}) - A_g\Delta q)\right]^{3/2}$$

$$R_4 = \left[1 + A_g.\left[F_g + 2.q_{ia}\right]\right]^{3/2}.\left\{\left[1 + \frac{A_g\Delta q}{1 + A_g.\left[F_g + 2.q_{ia}\right]}\right]^{3/2} - \left[1 - \frac{A_g\Delta q}{1 + A_g.\left[F_g + 2.q_{ia}\right]}\right]^{3/2}\right\}$$

Since, $(1+x)^{3/2} - (1-x)^{3/2} = (1+x).\sqrt{1+x} - (1-x).\sqrt{1-x} = \left(\sqrt{1+x} - \sqrt{1-x}\right) + x.\left(\sqrt{1+x} + \sqrt{1-x}\right)$

Also, $(1 - B_g) = [1 + A_g.(F_g + 2.q_{ia}) - A_g.H_q]$

$$\boxed{\begin{aligned}\frac{2.R_4}{3}.(1 - B_g) = {} & \frac{2}{3}.[1 + A_g.(F_g + 2.q_{ia}) - A_g.H_q]\left[1 + A_g.\left[F_g + 2.q_{ia}\right]\right]^{3/2} \\ & .\left\{\sqrt{1 + \frac{A_g\Delta q}{1 + A_g.\left[F_g + 2.q_{ia}\right]}} - \sqrt{1 - \frac{A_g\Delta q}{1 + A_g.\left[F_g + 2.q_{ia}\right]}} + \left(\frac{A_g\Delta q}{1 + A_g.\left[F_g + 2.q_{ia}\right]}\right)\right. \\ & \left..\left(\sqrt{1 + \frac{A_g\Delta q}{1 + A_g.\left[F_g + 2.q_{ia}\right]}} + \sqrt{1 - \frac{A_g\Delta q}{1 + A_g.\left[F_g + 2.q_{ia}\right]}}\right)\right\}\end{aligned}} \tag{D.2}$$

$$\boxed{R_5 = \left[1 + A_g.\left[F_g + 2.(q_{ia} + \frac{\Delta q}{2})\right]\right] - \left[1 + A_g.\left[F_g + 2.(q_{ia} - \frac{\Delta q}{2})\right]\right]} \tag{E.1}$$

Simplification:

$$R_5 = \left[1 + A_g.(F_g + 2.q_{ia}) + A_g\Delta q)\right] - \left[1 + A_g.(F_g + 2.q_{ia}) - A_g\Delta q)\right]$$

$$R_5 = \left[1 + A_g.\left[F_g + 2.q_{ia}\right]\right].\left\{\left[1 + \frac{A_g \Delta q}{1 + A_g.\left[F_g + 2.q_{ia}\right]}\right] - \left[1 - \frac{A_g \Delta q}{1 + A_g.\left[F_g + 2.q_{ia}\right]}\right]\right\}$$

Since, $(1 + x) - (1 - x) = 2x$

$$R_5 = \left[1 + A_g.\left[F_g + 2.q_{ia}\right]\right].\left\{\frac{2.A_g.\Delta q}{1 + A_g.\left[F_g + 2.q_{ia}\right]}\right\} = 2.A_g.\Delta q$$

Also, $(1 - B_g) = [1 + A_g.(F_g + 2.q_{ia}) - A_g.H_q]$

$$\boxed{\frac{R_5}{2}.(1 - B_g) = A_g.\Delta q.\left[1 + A_g.(F_g + 2.q_{ia}) - A_g.H_q\right]} \tag{E.2}$$

To obtain the final simplified expression of 14, we first only consider terms (A.2), (C.2) and (E.2) of $\Delta q_G$.

$$\widetilde{\Delta q_G} = \frac{1}{A_g^3.\Delta q.H_q}\left\{\frac{R_1}{6} + B_g.\frac{R_3}{4} - \frac{R_5}{2}.(1 - B_g)\right\} \tag{15}$$

$$\begin{aligned}\widetilde{\Delta q_G} = \frac{1}{A_g^3.\Delta q.H_q}&\left\{\left(A_g \Delta q\left[1 + A_g.\left[F_g + 2.q_{ia}\right]\right]^2 + \frac{(A_g \Delta q)^3}{3}\right)\right.\\ &+ \left(A_g^2.\Delta q.\left[H_q - (F_g + 2.q_{ia})\right]\left[1 + A_g.\left[F_g + 2.q_{ia}\right]\right]\right)\\ &\left. - \left(A_g.\Delta q.\left[1 + A_g.(F_g + 2.q_{ia}) - A_g.H_q\right]\right)\right\}\end{aligned}$$

$$\begin{aligned}\widetilde{\Delta q_G} = &\left(\frac{1}{A_g^2.H_q}\left[1 + A_g.\left[F_g + 2.q_{ia}\right]\right]^2 + \frac{1}{3}.\frac{\Delta q^2}{H_q}\right)\\ &+ \left(\frac{1}{A_g^2.H_q}.\left[A_g.H_q - A_g.(F_g + 2.q_{ia})\right]\left[1 + A_g.\left[F_g + 2.q_{ia}\right]\right]\right)\\ &- \left(\frac{1}{A_g^2.H_q}.\left[1 + A_g.(F_g + 2.q_{ia}) - A_g.H_q\right]\right)\end{aligned}$$

$$\widetilde{\Delta q_G} = \frac{1}{A_g^2.H_q}\Bigg\{\Big[1 + A_g.[F_g + 2.q_{ia}]\Big]^2 + \big[A_g.H_q - A_g.(F_g + 2.q_{ia})\big]\Big[1 + A_g.[F_g + 2.q_{ia}]\Big] - \Big[1 + A_g.(F_g + 2.q_{ia}) - A_g.H_q\Big]\Bigg\} + \frac{1}{3}.\frac{\Delta q^2}{H_q}$$

$$\widetilde{\Delta q_G} = \frac{1 + A_g.[F_g + 2.q_{ia}]}{A_g^2.H_q}\Bigg\{\Big[1 + A_g.[F_g + 2.q_{ia}]\Big] + \big[A_g.H_q - A_g.(F_g + 2.q_{ia})\big] - \Big[1 - \frac{A_g.H_q}{1 + A_g.(F_g + 2.q_{ia})}\Big]\Bigg\} + \frac{1}{3}.\frac{\Delta q^2}{H_q}$$

$$\widetilde{\Delta q_G} = \frac{1 + A_g.[F_g + 2.q_{ia}]}{A_g^2.H_q}\Bigg\{\Big[1 + A_g.[F_g + 2.q_{ia}]\Big] - \Big[1 + A_g.[F_g + 2.q_{ia}]\Big] + \Big[A_g.H_q + \frac{A_g.H_q}{1 + A_g.(F_g + 2.q_{ia})}\Big]\Bigg\} + \frac{1}{3}.\frac{\Delta q^2}{H_q}$$

$$\widetilde{\Delta q_G} = \frac{1 + A_g.[F_g + 2.q_{ia}]}{A_g^2.H_q}\Bigg\{A_g.H_q + \frac{A_g.H_q}{1 + A_g.(F_g + 2.q_{ia})}\Bigg\} + \frac{1}{3}.\frac{\Delta q^2}{H_q}$$

$$\widetilde{\Delta q_G} = \frac{1 + A_g.[F_g + 2.q_{ia}]}{A_g}\Bigg\{1 + \frac{1}{1 + A_g.(F_g + 2.q_{ia})}\Bigg\} + \frac{1}{3}.\frac{\Delta q^2}{H_q}$$

$$\boxed{\widetilde{\Delta q_G} = \frac{2 + A_g.(F_g + 2.q_{ia})}{A_g} + \frac{1}{3}.\frac{\Delta q^2}{H_q}} \tag{15.1}$$

Now, (14) can be written as:

$$\Delta q_G = \widetilde{\Delta q_G} + \frac{1}{A_g^3.\Delta q.H_q}\Big[2.\frac{R_4}{3}.(1 - B_g) - \frac{2.R_2}{5}\Big] \tag{16}$$

Let:

$$\frac{1}{A_g^3.\Delta q.H_q}\Big[2.\frac{R_4}{3}.(1 - B_g)\Big] = D \qquad \& \qquad \frac{1}{A_g^3.\Delta q.H_q}\Big[\frac{2.R_2}{5}\Big] = B$$

so that,

$$\Delta q_G = \widetilde{\Delta q_G} + D - B \tag{17}$$

We now simplify $D$ and $B$, and then obtain a simplified expression for $(D-B)$:

$$D=\frac{1}{A_g^3.\Delta q.H_q}\left[2.\frac{R_4}{3}.(1-B_g)\right] \tag{17.1}$$

$$\begin{aligned}\frac{2.R_4}{3}.(1-B_g)=&\frac{2}{3}.[1+A_g.(F_g+2.q_{ia})-A_g.H_q]\left[1+A_g.\left[F_g+2.q_{ia}\right]\right]^{3/2}\\&.\left\{\sqrt{1+\frac{A_g\Delta q}{1+A_g.\left[F_g+2.q_{ia}\right]}}-\sqrt{1-\frac{A_g\Delta q}{1+A_g.\left[F_g+2.q_{ia}\right]}}\right.\\&+\left(\frac{A_g\Delta q}{1+A_g.\left[F_g+2.q_{ia}\right]}\right).\left(\sqrt{1+\frac{A_g\Delta q}{1+A_g.\left[F_g+2.q_{ia}\right]}}\right.\\&\left.\left.+\sqrt{1-\frac{A_g\Delta q}{1+A_g.\left[F_g+2.q_{ia}\right]}}\right)\right\}\end{aligned}$$

$$\begin{aligned}D=&\frac{2}{3.A_g^3.\Delta q.H_q}.\left[1+A_g.(F_g+2.q_{ia})-A_g.H_q\right]\left[1+A_g.\left[F_g+2.q_{ia}\right]\right]^{3/2}\\&.\left\{\sqrt{1+\frac{A_g\Delta q}{1+A_g.\left[F_g+2.q_{ia}\right]}}-\sqrt{1-\frac{A_g\Delta q}{1+A_g.\left[F_g+2.q_{ia}\right]}}\right.\\&+\left(\frac{A_g\Delta q}{1+A_g.\left[F_g+2.q_{ia}\right]}\right).\left(\sqrt{1+\frac{A_g\Delta q}{1+A_g.\left[F_g+2.q_{ia}\right]}}\right.\\&\left.\left.+\sqrt{1-\frac{A_g\Delta q}{1+A_g.\left[F_g+2.q_{ia}\right]}}\right)\right\}\end{aligned}$$

$$\begin{aligned}D=&\frac{2}{3.A_g^3.\Delta q.H_q}.\left[1+A_g.(F_g+2.q_{ia})-A_g.H_q\right]\\&\times\frac{\sqrt{1+A_g.\left[F_g+2.q_{ia}\right]}}{\sqrt{1+\frac{A_g\Delta q}{1+A_g.\left[F_g+2.q_{ia}\right]}}+\sqrt{1-\frac{A_g\Delta q}{1+A_g.\left[F_g+2.q_{ia}\right]}}}\\&.\left\{(2.A_g.\Delta q)+A_g.\Delta q\left(2+2.\sqrt{1-\frac{(A_g\Delta q)^2}{\left(1+A_g.\left[F_g+2.q_{ia}\right]\right)^2}}\right)\right\}\end{aligned}$$

$$D = \frac{4}{3.A_g^2.H_q}.\left[1 + A_g.(F_g + 2.q_{ia}) - A_g.H_q\right]\frac{\sqrt{1 + A_g.\left[F_g + 2.q_{ia}\right]}}{\sqrt{1 + \frac{A_g\Delta q}{1+A_g.\left[F_g+2.q_{ia}\right]}} + \sqrt{1 - \frac{A_g\Delta q}{1+A_g.\left[F_g+2.q_{ia}\right]}}} .\left\{2 + \sqrt{1 - \frac{(A_g\Delta q)^2}{\left(1 + A_g.\left[F_g + 2.q_{ia}\right]\right)^2}}\right\} \tag{17.2}$$

$$B = \frac{1}{A_g^3.\Delta q.H_q}\left[\frac{2.R_2}{5}\right] \tag{17.3}$$

$$\begin{aligned}\frac{2.R_2}{5} = \frac{2}{5}.\left[1 + A_g.\left[F_g + 2.q_{ia}\right]\right]^{5/2} &.\left\{\left[1 + \frac{(A_g\Delta q)^2}{(1 + A_g.\left[F_g + 2.q_{ia}\right])^2}\right].\left(\sqrt{1 + \frac{A_g\Delta q}{1 + A_g.\left[F_g + 2.q_{ia}\right]}} - \sqrt{1 - \frac{A_g\Delta q}{1 + A_g.\left[F_g + 2.q_{ia}\right]}}\right)\right.\\ &\left. + 2.\left(\frac{A_g\Delta q}{1 + A_g.\left[F_g + 2.q_{ia}\right]}\right).\left(\sqrt{1 + \frac{A_g\Delta q}{1 + A_g.\left[F_g + 2.q_{ia}\right]}} + \sqrt{1 - \frac{A_g\Delta q}{1 + A_g.\left[F_g + 2.q_{ia}\right]}}\right)\right\}\end{aligned}$$

$$\begin{aligned}B = \frac{2}{5}.\frac{1}{A_g^3.\Delta q.H_q}.\left[1 + A_g.\left[F_g + 2.q_{ia}\right]\right]^{5/2} &.\left\{\left[1 + \frac{(A_g\Delta q)^2}{(1 + A_g.\left[F_g + 2.q_{ia}\right])^2}\right].\left(\sqrt{1 + \frac{A_g\Delta q}{1 + A_g.\left[F_g + 2.q_{ia}\right]}} - \sqrt{1 - \frac{A_g\Delta q}{1 + A_g.\left[F_g + 2.q_{ia}\right]}}\right)\right.\\ &\left. + 2.\left(\frac{A_g\Delta q}{1 + A_g.\left[F_g + 2.q_{ia}\right]}\right).\left(\sqrt{1 + \frac{A_g\Delta q}{1 + A_g.\left[F_g + 2.q_{ia}\right]}} + \sqrt{1 - \frac{A_g\Delta q}{1 + A_g.\left[F_g + 2.q_{ia}\right]}}\right)\right\}\end{aligned}$$

$$\begin{aligned}B = \frac{2}{5.A_g^3.\Delta q.H_q}.\frac{\left(1 + A_g.\left[F_g + 2.q_{ia}\right]\right)^{3/2}}{\sqrt{1 + \frac{A_g\Delta q}{1+A_g.\left[F_g+2.q_{ia}\right]}} + \sqrt{1 - \frac{A_g\Delta q}{1+A_g.\left[F_g+2.q_{ia}\right]}}}\\ .\left\{\left[1 + \frac{(A_g\Delta q)^2}{(1 + A_g.\left[F_g + 2.q_{ia}\right])^2}\right].(2.A_g.\Delta q) + 2.A_g\Delta q.\left(2 + 2.\sqrt{1 - \frac{(A_g\Delta q)^2}{\left(1 + A_g.\left[F_g + 2.q_{ia}\right]\right)^2}}\right)\right\}\end{aligned}$$

$$\begin{aligned}B = \frac{4}{5.A_g^2.H_q}.\frac{\left(1 + A_g.\left[F_g + 2.q_{ia}\right]\right)^{3/2}}{\sqrt{1 + \frac{A_g\Delta q}{1+A_g.\left[F_g+2.q_{ia}\right]}} + \sqrt{1 - \frac{A_g\Delta q}{1+A_g.\left[F_g+2.q_{ia}\right]}}}\\ .\left\{1 + \frac{(A_g\Delta q)^2}{\left(1 + A_g.\left[F_g + 2.q_{ia}\right]\right)^2} + 2 + 2.\sqrt{1 - \frac{(A_g\Delta q)^2}{\left(1 + A_g.\left[F_g + 2.q_{ia}\right]\right)^2}}\right\}\end{aligned}$$

$$B = \frac{4}{5.A_g^2.H_q} \cdot \frac{\left(1 + A_g.\left[F_g + 2.q_{ia}\right]\right)^{3/2}}{\sqrt{1 + \frac{A_g\Delta q}{1+A_g.\left[F_g+2.q_{ia}\right]}} + \sqrt{1 - \frac{A_g\Delta q}{1+A_g.\left[F_g+2.q_{ia}\right]}}} \cdot \left\{3 + \frac{(A_g\Delta q)^2}{\left(1 + A_g.\left[F_g + 2.q_{ia}\right]\right)^2} + 2.\sqrt{1 - \frac{(A_g\Delta q)^2}{\left(1 + A_g.\left[F_g + 2.q_{ia}\right]\right)^2}}\right\} \tag{17.4}$$

Let, $\alpha = \frac{1}{A_g^2.H_q} \cdot \frac{\sqrt{1+A_g.\left[F_g+2.q_{ia}\right]}}{\sqrt{1+\frac{A_g\Delta q}{1-A_g.\left[F_g+2.q_{ia}\right]}} + \sqrt{1-\frac{A_g\Delta q}{1-A_g.\left[F_g+2.q_{ia}\right]}}}$

Now,

$$D - B = \alpha.\left\{\frac{4}{3}.\left[1 + A_g.(F_g + 2.q_{ia}) - A_g.H_q\right].\left(2 + \sqrt{1 - \frac{(A_g\Delta q)^2}{\left(1 + A_g.\left[F_g + 2.q_{ia}\right]\right)^2}}\right) - \frac{4}{5}.\left(1 + A_g.\left[F_g + 2.q_{ia}\right]\right)\left[3 + \frac{(A_g\Delta q)^2}{\left(1 + A_g.\left[F_g + 2.q_{ia}\right]\right)^2} + 2.\sqrt{1 - \frac{(A_g\Delta q)^2}{\left(1 - A_g.\left[F_g + 2.q_{ia}\right]\right)^2}}\right]\right\}$$

$$\begin{aligned} D - B = \alpha.\Bigg\{&\frac{8}{3}.\left[1 + A_g.(F_g + 2.q_{ia})\right] \\ &+ \frac{4}{3}.\left[1 + A_g.(F_g + 2.q_{ia})\right]\left(\sqrt{1 - \frac{(A_g\Delta q)^2}{\left(1 + A_g.\left[F_g + 2.q_{ia}\right]\right)^2}}\right) \\ &- \frac{8}{3}.(A_g.H_q) - \frac{4}{3}.(A_g.H_q).\left(\sqrt{1 - \frac{(A_g\Delta q)^2}{\left(1 + A_g.\left[F_g + 2.q_{ia}\right]\right)^2}}\right) \\ &- \frac{12}{5}.\left(1 + A_g.\left[F_g + 2.q_{ia}\right]\right) - \frac{4}{5}.\frac{(A_g\Delta q)^2}{\left(1 + A_g.\left[F_g + 2.q_{ia}\right]\right)} \\ &- \frac{8}{5}.\left(1 + A_g.\left[F_g + 2.q_{ia}\right]\right)\sqrt{1 - \frac{(A_g\Delta q)^2}{\left(1 - A_g.\left[F_g + 2.q_{ia}\right]\right)^2}}\,\Bigg]\Bigg\} \end{aligned}$$

$$\begin{aligned} D - B = \alpha.\Bigg\{&\frac{4}{15}.\left[1 + A_g.(F_g + 2.q_{ia})\right] \\ &- \frac{4}{15}.\left[1 + A_g.(F_g + 2.q_{ia})\right]\left(\sqrt{1 - \frac{(A_g\Delta q)^2}{\left(1 + A_g.\left[F_g + 2.q_{ia}\right]\right)^2}}\right) \end{aligned}$$

$$- \frac{8}{3}.(A_g.H_q) - \frac{4}{3}.(A_g.H_q).\left(\sqrt{1 - \frac{(A_g\Delta q)^2}{\left(1 + A_g.\left[F_g + 2.q_{ia}\right]\right)^2}}\right)$$
$$- \frac{4}{5}.\frac{(A_g\Delta q)^2}{\left(1 + A_g.\left[F_g + 2.q_{ia}\right]\right)}\Bigg\}$$

$$\boxed{\begin{aligned} D - B = \alpha.\Bigg\{&\frac{4}{15}.\left[1 + A_g.(F_g + 2.q_{ia})\right]\left(1 - \sqrt{1 - \frac{(A_g\Delta q)^2}{\left(1 + A_g.\left[F_g + 2.q_{ia}\right]\right)^2}}\right) \\ &- \frac{4}{3}.(A_g.H_q)\left(2 + \sqrt{1 - \frac{(A_g\Delta q)^2}{\left(1 + A_g.\left[F_g + 2.q_{ia}\right]\right)^2}}\right) \\ &- \frac{4}{5}.\frac{(A_g\Delta q)^2}{\left(1 + A_g.\left[F_g + 2.q_{ia}\right]\right)}\Bigg\} \end{aligned}} \tag{17.5}$$

Now, with this simplification (17) reads as follows:

$$\boxed{\Delta q_G = \frac{2 + A_g.(F_g + 2.q_{ia})}{A_g} + \frac{1}{3}.\frac{\Delta q^2}{H_q} + (D - B)} \tag{18}$$

where, $(D - B)$ is given by (17.5) and,

$$\alpha = \frac{1}{A_g^2.H_q}.\frac{\sqrt{1 + A_g.\left[F_g + 2.q_{ia}\right]}}{\sqrt{1 + \frac{A_g\Delta q}{1+A_g.\left[F_g+2.q_{ia}\right]}} + \sqrt{1 - \frac{A_g\Delta q}{1+A_g.\left[F_g+2.q_{ia}\right]}}}$$

$$B_g = A_g.\left[H_q - (F_g + 2.q_{ia})\right]; \quad F_g = (v_g - v_{fb} - \psi_p);$$

$$A_g = \left(\frac{2}{\gamma_g'}\right)^2; \quad (\gamma_g')^2 = (\gamma_g)^2 / V_t$$

$$q_{ia} = \frac{q_s + q_d}{2}; \quad \Delta q = q_s - q_d; \quad H_q = (1 + 2q_{ia}/D_{vsat}) = 1 + \frac{q_s + q_d}{D_{vsat}}$$

Now, (5) reads as follows:

$$q_B = q_{B1} - \Delta q_G$$

where $q_{B1} = (v_g - v_{fb} - \psi_p) - 2.(n_q - 1).\int_0^1 q\, d\xi$

$$q_{B1} = F_g - \left\{(n_q - 1)\left(q_s + q_d + \frac{1}{3}.\frac{(q_s - q_d)^2}{1 + q_s + q_d}\right)\right\} \tag{19}$$

$$q_{B1} = F_g - 2.(n_q - 1)\left(q_{ia} + \frac{1}{6}.\frac{(\Delta q)^2}{1 + q_s + q_d}\right) \tag{20}$$

Using this expression for $q_{B1}$ in (18):

$$\boxed{\begin{aligned} q_B = F_g - 2.(n_q - 1)\left(q_{ia} + \frac{1}{6}.\frac{(\Delta q)^2}{1 + q_s + q_d}\right) \\ - \frac{[2 + A_g.(F_g + 2.q_{ia})]}{A_g} - \frac{1}{3}.\frac{\Delta q^2}{H_q} - (D - B) \end{aligned}} \tag{21}$$

Upon rearranging:

$$\begin{aligned} q_B = F_g - 2.n_q.\left(q_{ia} + \frac{1}{6}.\frac{(\Delta q)^2}{H_q}\right) + 2.\left(q_{ia} + \frac{1}{6}.\frac{(\Delta q)^2}{H_q}\right) \\ - \frac{2}{A_g} - F_g - 2.q_{ia} - \frac{1}{3}.\frac{\Delta q^2}{H_q} - (D - B) \end{aligned}$$

$$q_B = -2.n_q.\left(q_{ia} + \frac{1}{6}.\frac{(\Delta q)^2}{H_q}\right) - \frac{2}{A_g} - (D - B) \tag{22}$$

$$\begin{aligned} q_B = -2.n_q.\left(q_{ia} + \frac{1}{6}.\frac{(\Delta q)^2}{H_q}\right) - \frac{2}{A_g} \\ - \alpha.\left\{\frac{4}{15}.\left[1 + A_g.(F_g + 2.q_{ia})\right]\left(1 - \sqrt{1 - \frac{(A_g \Delta q)^2}{(1 + A_g.[F_g + 2.q_{ia}])^2}}\right)\right. \\ \left. - \frac{4}{3}.(A_g.H_q)\left(2 + \sqrt{1 - \frac{(A_g \Delta q)^2}{(1 + A_g.[F_g + 2.q_{ia}])^2}}\right) - \frac{4}{5}.\frac{(A_g \Delta q)^2}{(1 + A_g.[F_g + 2.q_{ia}])}\right\} \end{aligned}$$

$$q_B = -2.n_q.\left(q_{ia} + \frac{1}{6}.\frac{(\Delta q)^2}{H_q}\right) - \frac{2}{A_g}$$
$$-\alpha.\left\{\frac{4}{15}.\left[1 + A_g.(F_g + 2.q_{ia})\right]\left(1 - \sqrt{1 - \frac{(A_g\Delta q)^2}{(1 + A_g.[F_g + 2.q_{ia}])^2}}\right)\right.$$
$$\left.-\frac{4}{3}.(A_g.H_q)\left(2 + \sqrt{1 - \frac{(A_g\Delta q)^2}{(1 + A_g.[F_g + 2.q_{ia}])^2}}\right) - \frac{4}{5}.\frac{(A_g\Delta q)^2}{(1 + A_g.[F_g + 2.q_{ia}])}\right\}$$

$$q_B = -n_q.\left(q_s + q_d + \frac{1}{3}.\frac{(q_s - q_d)^2}{1 + q_s + q_d}\right) - \frac{2}{A_g}$$
$$-\alpha.\left\{\frac{4}{15}.\left[1 + A_g.(F_g + 2.q_{ia})\right]\left(1 - \sqrt{1 - \frac{(A_g\Delta q)^2}{(1 + A_g.[F_g + 2.q_{ia}])^2}}\right)\right.$$
$$\left.-\frac{4}{3}.(A_g.H_q)\left(2 + \sqrt{1 - \frac{(A_g\Delta q)^2}{(1 + A_g.[F_g + 2.q_{ia}])^2}}\right) - \frac{4}{5}.\frac{(A_g\Delta q)^2}{(1 + A_g.[F_g + 2.q_{ia}])}\right\} \tag{23}$$

where,

$$\alpha = \frac{1}{A_g^2.H_q}.\frac{\sqrt{1 + A_g.[F_g + 2.q_{ia}]}}{\sqrt{1 + \frac{A_g\Delta q}{1 - A_g.[F_g + 2.q_{ia}]}} + \sqrt{1 - \frac{A_g\Delta q}{1 - A_g.[F_g + 2.q_{ia}]}}}$$

---

Now we proceed by defining following:

$$\delta_s = \frac{A_g}{4}.(F_g + 2.q_s) = \frac{v_g - v_{fb} - \psi_p + 2.q_s}{(\gamma_g')^2} = \frac{v_g - v_{fb} - \psi_p + q_s - q_d + q_s + q_d}{(\gamma_g')^2}$$

$$\delta_s = \frac{v_g - v_{fb} - \psi_p + \Delta q + 2.q_{ia}}{(\gamma_g')^2} \quad \Longrightarrow \quad \boxed{\delta_s = \frac{A_g}{4}.(F_g + 2.q_{ia} + \Delta q)}$$

$$\delta_d = \frac{v_g - v_{fb} - \psi_p + 2.q_{ia} - \Delta q}{(\gamma_g')^2} \quad \Longrightarrow \quad \boxed{\delta_d = \frac{A_g}{4}.(F_g + 2.q_{ia} - \Delta q)}$$

$$\boxed{x = \frac{1}{4} + \delta_s} \quad \text{and} \quad \boxed{y = \frac{1}{4} + \delta_d}$$

---

In pursuit of expressing (23) in terms of x and y, we evaluate following terms:

$$\left(1 + A_g.\left[F_g + 2.q_{ia}\right]\right) = 1 + 2.(\delta_s + \delta_d) = 2.\left[\frac{1}{2} + (\delta_s + \delta_d)\right]$$
$$= 2.\left[\frac{1}{4} + \delta_s + \frac{1}{4} + \delta_d\right] = 2.(x + y)$$

$$1 + \frac{A_g \Delta q}{1 + A_g.\left[F_g + 2.q_{ia}\right]} = \frac{1 + A_g.\left[F_g + 2.q_{ia} + \Delta q\right]}{1 + A_g.\left[F_g + 2.q_{ia}\right]} = \frac{1 + 4.\delta_s}{1 + 2.(\delta_s + \delta_d)}$$
$$= \frac{\frac{1}{4} + \delta_s}{\frac{1}{4} + \frac{1}{2}.(\delta_s + \delta_d)} = 2.\frac{x}{(x + y)}$$

$$1 - \frac{A_g \Delta q}{1 + A_g.\left[F_g + 2.q_{ia}\right]} = \frac{1 + A_g.\left[F_g + 2.q_{ia} - \Delta q\right]}{1 + A_g.\left[F_g + 2.q_{ia}\right]} = \frac{1 + 4.\delta_d}{1 + 2.(\delta_s + \delta_d)}$$
$$= \frac{\frac{1}{4} + \delta_d}{\frac{1}{4} + \frac{1}{2}.(\delta_s + \delta_d)} = 2.\frac{y}{(x + y)}$$

$$\Delta q = \frac{2}{A_g}.(\delta_s - \delta_d) = \frac{2}{A_g}.(x - y) \quad or \quad A_g = \frac{2.(x - y)}{\Delta q}$$

Now, (23) reads as follows:

$$\begin{aligned} q_B = & -n_q.\left(q_s + q_d + \frac{1}{3}.\frac{(q_s - q_d)^2}{1 + q_s + q_d}\right) - \frac{\Delta q}{(x - y)} \\ & - \alpha.\left\{\frac{8}{15}.[x + y]\left(1 - \sqrt{\frac{2.x}{x + y}.\frac{2.y}{x + y}}\right)\right. \\ & \left. - \frac{4}{3}.\left(\frac{2.(x - y)}{\Delta q}.H_q\right)\left(2 + \sqrt{\frac{2.x}{x + y}.\frac{2.y}{x + y}}\right) - \frac{4}{5}.\frac{[2.(x - y)]^2}{2.(x + y)}\right\} \end{aligned} \tag{24}$$

where, $\alpha = \frac{1}{\left[\frac{2.(x-y)}{\Delta q}\right]^2.H_q}.\frac{\sqrt{2.(x+y)}}{\sqrt{2.\frac{x}{(x+y)}} + \sqrt{2.\frac{y}{(x+y)}}} = \frac{1}{\left[\frac{2.(x-y)}{\Delta q}\right]^2.H_q}.\frac{\sqrt{(x+y)}}{\sqrt{\frac{x}{(x+y)}} + \sqrt{\frac{y}{(x+y)}}}$

Let, $\widetilde{q_B} = -n_q.\left(q_s + q_d + \frac{1}{3}.\frac{(q_s - q_d)^2}{1 + q_s + q_d}\right) - \frac{\Delta q}{(x - y)}$

Then we can write:

$$q_B = \widetilde{q_B} - \frac{1}{4.[\frac{(x-y)}{\Delta q}]^2.H_q} \cdot \frac{\sqrt{(x+y)}}{\sqrt{\frac{x}{(x+y)}} + \sqrt{\frac{y}{(x+y)}}} \cdot \left\{ \frac{8}{15}.[x+y]\left(1 - 2.\sqrt{\frac{x.y}{(x+y)^2}}\right) - \frac{8}{3}.\left(\frac{2.(x-y)}{\Delta q}.H_q\right)\left(1 + \sqrt{\frac{x.y}{(x+y)^2}}\right) - \frac{8}{5}.\frac{[(x-y)]^2}{(x+y)} \right\} \tag{25}$$

Simplification:

$$q_B = \widetilde{q_B} - \frac{(\Delta q)^2(x+y)}{4.(x-y)^2.H_q(\sqrt{x}+\sqrt{y})} \cdot \left\{ \frac{8}{15}.[x+y]\left(1 - 2.\frac{\sqrt{x.y}}{x+y}\right) - \frac{16}{3}.\left(\frac{(x-y)}{\Delta q}.H_q\right)\left(1 + \frac{\sqrt{x.y}}{x+y}\right) - \frac{8}{5}.\frac{[(x-y)]^2}{(x+y)} \right\}$$

$$q_B = \widetilde{q_B} - \frac{(\Delta q)^2(x+y)}{(x-y)^2.H_q(\sqrt{x}+\sqrt{y})} \cdot \left\{ \frac{2}{15}.\left(x+y-2.\sqrt{x.y}\right) - \left(\frac{4.(x-y).H_q}{3.\Delta q}\right)\left(\frac{x+y+\sqrt{x.y}}{x+y}\right) - \frac{2}{5}.\frac{[(x-y)]^2}{(x+y)} \right\}$$

$$q_B = \widetilde{q_B} - \frac{(\Delta q)^2(x+y)}{(x-y)^2.H_q(\sqrt{x}+\sqrt{y})} \cdot \left\{ \frac{2}{15}.\left(x+y-2.\sqrt{x.y}\right) - \frac{2}{5}.\frac{[(x-y)]^2}{(x+y)} \right\} + \left(\frac{4.\Delta q}{3.(x-y)}\right)\left(\frac{x+y+\sqrt{x.y}}{(\sqrt{x}+\sqrt{y})}\right)$$

$$q_B = \widetilde{q_B} - \frac{(\Delta q)^2}{(x-y)^2.H_q(\sqrt{x}+\sqrt{y})} \cdot \frac{2}{15} \cdot \left\{ \left(x^2+y^2+2.x.y-2.(x+y).\sqrt{x.y}\right) - 3.[x^2+y^2-2.x.y] \right\} + \frac{4.\Delta q}{3.(x-y)} \cdot \left(\frac{x+y+\sqrt{x.y}}{(\sqrt{x}+\sqrt{y})}\right)$$

$$q_B = \widetilde{q_B} + \frac{(\Delta q)^2}{(x-y)^2.H_q(\sqrt{x}+\sqrt{y})} \cdot \frac{4}{15} \cdot \left\{ x^2+y^2-4.x.y+(x+y).\sqrt{x.y} \right\} + \frac{4.\Delta q}{3.(x-y)} \cdot \left(\frac{x+y+\sqrt{x.y}}{(\sqrt{x}+\sqrt{y})}\right)$$

$$q_B = \widetilde{q_B} + \frac{(\Delta q)^2}{(x-y)^2.H_q(\sqrt{x}+\sqrt{y})} \cdot \frac{4}{15} \cdot \left\{ (x-y)^2-2.x.y+(x+y).\sqrt{x.y} \right\} + \frac{4.\Delta q}{3.(x-y)} \cdot \left(\frac{x+y+\sqrt{x.y}}{(\sqrt{x}+\sqrt{y})}\right)$$

$$q_B = -n_q.\left(q_s + q_d + \frac{1}{3}.\frac{(q_s - q_d)^2}{1 + q_s + q_d}\right) - \frac{\Delta q}{(x-y)}$$
$$+ \frac{(\Delta q)^2}{(x-y)^2.H_q(\sqrt{x}+\sqrt{y})}.\frac{4}{15}.\left\{(x-y)^2 - 2.x.y + (x+y).\sqrt{x.y}\right\}$$
$$+ \frac{4.\Delta q}{3.(x-y)}.\left(\frac{x + y + \sqrt{x.y}}{(\sqrt{x}+\sqrt{y})}\right)$$

$$q_B = -n_q.\left(q_s + q_d + \frac{1}{3}.\frac{(q_s - q_d)^2}{1 + q_s + q_d}\right) - \frac{2}{A_g}$$
$$+ \frac{(\Delta q)^2}{(x-y)^2.H_q(\sqrt{x}+\sqrt{y})}.\frac{4}{15}.\left\{x^2 + y^2 - 4.x.y + (x+y)\sqrt{x.y}\right\}$$
$$+ \frac{4.\Delta q}{3.(x-y)}.\left(\frac{x + y + \sqrt{x.y}}{(\sqrt{x}+\sqrt{y})}\right) \tag{26}$$

Let $\hat{q_B} = -n_q.\left(q_s + q_d + \frac{1}{3}.\frac{(q_s-q_d)^2}{1+q_s+q_d}\right)$ so that (26) reads:

$$q_B = -\frac{2}{A_g} + \frac{4.\Delta q}{3.(x-y)}.\left(\frac{x + y + \sqrt{x.y}}{(\sqrt{x}+\sqrt{y})}\right)$$
$$+ \frac{(\Delta q)^2}{(x-y)^2.H_q(\sqrt{x}+\sqrt{y})}.\frac{4}{15}.\left\{x^2 + y^2 - 4.x.y + (x+y)\sqrt{x.y}\right\} + \hat{q_B}$$

$$q_B = -\frac{2}{A_g} + \frac{2}{A_g}\frac{4}{3}.\left(\frac{x + y + \sqrt{x.y}}{(\sqrt{x}+\sqrt{y})}\right)$$
$$+ \frac{(\Delta q)^2}{(x-y)^2.H_q(\sqrt{x}+\sqrt{y})}.\frac{4}{15}.\left\{\left[x + y + 3\sqrt{x.y}\right].\left[x + y - 2\sqrt{x.y}\right]\right\} + \hat{q_B} \tag{27}$$

Now, let $\Theta = -\frac{2}{A_g} + \frac{2}{A_g}\frac{4}{3}.\left(\frac{x+y+\sqrt{x.y}}{(\sqrt{x}+\sqrt{y})}\right)$

Rearranging:

$$\Theta = -\frac{2}{A_g} + \frac{2}{A_g}\frac{1}{3}.\left(\frac{4x + 4y + 4\sqrt{x.y}}{(\sqrt{x}+\sqrt{y})}\right)$$
$$= -\frac{2}{A_g} + \frac{2}{A_g}\frac{1}{3}.\left(\frac{3x + 3y + 6\sqrt{x.y} + x + y - 2\sqrt{x.y}}{(\sqrt{x}+\sqrt{y})}\right)$$

$$\Theta = -\frac{2}{A_g} + \frac{2}{A_g}\frac{1}{3}.\left(\frac{3.(\sqrt{x}+\sqrt{y})^2 + (\sqrt{x}-\sqrt{y})^2}{(\sqrt{x}+\sqrt{y})}\right)$$
$$= -\frac{2}{A_g} + \frac{2}{A_g}.\left((\sqrt{x}+\sqrt{y}) + \frac{(\sqrt{x}-\sqrt{y})^2}{3.(\sqrt{x}+\sqrt{y})}\right)$$

$$\Theta = \frac{2}{A_g}.\left\{\sqrt{x}+\sqrt{y}-1\right\} + \frac{2}{A_g}.\left(\frac{(\sqrt{x}-\sqrt{y})^2}{3.(\sqrt{x}+\sqrt{y})}\right)$$
$$= \frac{2}{A_g}.\left\{\sqrt{x}+\sqrt{y}-1\right\} + \frac{2}{A_g}.\left(\frac{(\sqrt{x}+\sqrt{y})^2.(\sqrt{x}-\sqrt{y})^2}{3.(\sqrt{x}+\sqrt{y})^3}\right)$$

$$\Theta = \frac{2}{A_g}.\left\{\sqrt{x}+\sqrt{y}-1\right\} + \frac{2}{A_g}.\left(\frac{(x-y)^2}{3.(\sqrt{x}+\sqrt{y})^3}\right)$$
$$= \frac{2}{A_g}.\left\{\sqrt{x}+\sqrt{y}-1\right\} + \frac{1}{3.(\sqrt{x}+\sqrt{y})^3}.(\frac{2}{A_g})^2.(x-y)^2.\frac{A_g}{2}$$

$$\boxed{\Theta = \frac{2}{A_g}.\left\{\sqrt{x}+\sqrt{y}-1\right\} + \frac{1}{3.(\sqrt{x}+\sqrt{y})^3}.(\frac{2}{A_g})^2.(x-y)^2.\frac{A_g}{2}} \quad (28)$$

Now,

$$\frac{2}{A_g}.\left((\sqrt{x}+\sqrt{y})-1\right) = \frac{1}{A_g}.\left(2.\sqrt{x}-1+2.\sqrt{y}-1\right)$$
$$= \frac{1}{A_g}.\left(\frac{(1+2.\sqrt{x}).(2.\sqrt{x}-1).(1+2.\sqrt{y}) + (1+2.\sqrt{x}).(2.\sqrt{y}-1).(1+2.\sqrt{y})}{(1+2.\sqrt{x}).(1+2.\sqrt{y})}\right)$$

$$\frac{2}{A_g}.\left((\sqrt{x}+\sqrt{y})-1\right) = \frac{1}{A_g}.\left(\frac{(4x-1).(1+2.\sqrt{y}) + (1+2.\sqrt{x}).(4y-1)}{(1+2.\sqrt{x}).(1+2.\sqrt{y})}\right)$$
$$= \frac{1}{A_g}.\left(\frac{4.\delta_s.(1+2.\sqrt{y}) + (1+2.\sqrt{x}).4.\delta_d}{(1+2.\sqrt{x}).(1+2.\sqrt{y})}\right)$$

$$\frac{2}{A_g}.\left((\sqrt{x}+\sqrt{y})-1\right) = \frac{1}{A_g}.\left(\frac{4.\delta_s}{(1+2.\sqrt{x})} + \frac{4.\delta_d}{(1+2.\sqrt{y})}\right)$$
$$= \left(\frac{F_g+2.q_s}{(1+2.\sqrt{x})} + \frac{F_g+2.q_d}{(1+2.\sqrt{y})}\right)$$

$$\boxed{\frac{2}{A_g}.\left((\sqrt{x}+\sqrt{y})-1\right) = \left(\frac{F_g+2.q_s}{(1+2.\sqrt{x})} + \frac{F_g+2.q_d}{(1+2.\sqrt{y})}\right) = A+B} \quad (29)$$

$$\boxed{\Theta \quad = A + B + \frac{1}{3.(\sqrt{x}+\sqrt{y})^3}.(\frac{2}{A_g})^2.(x-y)^2.\frac{A_g}{2}} \tag{30}$$

Using (30), Eq. (27) reads:

$$q_B = A + B + \frac{1}{3.(\sqrt{x}+\sqrt{y})^3}.(\frac{2}{A_g})^2.(x-y)^2.\frac{A_g}{2} + \frac{(\Delta q)^2}{(x-y)^2.H_q(\sqrt{x}+\sqrt{y})}.\frac{4}{15}.\left\{[x+y+3\sqrt{x.y}].[x+y-2\sqrt{x.y}]\right\} + \hat{q_B} \tag{31}$$

Rearranging:

$$q_B = A + B + \frac{1}{3.(\sqrt{x}+\sqrt{y})^3}.(\Delta q)^2.\frac{A_g}{2} + \frac{(\Delta q)^2}{(x-y)^2.H_q(\sqrt{x}+\sqrt{y})}.\frac{4}{15}.\left\{(x-y)^2 - 2.x.y + (x+y).\sqrt{x.y}\right\} + \hat{q_B}$$

$$q_B = A + B + \frac{1}{3.(\sqrt{x}+\sqrt{y})^3}.(\Delta q)^2.\frac{A_g}{2} + \frac{(\Delta q)^2}{H_q(\sqrt{x}+\sqrt{y})}.\frac{4}{15}.\left\{1 - \frac{2.x.y}{(x-y)^2} + \frac{(x+y).\sqrt{x.y}}{(x-y)^2}\right\} + \hat{q_B}$$

$$q_B = A + B + \frac{1}{3.(\sqrt{x}+\sqrt{y})^3}.(\Delta q)^2.\frac{A_g}{2} + \frac{(\Delta q)^2}{H_q(\sqrt{x}+\sqrt{y})^3}.\frac{4}{15}.\left\{(\sqrt{x}+\sqrt{y})^2 - \frac{2.x.y}{(\sqrt{x}-\sqrt{y})^2} + \frac{(x+y).\sqrt{x.y}}{(\sqrt{x}-\sqrt{y})^2}\right\} + \hat{q_B}$$

$$q_B = A + B + \frac{1}{3.(\sqrt{x}+\sqrt{y})^3}.(\Delta q)^2.\frac{A_g}{2} + \frac{(\Delta q)^2}{H_q(\sqrt{x}+\sqrt{y})^3}.\frac{4}{15}.\left\{\frac{(x-y)^2 - 2.x.y + (x+y)\sqrt{x.y}}{(\sqrt{x}-\sqrt{y})^2}\right\} + \hat{q_B}$$

$$q_B = A + B + \frac{1}{3.(\sqrt{x}+\sqrt{y})^3}.(\Delta q)^2.\frac{A_g}{2} + \frac{(\Delta q)^2}{H_q(\sqrt{x}+\sqrt{y})^3}.\frac{4}{15}.\left\{\frac{(x+y+3\sqrt{x.y}).(\sqrt{x}-\sqrt{y})^2}{(\sqrt{x}-\sqrt{y})^2}\right\} + \hat{q_B}$$

$$q_B = A + B + \frac{1}{3.(\sqrt{x}+\sqrt{y})^3}.(\Delta q)^2.\frac{A_g}{2} + \frac{(\Delta q)^2}{H_q(\sqrt{x}+\sqrt{y})^3}.\frac{4}{15}.\left\{(x+y+3\sqrt{x.y})\right\} + \hat{q_B}$$

$$q_B = A + B + \frac{(\Delta q)^2}{3.(\sqrt{x}+\sqrt{y})^3}.\left\{\frac{4}{5}.(x+y+3\sqrt{x.y}).\frac{1}{H_q} + \frac{A_g}{2}\right\} + \hat{q_B}$$

$$q_B = A + B + \frac{(\Delta q)^2}{3.(\sqrt{x}+\sqrt{y})^3}.\left\{\frac{4}{5}.(x+y+3\sqrt{x.y}).\frac{1}{H_q} + \frac{2}{\gamma_g')^2}\right\} + \hat{q_B}$$

$$q_B = A + B + \frac{(\Delta q)^2}{3.(\sqrt{x}+\sqrt{y})^3}.\left\{\frac{4}{5}.\left((\sqrt{x}+\sqrt{y})^2 + \sqrt{x.y}\right).\frac{1}{H_q} + \frac{2}{\gamma_g')^2}\right\} + \hat{q_B}$$

$$q_B = A + B + \frac{(\Delta q)^2}{3.(\sqrt{x}+\sqrt{y})^3}.\left\{\frac{4}{5}.\left((\sqrt{x}+\sqrt{y})^2 + \sqrt{x.y}\right).\frac{1}{H_q} + \frac{2}{(\gamma_g')^2}\right\} - n_q.\left(q_s + q_d + \frac{1}{3}.\frac{(q_s-q_d)^2}{1+q_s+q_d}\right) \quad (32)$$

$$\boxed{q_B = A + B + \frac{1}{3}.\frac{(\Delta q)^2}{C^3}.\left\{\frac{8}{10}\left(C^2 + P.Q\right).\frac{1}{1+q_s+q_d} + \frac{2}{(\gamma_g')^2}\right\} - n_q.\left(q_s + q_d + \frac{1}{3}.\frac{(q_s-q_d)^2}{1+q_s+q_d}\right)} \quad (33)$$

where

$$A = \frac{v_g - v_{fb} - \psi_p + 2.q_s}{1+2.\sqrt{x}} \qquad B = \frac{v_g - v_{fb} - \psi_p + 2.q_d}{1+2.\sqrt{y}}$$

$$x = \frac{1}{4} + \frac{v_g - v_{fb} - \psi_p + 2.q_s}{(\gamma_g')^2} \qquad y = \frac{1}{4} + \frac{v_g - v_{fb} - \psi_p + 2.q_d}{(\gamma_g')^2}$$

$$P = \sqrt{x} \qquad Q = \sqrt{y} \qquad C = P + Q$$

# References


[1] Soo-Young Oh, D.E. Ward, R.W. Dutton, Transient analysis of MOS transistors, IEEE Transactions on Electron Devices 27 (8) (1980) 1571–1578.

[2] W. Liu, X. Jin, Y. King, C. Hu, An efficient and accurate compact model for thin-oxide-MOSFET intrinsic capacitance considering the finite charge layer thickness, IEEE Transactions on Electron Devices 46 (5) (1999) 1070–1072.

[3] Y.C. King, H. Fujioka, S. Kamohara, K. Chen, C. Hu, Dc electrical oxide thickness model for quantization of the inversion layer in MOSFETs, Semiconductor Science and Technology 13 (1998) 963–966.

[4] R. Shrivastava, K. Fitzpatrick, A simple model for the overlap capacitance of a VLSI MOS device, IEEE Transactions on Electron Devices 29 (12) (1982) 1870–1875.

[5] W. Liu, C. Hu, BSIM4 and MOSFET Modeling for IC Simulation, World Scientific, 2011.

[6] W. Liu, MOSFET Models for SPICE Simulation Including BSIM3v3 and BSIM4, Wiley, 2001.

[7] Y. Tsividis, C. McAndrew, Operation and Modeling of the MOS Transistor, third ed., Oxford University Press, 2010.

[8] Y.S. Chauhan, et al., BSIM6: analog and RF compact model for bulk MOSFET, IEEE Transactions on Electron Devices 61 (2) (2014) 234–244, https://doi.org/10.1109/TED.2013.2283084.

CHAPTER

# 6

# Noise and RF Modeling

## 6.1 Introduction

Like any other semiconductor device, bulk MOSFET, too, produces noise. There are primarily the following types of device noises that are critical for circuit designers: thermal noise (a.k.a. white noise), flicker noise, shot noise, and gate-induced noise. The power spectral density of the flicker noise varies inversely with the frequency, while that of thermal noise is independent of the frequency. Gate-induced noise is partially correlated with channel thermal noise and is important at high frequencies.

Fluctuations observed in the transistor's current or voltages due to noise have zero average value. Therefore it is standard practice to rather determine the average of the square of current/voltage noise. Mathematically,

$$v_n^2 = S_v(f) * \Delta f \tag{6.1}$$

$$i_n^2 = S_i(f) * \Delta f \tag{6.2}$$

here, $\Delta f$ is the bandwidth over which noise is measured. $S_v(f)$ and $S_i(f)$ are the voltage and current noise power spectral density at 1 Hz bandwidth, respectively. Noise models in a compact model are developed by determining correct $S_v(f)$ (or $S_i(f)$) based on the physical origin of the noise sources. It can be bias/temperature/frequency dependent. In the next section, we discuss modeling of $S_v(f)$ (or $S_i(f)$) for different noise sources.

## 6.2 Channel and gate thermal noise

Thermal noise is due to the random thermal motion of the carriers in a semiconductor. For a semiconductor bar of conductance G, $S_i(f)$ due to

BSIM-Bulk MOSFET Model for IC Design - Digital, Analog, RF and High-Voltage
https://doi.org/10.1016/B978-0-32-385677-5.00010-X

thermal noise is given as:

$$S_{i,th}(f) = 4kT * G \tag{6.3}$$

here k is the Boltzmann constant, and T is the temperature. In the BSIM-BULK compact model, there are two channel thermal noise models. One is a charge-based model (default model) similar to that used in BSIM3v3.2 and BSIM4.7.0 (TNOIMOD=0). The other is the holistic model similar to BSIM4.7.0 (TNOIMOD=2). These two models can be selected through the model selector TNOIMOD.

### 6.2.1 Charge based noise model, TNOIMOD=0

This is the default thermal noise model of the BSIM-BULK compact model. (6.3) is used to estimate thermal noise in the MOSFETs. Total resistance of the MOS transistor biased in linear region is given as:

$$R_{MOS} = R_{ch} + R_{dsi} \tag{6.4}$$

here $R_{ch}$ and $R_{dsi}$ represent channel resistance and LDD resistance, respectively. It can be shown that the channel resistance is given by

$$R_{ch} = \frac{L_{eff}^2}{\mu_{eff} * Q_{inv}}; \tag{6.5}$$

Thus

$$R_{MOS} = \frac{L_{eff}^2}{\mu_{eff} * Q_{inv}} + R_{dsi} \tag{6.6}$$

$$G_{tnoi} = \frac{1}{R_{MOS}} = \frac{T0}{R_{dsi} * T0 + L_{eff}^2} \tag{6.7}$$

where

$$T0 = \mu_{eff} * Q_{inv} \tag{6.8}$$

$$Q_{inv} = Q_{Si} + Q_{Di} \tag{6.9}$$

$$Q_{Si} = -NF * W_{eff} * L_{eff} * C_{ox} * V_t * Q_s \tag{6.10}$$

$$Q_{Di} = -NF * W_{eff} * L_{eff} * C_{ox} * V_t * Q_d \tag{6.11}$$

Thermal noise current power spectral density is then given as:

$$S_{i,th} = 4kT * G_{tnoi} * NTNOI \tag{6.12}$$

here the parameter NTNOI is introduced for the thermal noise model flexibility, especially for the short-channel transistors. Thermal noise due to

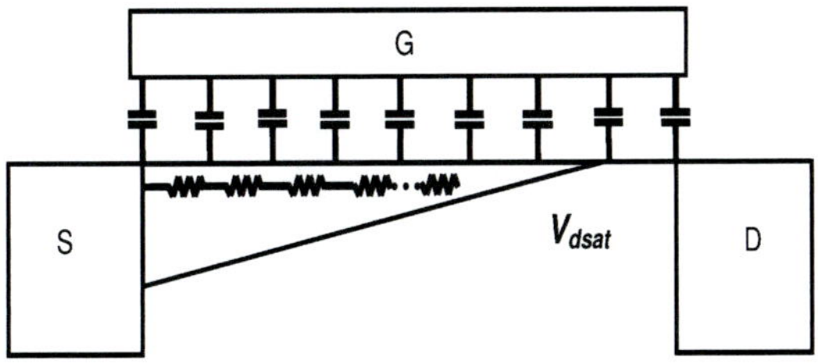


FIGURE 6.1 Distributed channel representation for TNOIMOD=1.

series source/drain resistance is separately calculated in the BSIM-BULK model using (6.3). Therefore, for RDSMOD=1, where the LDD resistances are external, $R_{dsi}$ takes zero value to avoid double counting of noise due to series resistance.

### 6.2.2 Correlated noise model, TNOIMOD=1

The main limitation of the TNOIMOD=0 model is that it does not model induced gate noise since it does not take distributed nature of the MOS channel into consideration (see Fig. 6.1). As a result, induced gate noise is not captured. TNOIMOD=1 is the correlated thermal noise model, which is holistic in nature. It is activated by setting TNOIMOD=1. The channel of a transistor is strongly coupled with the gate through the oxide capacitance. Therefore fluctuations in the channel current causing thermal noise in the drain current also lead to the gate current noise. This is termed as induced gate noise, as the origin of the noise source is the channel. In this thermal noise model, all the short-channel effects and velocity saturation effects incorporated in the IV model are included, and both the gate and the drain noise are implemented as current noise sources. The drain current noise flows from the drain to the source, whereas the induced gate current noise flows from the gate to the source. The correlation between the two noise sources is independently controllable and can be tuned using the parameter RNOIC, although the use of a default value of 0.395 is recommended when measured data is not available.

The thermal noise model in BSIM-BULK is derived using the Langevin method [1], where the Langevin noise source is added, and the small signal noise current is calculated. It can be shown that the drain current thermal noise power spectral density comes out to be [2]:

$$S_{id}(f) = \frac{4kT}{I_{ds}L_{eff}^2} \int_0^{V_{ds}} g_0^2 dV \tag{6.13}$$

Induced gate noise is expressed as [2]:

$$S_{ig}(f) = \frac{\omega W_{eff}^2}{L_{eff}^2}\frac{4kT}{I_{ds}^5}C_{ox}^2\int_0^{V_{ds}} g_0^2(v)dV\left[\int_0^{V_{ds}}(u-v)g_0(u)du\right]^2 dv \quad (6.14)$$

And the correlation between drain and gate noise [2]:

$$S_{ig,id}(f) = 4kT\frac{j\omega W_{eff}C_{ox}}{I_{ds}^3 L_{eff}^2}\int_0^{V_{ds}} g_0^2(v)dV\int_0^{V_{ds}}(V-U)g_0(U)dUdV \quad (6.15)$$

The MOSFET channel conductance can be expressed in terms of inversion charge density as:

$$g_0(V) = \mu C_{ox} W V_t q_i \quad (6.16)$$

Using (6.16) and evaluating (6.13), (6.14), and (6.15) leads to:

$$S_{id} = 4KT\cdot NF\mu C_{ox}\frac{W_{eff}}{L_{vsat}}V_t D_{ptwg}M_{oc}\left[q_{ia}\cdot\left(1+\frac{\beta_{lowId}^2}{TNOIK2+q_{ia}}\cdot\frac{V_{dseff}}{V_{dsat}}\right)\right.$$
$$\left.+(3\cdot\beta_{tnoi}^2)\cdot\frac{(q_s-q_{deff})^2}{12\left(\frac{1+q_s+q_{deff}}{2}\right)}\right] \quad (6.17)$$

$$mig = \frac{1}{12\cdot NF\cdot W_{eff}\mu_{eff}\cdot D_{ptwg}M_{oc}C_{ox}\cdot V_t}\frac{L_{vsat}^3}{L_{eff}^2}\cdot\left[\frac{q_{ia}}{\left(\frac{1+q_s+q_{deff}}{2}\right)^2}\right.$$
$$\left.-\frac{\left(6q_{ia}+\frac{1}{2}\right)q_{sd}^2}{60\left(\frac{1+q_s+q_{deff}}{2}\right)^4}+\frac{q_{sd}^4}{144\left(\frac{1+q_s+q_{deff}}{2}\right)^5}\right]\cdot\frac{15}{4}\cdot\theta_{tnoi}^2 \quad (6.18)$$

$$S_{ig} = \frac{4KT\cdot[\omega\cdot C_{ox}\cdot W\cdot NF\cdot L]^2\cdot mig}{1+(\omega\cdot C_{0x}\cdot L\cdot W\cdot NF\cdot mig)^2} \quad (6.19)$$

$$S_{ig,id} = -j\omega\cdot 4KT\cdot\mu C_{ox}D_{ptwg}M_{oc}V_t\left(\frac{L_{vsat}}{L_{eff}}\right)\cdot$$
$$\left[\frac{(q_s-q_{deff})}{12\left(\frac{1+q_s+q_{deff}}{2}\right)}-\frac{(q_s-q_{deff})^3}{144\left(\frac{1+q_s+q_{deff}}{2}\right)^3}\right]\cdot\frac{c_{tnoi}}{0.395} \quad (6.20)$$

$$c = \frac{S_{ig,id}}{\sqrt{S_{ig}}\cdot\sqrt{S_{id}}} \quad (6.21)$$

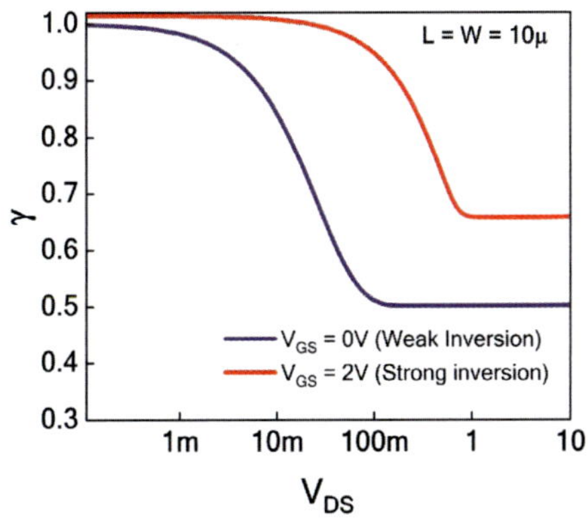


FIGURE 6.2 White noise gamma test [3]: plot of simulated $\gamma$ vs. $V_{ds}$ in strong and weak inversion.

where

$$\beta_{tnoi} = RNOIA \cdot \left[1.0 + TNOIA \cdot L_{eff} \cdot \left(\frac{q_{ia}}{E_{sat,noi} L_{eff}}\right)^2\right] \tag{6.22}$$

$$\theta_{tnoi} = RNOIB \cdot \left[1.0 + TNOIB \cdot L_{eff} \cdot \left(\frac{q_{ia}}{E_{sat,noi} L_{eff}}\right)^2\right] \tag{6.23}$$

$$c_{tnoi} = RNOIC \cdot \left[1.0 + TNOIC \cdot L_{eff} \cdot \left(\frac{q_{ia}}{E_{sat,noi} L_{eff}}\right)^2\right] \tag{6.24}$$

In Fig. 6.2, simulated $\gamma = S_{id}/4kTg_{ds0}$ [3] is plotted versus $V_{ds}$ in weak and strong region of operation for TNOIDMOD=1. Theoretically, $\gamma$ should be 1 at $V_{ds} = 0\,\text{V}$, and at high $V_{ds}$, it approaches $1/2$ and $2/3$ in weak inversion and strong inversion, respectively. As seen in Fig. 6.2, the BSIM-BULK model correctly simulates the thermal noise behavior.

## 6.3 Series resistance thermal noise

For RDSMOD=0 and 1, source and drain resistances are created in series with the MOS channel. Thermal noises due to these resistances are modeled as follows:

$$i_{s,si}^2 = \frac{4kT}{R_{source}} \Delta f \tag{6.25}$$

$$i_{d,di}^2 = \frac{4kT}{R_{drain}} \Delta f \tag{6.26}$$

Series resistances $R_{source}$ and $R_{drain}$ are discussed in Chapter 3.

## 6.4 Gate current shot noise

Gate current shot noise is associated with quantum mechanical tunneling. For IGCMOD=1, it is modeled as follows:

$$i_{gs}^2 = 2q(I_{gcs} + I_{gs})\Delta f \tag{6.27}$$

$$i_{gd}^2 = 2q(I_{gcd} + I_{gd})\Delta f \tag{6.28}$$

For IGBMOD=1, it is modeled as follows:

$$i_{gb}^2 = 2qI_{gb}\Delta f \tag{6.29}$$

Current components $I_{gcs}$, $I_{gs}$, $I_{gcd}$, $I_{gd}$ and $I_{gb}$ are discussed in Chapter 4.

## 6.5 Flicker noise

The BSIM-BULK compact model offers two flicker noise models: one for low pocket (halo implants) dose/uniformly doped transistors and the other for strong pocket dose transistors. The former is activated by setting FNOIMOD=0, while the latter by FNOIMOD=1. We will first discuss the FNOIMOD=0 model, followed by FNOIMOD=1 in the next section.

### 6.5.1 FNOIMOD=0

The origin of the flicker noise (a.k.a. 1/f noise) in semiconductor devices is widely debated. One school of thought attributes it to inversion charge carrier number fluctuations, while the other to carrier mobility fluctuations [4,5]. The unified theory of 1/f noise is based on carrier trapping in the oxide, leading to both carrier number as well as mobility fluctuations. The unified model was developed by the BSIM group, and it is used in all the industry standard compact models [6,7].

Starting from the drift-diffusion current model equation, the drain current flicker noise power spectral density can be expressed as [7]:

$$S_{ID} = \frac{kTI_{DS}^2}{\gamma f W L^2}\int_0^L N_t^*(E_{F_n})\frac{R^2}{N^2}dx \tag{6.30}$$

$$S_{ID} = \frac{kTqI_{DS}\mu_{eff}}{\gamma f L^2}\int_0^{V_d} N_t^*(E_{F_n})\frac{R^2}{N}dV \tag{6.31}$$

here $\gamma$ is the attenuation coefficient of the electron wave function in the oxide. R represents the ratio of the fluctuations in carrier number ($\delta\Delta N$) to fluctuations in occupied trap number ($\delta\Delta N_t$):

$$R = \frac{\delta\Delta N}{\delta\Delta N_t} = -\frac{C_i}{C_i + C_d + C_{ox} + C_{it}} = -\frac{N}{N+N^*} \tag{6.32}$$

where $C_i, C_d, C_{it}$ are the inversion layer, depletion layer, and interface trap capacitances, respectively, and $N^*$ is defined as follows:

$$N^* = \frac{kT}{q^2} * (C_d + C_{ox} + C_{it}) \tag{6.33}$$

The above integral assumes that the major contribution to noise is due to traps located around the electron quasi-Fermi level ($E_{F_n}$). $N_t^*$ is the effective trap charge density, which is approximated as a parabolic function of carrier number (N) as follows [7]:

$$N_t^* = NOIA + NOIB * N + NOIC * N^2 \tag{6.34}$$

NOIA, NOIB, and NOIC are the model parameters. On solving (6.31) in the linear region, we get

$$S_{si} = \frac{q^3 * V_t * i_{ds} * \mu_{eff}}{\gamma * C_{ox} * L_{effnoi}^2} * (T1 + T2 + T3) \tag{6.35}$$

where

$$T1 = NOIA * ln\left(\frac{N_0 + N^*}{N_L + N^*}\right) \tag{6.36}$$

$$T2 = NOIB * (N_0 - N_L) \tag{6.37}$$

$$T3 = 0.5 * NOIC * (N_0^2 - N_L^2) \tag{6.38}$$

$$L_{effnoi} = L_{eff} - 2 * LINTNOI \tag{6.39}$$

$$N_0 = 2 * n_q * C_{ox} * V_t * q_s * Mnud1 * Mnud/q \tag{6.40}$$

$$N_L = 2 * n_q * C_{ox} * V_t * q_{d,eff} * Mnud1 * Mnud/q \tag{6.41}$$

$$N^* = V_t/q * (C_{ox} + C_{dep} + CIT) \tag{6.42}$$

$$T0 = \gamma * C_{ox} * L_{effnoi}^2 \tag{6.43}$$

$$T4 = \gamma * L_{effnoi}^2 * W_{eff} * NF \tag{6.44}$$

here $N_0$ and $N_L$ represent inversion charge density at the source and drain ends, respectively.

In the saturation region, the channel is composed of a triode region and a pinch-off region. The noise contribution of the triode region is the same as in (6.35), except that $V_d$ is replaced by $V_{d,sat}$. The flicker noise contribution of the pinch-off region is obtained by evaluating the integral (6.31) at x=L and substituting dx by $\Delta L_{clm}$ (channel length modulation distance).

$$S_{ID,pinchoff} = T0b\Delta L_{clm} * \frac{T0c}{T0d} \tag{6.45}$$

where

$$\Delta L_{clm} = litl \cdot log\left(\frac{\frac{V_{ds}-V_{dseff}}{litl} + EM}{E_{sat}}\right)$$

$$E_{sat} = \frac{2 \cdot VSAT}{\mu_{eff}}; \quad litl = \sqrt{\frac{\epsilon_{Si}}{\epsilon_{ox}} * TOXE * XJ} \tag{6.46}$$

$$T0b = \frac{kTi_{ds}^2}{\gamma f W L_{effnoi}^2} \tag{6.47}$$

$$T0c = NOIA + NOIB * N_L + NOIC * N_L^2 \tag{6.48}$$

$$T0d = (N_L + N^*)^2 \tag{6.49}$$

Thus flicker noise power spectral density is strong inversion expressed as:

$$S_{si} = \frac{q^3 * Vt * i_{ds} * \mu_{eff}}{\gamma * Cox * L_{effnoi}^2} * (T1 + T2 + T3) + T0b * \Delta L_{clm} * \frac{T0c}{T0d} \tag{6.50}$$

In the subthreshold region, inversion charge density is much smaller, and under the assumption $N << N^*$ and $N_t^* = NOIA$, integral in (6.31) gives:

$$S_{wi} = i_{ds}^2 \frac{NOIA * kT}{W_{eff} * NF * L_{effnoi} * \gamma * N^{*2} * f} \tag{6.51}$$

Finally, $S_{id}$ is expressed as:

$$S_{id} = \frac{S_{si} * S_{wi}}{S_{si} + S_{wi}} \tag{6.52}$$

## 6.5.2 FNOIMOD=1, flicker noise in strong halo-doped transistors

Bulk transistors employ halo (pocket) implants to combat short-channel-effects. The pocket regions also contribute to the flicker noise, which can be different from the channel noise due to several factors like different trap density, mobility, etc. Fig. 6.3 shows the measured median drain current ($I_D$) flicker (1/f) noise power spectral density ($S_{ID}$) normalized to channel width for long and short channel devices in 45 nm low power CMOS technology node [8]. Each point in this plot is the median of 21 samples. As can be seen in the figure, the strong pocket devices show no impact of channel length scaling for 1/f noise in the subthreshold region. Further, unlike the short-channel device, the long-channel device shows anomalous bias dependence through a sudden decrease in slope. Such behavior cannot be captured by the existing noise model as shown in Fig. 6.3. Although a noise model specific to the Halo devices was earlier developed by Wu et al. [9], it adds the noise contributions from the different

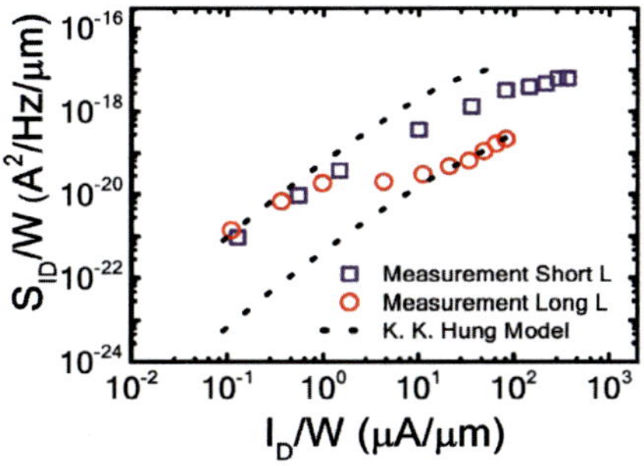


FIGURE 6.3 Drain current flicker noise power spectral density vs. drain current at $V_{DS} =$ 0.55 V, both normalized to the channel width. Measurements are from the same technology as in [8] and show complex dependency on the gate voltage, especially for long-channel device. In weak inversion, there is no impact of length scaling on $1/f$ noise. The dotted lines show the unified model [6] result for the long-channel device, and it is clearly seen that it cannot model the noise characteristics of the Halo devices.

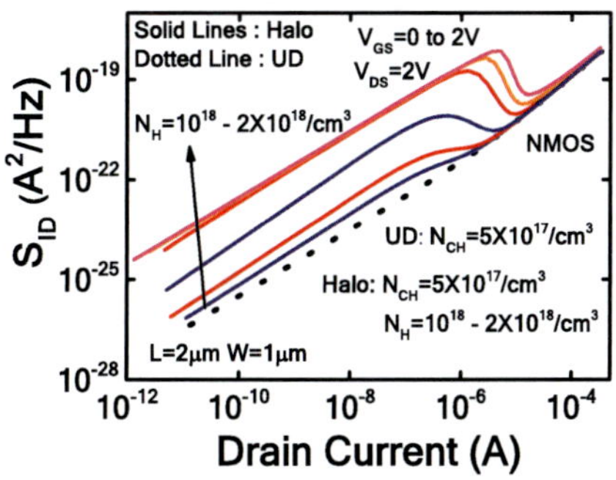


FIGURE 6.4 Comparison of TCAD $S_{ID}$ vs. drain current of UD and Halo for different doping conditions at $V_{DS} = 2\,\text{V}$ [10]. In weak inversion, $S_{ID}$ of the Halo device is significantly higher than that of UD. In strong inversion, all have the same $S_{ID}$. As doping of the halo region increases, nonmonotonicity is observed. Such behavior is also observed in the real device measurements [11] and therefore establishes the qualitative correctness of the TCAD simulation environment. Gate oxide thickness $T_{ox} = 4\,\text{nm}$, channel doping $N_{CH}$=5X10$^{17}$/cm$^3$, channel length L=2 μm, and width W=1 μm.

regions with equal proportions. Such formulation might lead to the excess contribution of noise from the pocket part in strong inversion or excess contribution from the channel part at lower biases.

Fig. 6.4 compares TCAD $S_{ID}$ versus $I_D$ of the Halo and the uniformly doped (UD) devices for various doping concentrations of the pocket regions at $V_{DS}$=2 V. Halo device shows a significant departure from the conventional UD behavior at small current levels (typically for $V_{GS} < V_{TH}$); however, in strong inversion, it asymptotically follows UD. Furthermore, at higher doping levels of the halo region, $S_{ID}$ has a nonmonotonic trend as it initially increases, then decreases and again increases with $I_D$. These

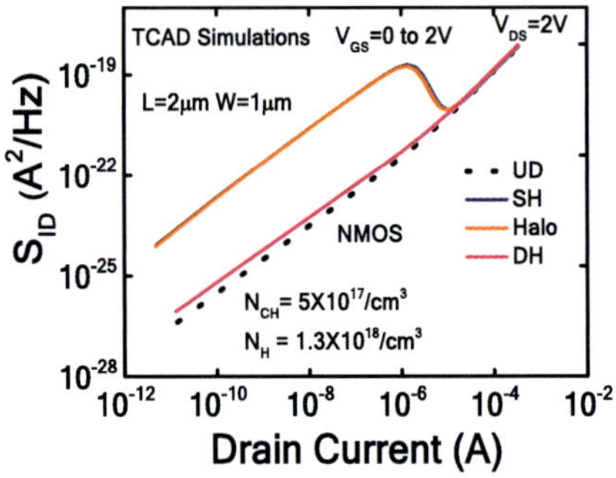


FIGURE 6.5 Comparison of $S_{ID}$ vs. drain current of UD, SH, DH and Halo at $V_{DS} = 2$ V. $N_{CH}$=5X10$^{17}$/cm$^3$ and $N_H$=1.3X10$^{18}$/cm$^3$. Since DH does not depicts unconventional bias dependency, it can be inferred that the drain side halo region has little impact on the overall noise behavior of Halo devices [10]. L=2 μm and W=1 μm.

TCAD results are in qualitative agreement with the real device measurements reported in [11]. To investigate the individual effect of source and drain side halo regions, $S_{ID}$ vs. $I_D$ of source side halo (SH), drain side halo (DH), Halo (i.e., pocket implants at both sides) and UD at $V_{DS} = 2$ V are plotted in Fig. 6.5. At a given current level, $S_{ID}$ of the Halo and SH is significantly higher than that of UD in weak inversion. Characteristics of SH and Halo are similar, and interestingly, DH does not show unconventional bias dependency. From these results, it can be inferred that the anomalous behavior observed in Halo devices is due to the source side halo region, and drain side halo region has little role to play in saturation. This is a key feature that needs to be understood because traditionally source side halo region is overlooked when modeling Halo devices. We offer our understanding of this effect, which is as follows: equivalent conductance of the MOSFET with halo implants can be expressed as [12],

$$\frac{1}{G_{eq}} = \frac{1}{G_{eq}|_{SH}} + \frac{1}{G_{eq}|_{CH}} + \frac{1}{G_{eq}|_{DH}} \tag{6.53}$$

$G_{eq}|_{SH}$, $G_{eq}|_{DH}$ and $G_{eq}|_{CH}$ represent equivalent conductances of the source side halo, drain side halo, and channel region, respectively, and are proportional to the ratio of inversion charge density and length of the region. In saturation, drain side barrier (between lightly doped channel and drain side pocket region) is lowered due to high drain voltage, which increases effective conductance (as it exponentially depends on the surface potential in weak inversion) of the drain side halo region. For $L_{ch} >> L_h$, $G_{eq} \approx G_{eq}|_{SH}$ from (6.53), the overall conduction in the device is limited by the source side halo region. Since overall noise is a function of equivalent conductance (or resistance) of each region [13], the drain side halo region does not contribute significantly to the overall noise performance. At low $V_{DS}$ operations, $G_{eq}|_{DH} \approx G_{eq}|_{SH}$, and one can write $G_{eq}|_{SH,eff} \approx 0.5 * G_{eq}|_{SH}$. Therefore, by parameterizing $G_{eq}|_{SH}$ and neglecting $G_{eq}|_{DH}$ (thereby reducing computational time), it is possible to accurately model noise behavior both in linear and saturation region of operation.

#### 6.5.2.1 Compact model formulation

Starting with the earlier Langevin method for noise modeling, other approaches exist, like the equivalent circuit method, impedance field method, Klaassen Prin (KP) approach, etc. [14–16]. The presented formulation is based on a small signal approach similar to the impedance field method. Halo-doped devices, in principle, can be modeled by three transistor equivalent subcircuit, each representing the source/drain halo region and the channel region. This creates two extra nodes in the circuit, increasing the computer simulation time. In the BSIM-BULK model, we have developed a complete analytical model of flicker noise in strong-pocked devices [17]. The new model is activated by setting FNOIMOD=1.

Firstly, the I-V parameters of the family of devices under test are extracted. For the flicker noise modeling, it is now assumed that the transistor is composed of two transistors as shown in Fig. 6.6. The total length L is segmented into two parts: a region of higher doping of length $L_h$ with equivalent resistance $R_h$, noise PSD $S_{ID,h}$ and another low-doped region of length L-$L_h$ with resistance $R_{ch}$ and noise PSD $S_{ID,ch}$. Fig. 6.6(b) shows the equivalent circuit representation of Fig. 6.6(a). The individual contribution of the halo and the channel transistors to overall noise is obtained using small signal analysis and the principle of superposition as shown in Fig. 6.7. From Fig. 6.7(a), the drain current noise PSD due to halo transistor is obtained by assuming channel transistor to be noiseless. Using small signal analysis [17],

$$I_{n1} \simeq I_{n,h} \frac{g_{m,ch} + g_{d,ch}}{g_{m,ch} + g_{d,ch} + g_{d,h}} \tag{6.54}$$

$$S_{ID,1} = S_{ID,h} \left[ \frac{g_{m,ch} + g_{d,ch}}{g_{m,ch} + g_{d,ch} + g_{d,h}} \right]^2 \tag{6.55}$$

where $g_{m,ch}$, $g_{d,ch}$ are the transconductance and output conductance of the channel transistor, and $g_{d,h}$ is the transconductance of the halo transistor. Similarly, from Fig. 6.7(b), the noise PSD due to the channel transistor is

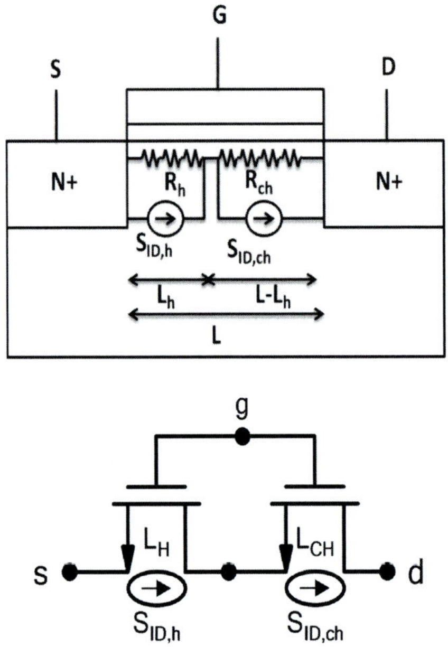


FIGURE 6.6 Representation of the halo-implanted MOSFET for noise modeling: (a) Channel length L can be divided into two parts - the halo region of length $L_h$ with doping $N_h$ and the channel region of length L-$L_h$ with doping $N_{ch}$ (b) Equivalent circuit representation. Source side electrostatics is captured by the high threshold voltage transistor of length $L = L_h$ [17].

expressed as,

$$I_{n2} \simeq I_{n,ch} \frac{g_{d,h}}{g_{m,ch} + g_{d,ch} + g_{d,h}} \tag{6.56}$$

$$S_{ID,2} = S_{ID,ch} \left[ \frac{g_{d,h}}{g_{m,ch} + g_{d,ch} + g_{d,h}} \right]^2 \tag{6.57}$$

Total drain current noise PSD becomes,

$$S_{ID} = S_{ID,1} + S_{ID,2} \tag{6.58}$$

$$= S_{ID,h} \left[ \frac{g_{m,ch} + g_{d,ch}}{g_{m,ch} + g_{d,ch} + g_{d,h}} \right]^2 + S_{ID,ch} \left[ \frac{g_{d,h}}{g_{m,ch} + g_{d,ch} + g_{d,h}} \right]^2 \tag{6.59}$$

$$S_{ID} = S_{ID,h}.CF_h + S_{ID,ch}.CF_{ch} \tag{6.60}$$

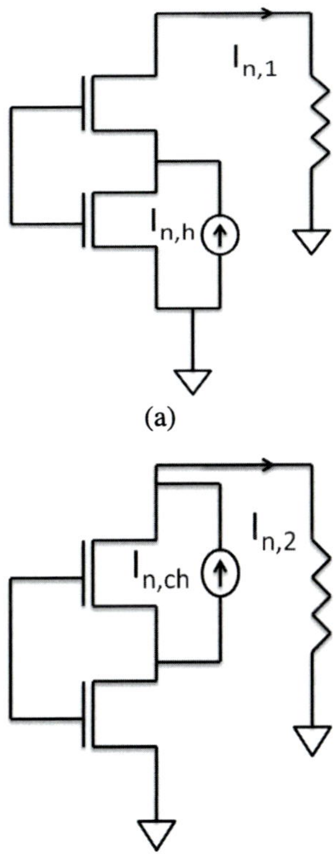


FIGURE 6.7 Small signal analysis of two transistor noise equivalent circuit: superposition principle is used to obtain the total noise (a) Noisy halo transistor, noiseless channel transistor (b) Noisy channel transistor, noiseless halo transistor [17].

We refer to the multiplying factors to $S_{ID,h}$ and $S_{ID,ch}$ in (6.59) as contribution factors (CF). Transconductance and output conductance are expressed as [18],

$$g_{d,ch} = 2n_q \mu C_{ox} \frac{W}{L - L_h} V_t q_{d,ch} \tag{6.61}$$

$$g_{d,h} = 2n_q \mu C_{ox} \frac{W}{L_h} V_t q_{d,h} \tag{6.62}$$

$$g_{m,ch} = 2\mu C_{ox} \frac{W}{L - L_h} V_t (q_{s,ch} - q_{d,ch}) \tag{6.63}$$

where $n_q$, $\mu$, $C_{ox}$, and $V_t$ are the slope factor, effective mobility, oxide capacitance per unit area, and thermal voltage, respectively. Note that all these quantities are already extracted during CV and IV extraction. $q_{s,ch}$, $q_{d,ch}$, $q_{s,h}$, and $q_{d,h}$ are the normalized inversion charge densities at the source and drain ends of the channel, and the halo transistor, respectively. $q_{sh}$ and $q_{d,ch}$ are obtained from the analytical solution of the BSIM-BULK core model equations [19] charge equation given by (6.64), where pinch-off potential ($\psi_p$, normalized to the thermal voltage $V_t$) is given by (6.65) and is calculated independently for the two transistors using source potential $v_{ch} = v_s$ and $v_{ch} = v_d$, respectively (interested readers may refer to Chapter 2 for solution of the core model).

$$\ln(q_i) + \ln\left[\frac{2n_q}{\gamma}\left(q_i\frac{2n_q}{\gamma} + 2\sqrt{\psi_p - 2q_i}\right)\right] + 2q_i = \psi_p - 2\phi_f - v_{ch} \tag{6.64}$$

$$\psi_p = \begin{cases} -\ln\left[1 - \psi_{p0} + \left(\frac{v_g - v_{fb} - \psi_{p0}}{\gamma}\right)^2\right] & \text{if } v_g - v_{fb} < 0 \\ 1 - e^{-\psi_{p0}} + \left[\sqrt{v_g - v_{fb} - 1 + e^{-\psi_{p0}} + \left(\frac{\gamma}{2}\right)^2} - \frac{\gamma}{2}\right]^2 & \text{otherwise} \end{cases} \tag{6.65}$$

Here $v_g$, $v_{fb}$, $\phi_f$ and $\gamma$ are the normalized gate voltage, flat band voltage, bulk potential, and body factor, respectively, and $\psi_{p0}$ is an approximation of the pinch-off potential when it is close to zero.

Now it is remaining to calculate $q_{d,h}$ and $q_{s,ch}$. For this, current continuity between the two transistors in the subcircuit is used. In BSIM6, drain current can be expressed as a function of inversion charge densities at the source and the drain ends, which gives,

$$i_h = \frac{I_{DS}}{2n_q\mu C_{ox}\frac{W}{L_h}V_t^2} = (q_{s,h}^2 + q_{s,h}) - (q_{d,h}^2 + q_{d,h}) \tag{6.66}$$

$$i_{ch} = \frac{I_{DS}}{2n_q\mu C_{ox}\frac{W}{L-L_h}V_t^2} = (q_{s,ch}^2 + q_{s,ch}) - (q_{d,ch}^2 + q_{d,ch}) \tag{6.67}$$

where $i_h$ and $i_{ch}$ are the normalized drain current of halo and channel transistor, respectively. Since drain current is already known from the DC fitting, one can write,

$$q_{d,h} = -\frac{1}{2} + \frac{1}{2}\sqrt{1 + 4(q_{s,h}^2 + q_{s,h} - i_h)} \tag{6.68}$$

$$q_{s,ch} = -\frac{1}{2} + \frac{1}{2}\sqrt{1 + 4(q_{d,ch}^2 + q_{d,ch} + i_{ch})} \tag{6.69}$$

Thus the inversion charge densities at the end point of the halo and channel region are explicitly known.

#### *6.5.2.2 The noise source*

The source of flicker noise in MOSFETs is attributed to mobility fluctuation and/or carrier number fluctuation [4,5,20]. There exist popular

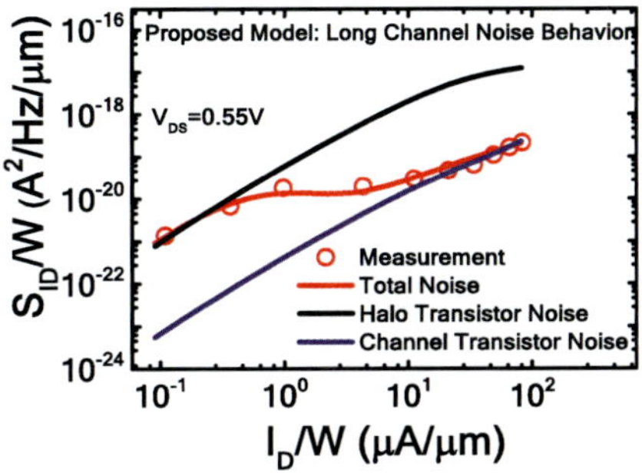


FIGURE 6.8 Model validation with experimental data of long channel device: $S_{ID}$ vs. $I_D$ at $V_{DS} = 0.55$ V. $S_{ID}$ asymptotically follows halo transistor noise in weak inversion and channel transistor noise in strong inversion. L=1 μm [17].

models that unify the two approaches [6,7,21]. Here we have used the unified model presented in [6] (which has been widely used in industry standard bulk MOSFET models [22–24]) for halo and channel transistors separately, where $S_{ID}$ is expressed as

$$S_{ID,h} = \frac{kTI_{DS}^2}{\gamma f W L_h^2} \int_0^{L_h} \frac{N_{t,h}^*(E_{F_n})}{N_h^2} dx \tag{6.70}$$

$$S_{ID,ch} = \frac{kTI_{DS}^2}{\gamma f W (L - L_h)^2} \int_{L_h}^{L} \frac{N_{t,ch}^*(E_{F_n})}{N_{ch}^2} dx \tag{6.71}$$

where apparent trap density $N_{t,ch(h)}^*(E_{F_n}) = A_{ch(h)} + B_{ch(h)} N_{ch(h)} + C_{ch(h)} N_{ch(h)}^2$, $A$, $B$, $C$ are the noise parameters, $\gamma$ is the tunneling parameter, $k$ is the Boltzmann constant, and $T$ is the temperature. It is important to note that $S_{ID,h}$ and $S_{ID,ch}$ are in explicit form since (6.70) and (6.71) can be expressed as a function of $q_{s,h}$, $q_{d,h}$ and $q_{s,ch}$ and $q_{d,ch}$, respectively [25].

Also note that $S_{ID}$ in the halo region might be locally higher than in the channel region due to increased trap density, which is captured by the noise parameters of the halo transistor. Hence formulation of $S_{ID}$ is based on the local trap density, inversion charge densities are specific to the region that generates the noise as well as the length of the region. Using (6.70) and (6.71) and substituting (6.68), (6.69), $q_{s,h}$, $q_{d,ch}$ in (6.59) gives the overall PSD. Figs. 6.8, 6.9, 6.10, 6.11 show the model results and validation with the experimental data.

## 6.6 BSIM-BULK RF models

Nowadays, low-power wireless applications, such as wireless sensor networks, wireless body area networks, and the Internet of Things (IoT),

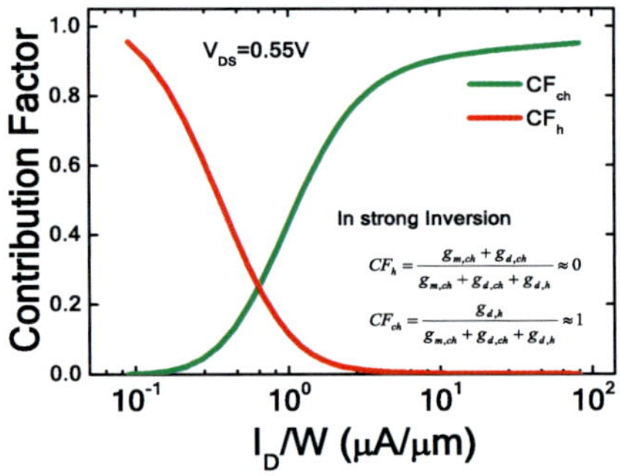


FIGURE 6.9 Contribution factor of the halo and the channel transistor vs. drain current. In weak inversion, $CF_h >> CF_{ch}$, and thus $S_{ID}$ is dominated by the noise from the halo region (see (6.60)). On the other hand, $CF_h$ falls rapidly in the strong inversion leading to the negligible contribution from halo region in $S_{ID}$. The role of CF in this model is similar to that of impedance field [14], which is responsible for noise propagation from a point in the channel to the drain terminal. Note that [9] adds halo and channel contributions with equal weights and therefore will overestimate noise in strong inversion.

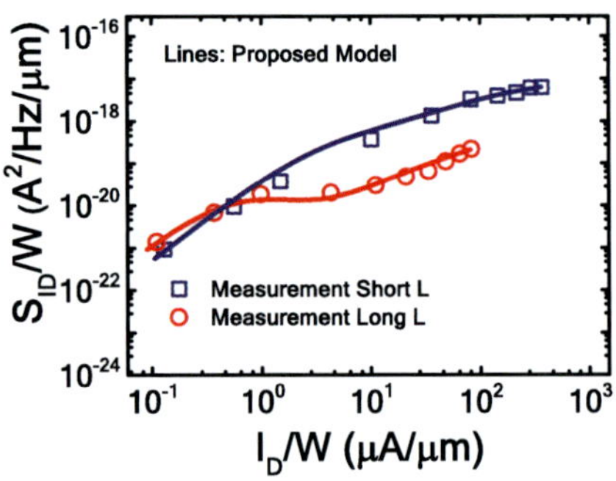


FIGURE 6.10 Model validation with experimental data of a short channel device: $S_{ID}$ vs. $I_{DS}$ at $V_{DS} = 0.55$ V. L=1 μm and 120 nm [26].

are setting stringent constraints on power consumption [27]. This can be partly addressed by taking advantage of the phenomenal RF performance obtained from the continuous downscaling of CMOS technology. Indeed, advanced nano-scale transistors achieve impressive peak transit frequency, extremely low thermal noise and minimum noise figure (NF) [28].

A considerable effort has already been put into characterizing and modeling the MOS transistor at the RF regime [29,30]. Due to the low-cost and low-power characteristics of CMOS technology, capturing the circuit performance operating near the device cut-off frequency and beyond is essential for accurate and reliable circuit design. Current research studies mainly focus on integrating the RF power amplifier(PA) with CMOS technology. Conventionally PA and other millimeter-wave circuits are widely

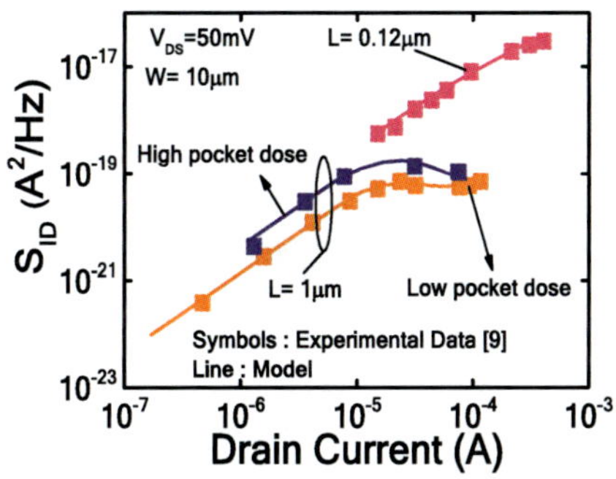


FIGURE 6.11 Model validation with experimental data at $V_{DS} = 50\,\text{mV}$ [11]. Unlike long-channel lengths, short-channel length device does not show a significant bias dependency in $S_{ID}$. The short-channel length device behaves similarly to a uniformly doped device as the halo regions from both ends merge. The compact model can also accurately model $S_{ID}$ in linear region, both for long- and short-channel length devices [10]. $V_{DS} = 50\,\text{mV}$.

implemented in SiGe or GaAs technology due to the high $f_t$ and low loss in the substrate [29]. However, these processes have limited integration capability. As a result, the CMOS, especially advanced nano-CMOS, tends to be more popular due to its potential for higher integration with digital signal processing blocks [29].

A complete compact RF model does not yet exist in commercial circuit simulators. Thus the most common RF modeling approach is to build subcircuits based on MOSFET models suitable for analog/digital applications. To further improve the model accuracy at high frequencies, the subcircuit, parasitic elements around the gate, source, drain, and substrate are added. An example of a subcircuit model for RF applications is given in Fig. 6.12. For accurate RF modeling, it is crucial to incorporate physical and scalable model equations for the parasitic elements at the source, drain, gate, and substrate. The scalability of the intrinsic device is taken care of by the core model, for example, BSIM-BULK. Here, we will discuss the modeling of the gate and substrate resistances.

## 6.7 BSIM-BULK small signal model

BSIM-BULK has a physical charge-based core model to accurately capture the terminal charges and drain current of MOSFET [31]. Distributed effects of MOSFET have been included in the BSIM-BULK small signal model as shown in Fig. 6.13. BSIM-BULK has models for the extrinsic series resistance $R_s$ and $R_d$, parasitic overlap capacitance $C_{gso}$ and $C_{gdo}$, fringing capacitance $C_f$, gate resistance $R_{geltd}$, channel reflected NQS resistance $R_{ch}$ (Modeling of gate resistance network is discussed in the next section), 5-resister substrate network (Modeling of the substrate network

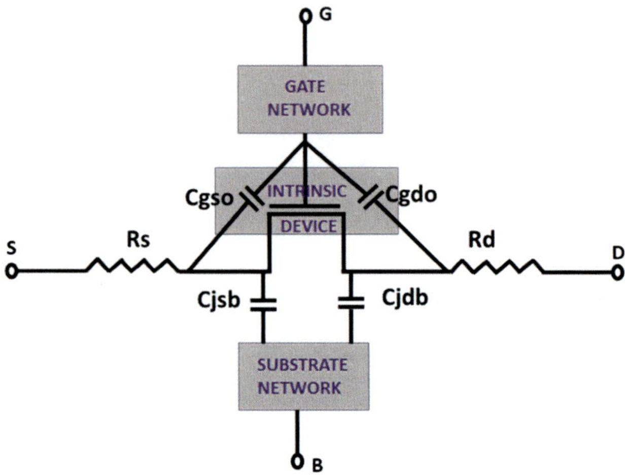


FIGURE 6.12 A subcircuit with parasitic elements added to an intrinsic MOSFET model of BSIM-BULK for RF circuit simulation.

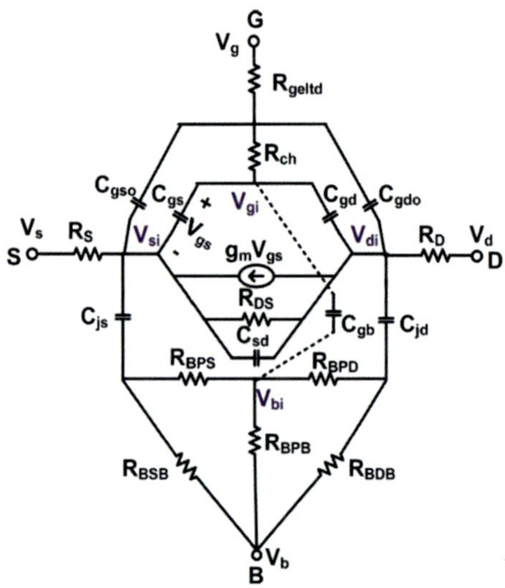


FIGURE 6.13 High-frequency small signal equivalent representation of MOSFET, used in BSIM-BULK Model, $R_s$ and $R_d$ represent the extrinsic source-drain series resistances, $C_{gso}$ and $C_{gdo}$ represent parasitic overlap capacitance, $R_{geltd}$ is the gate electrode resistance, $R_{ch}$ is channel-reflected NQS resistance, $R_{BPS}$, $R_{BPD}$, $R_{BPB}$, $R_{BSB}$, and $R_{BDB}$ represent the 5-resistor substrate network, and $C_{js}$ and $C_{jd}$ represent source-drain junction capacitance.

will be explained in the upcoming section) and junction capacitance $C_{js}$ and $C_{jd}$, that affect the device behavior considerably at high frequencies for the accurate modeling of RF device performance metrics [31]. From the analysis of the high-frequency small signal network in Fig. 6.13, simplified circuit equivalent expressions for the Y-parameters can be derived [29]. These derivations are based on the assumption that, for the frequency range below tera-hertz, all the frequency terms that are higher than second order can be neglected. We have used the same expressions derived in [29] to explain the high-frequency physics of MOSFET. The 2-port Y-parameters [port 1-gate, and port 2-drain with source and body terminals tied to ground] are as follows

$$Y_{gg} \approx \omega^2 C_{gg}^2 R_g + j\omega C_{gg} \tag{6.72}$$

$$Y_{gd} \approx \omega^2 (C_{gb} C_{jd} R_{sub} - C_{gd} C_{gg} R_g) - j\omega C_{gd} \tag{6.73}$$

$$Y_{dg} \approx g_m + \omega^2 (C_{gb} C_{jd} R_{sub} - C_{dg} C_{gg} R_g) - j\omega C_{dg} - j\omega C_{gg} g_m R_g \tag{6.74}$$

$$Y_{dd} \approx g_{ds} + \omega^2 (C_{jd}^2 R_{sub} + C_{gd} C_{dg} R_g) + j\omega (C_{gd} + C_{jd}) \tag{6.75}$$

### 6.7.1 BSIM-BULK RF parameter extraction (without NQS effect)

For accurate RF parameter extraction, the following requirements are essential:

1: Fit the DC measurement data.

2: RF measurement data, that is 2-port S-parameter data.

The S-parameter measurements need to be accurately de-embedded with well-defined procedures that give accuracy up to the frequency of interest and above [30]. Otherwise, layout and lead parasitics could have a chance to fold into the MOSFET model. These parasitics may significantly influence the high-frequency behavior and mask the actual MOSFET behavior, leading to the incorrect estimation of RF MOSFET components.

For the validation of the BSIM-BULK model in the RF operation, first, it is important to fit the transconductance ($G_m$), output conductance ($G_{DS}$), gate capacitance ($C_{gg}$) and gate resistance $R_{gg}$, as illustrated in Fig. 6.14 and Fig. 6.15. Now, once we extract the $G_m$, $G_{DS}$, $C_{gg}$ and $R_{gg}$, we can start the studies of de-embedded Y-parameters of a wide ($W_f$ = 0.6 μm), multi-finger (M = 4, $N_f$ = 16), and short (L = 70 nm) RF NMOS. The model is compared against measurements for frequencies up to 100 GHz and for different bias conditions, remaining always in strong inversion and saturation, i.e., $V_{GS} - V_{TH0} > 0\,V$ and $V_{DS} = 0.55\,V$. The results for the normalized Y-parameters versus frequency F are shown in Figs. 6.16–6.19.

For further verification, the input and output reflection coefficients, namely the $S_{11}$ and $S_{22}$ parameters, under the same bias conditions, are also shown in Fig. 6.20. These S-parameters provide a measure of the input and output mismatch loss, and thus they can be important from a designer's point of view. Overall, an adequate fitting between the BSIM-BULK model and the measurements for the RF parameters is observed.

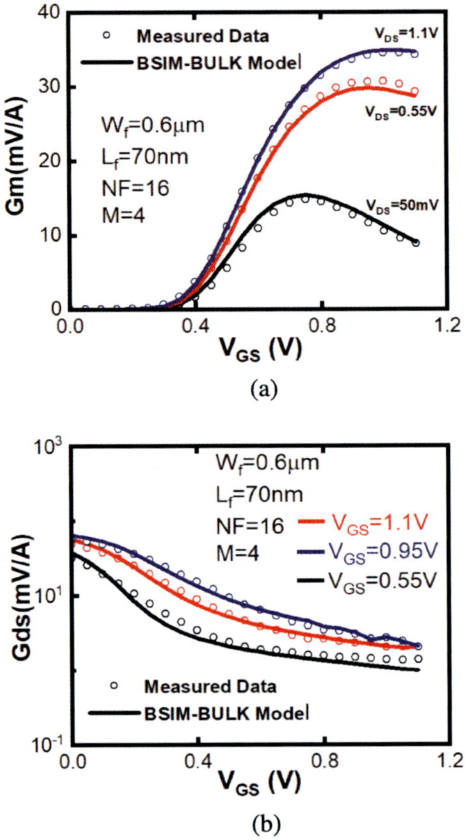


FIGURE 6.14 DC fitting: (a) Fitting of transconductance (b) Fitting of output conductance. Solid: Model, Symbols: data.

## 6.8 Modeling of the gate resistance

The gate electrode resistance influences the input impedance. The physical gate electrode resistance represents the resistance of the poly-silicon material or metallic gate material on top of the channel. It is given by [32]

$$R_{geltd} = n \cdot \frac{W_{eff}}{L_{eff}} \frac{R_{shg}}{NF} \tag{6.76}$$

where $R_{shg}$ is the sheet resistivity of the gate material, $W_{eff}$ represents the channel width, $L_{eff}$ is the channel length, and $NF$ is the number of fingers. The factor $n$ is a constant derived from a gate electrode distributed R-C transmission line model. Thus $n = 1/3$ for a single side gate contact, and $n = 1/12$ for double side gate contact.

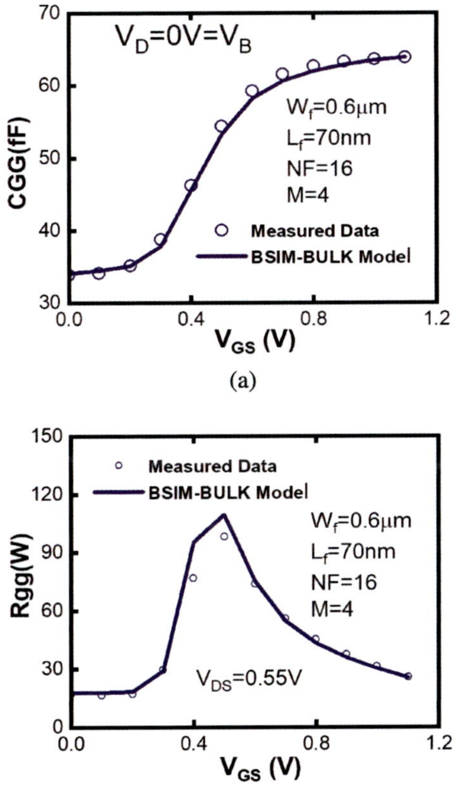


FIGURE 6.15 (a) Gate capacitance fitting (b) The gate electrode resistance $R_{geltd}$ and the channel-induced gate resistance $R_{ch}$ parameters are extracted from this curve. The channel-induced gate resistance is extracted using the BSIM-BULK NQS model. Solid: Model, Symbols: data.

This gate resistance is modeled in BSIM-BULK as [31]

$$R_{geltd} = \frac{RSHG \cdot (XGW + \frac{W_{eff}}{3 \cdot NGCON})}{NGCON \cdot (L_{drawn} - XGL) \cdot NF} \tag{6.77}$$

where, $RSHG$ is sheet resistance of the gate electrode, $XGW$ is the distance from the gate contact to the channel edge, $XGL$ is the offset of the gate length due to variations in patterning, and $NGCON$ is the number of gate contacts [31].

However, it has been reported in the literature that not only the distributed RC effect of the gate but also the nonquasi-static effect or the distributed RC effect of the channel must be accounted for the modeling of high-frequency behavior of the MOSFET. Therefore it is necessary to add

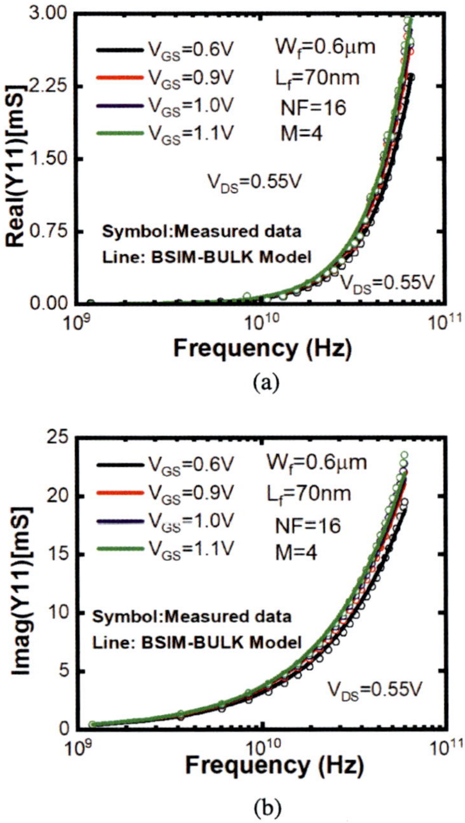


FIGURE 6.16 RF Model validation: (a) $Real(Y_{11})$. (b) $Imag(Y_{11})$. Device used in this RF study has $L = 70\,\text{nm}$, $W = 0.6\,\mu\text{m}$, $NF = 16$ and $M = 4$.

an additional component to the gate resistance that represents the channel distributed RC effect. When a MOSFET operates at high frequency, not only the distributed RC effect of the gate will show its contribution, but also the nonquasi-static effect or the distributed RC effect of the channel must be considered, as seen by the signal applied to the gate. Thus the effective gate resistance consists of two parts: the distributed gate electrode resistance ($R_{geltd}$) and the distributed channel resistance seen from the gate ($R_{ch}$), as shown in Fig. 6.21.

$$R_g = R_{geltd} + R_{ch} \tag{6.78}$$

Similar to the BSIM4 MOSFET model, BSIM-BULK has a channel-induced gate resistance to incorporate the NQS effect. In the channel-induced gate resistance model, a virtual gate resistance is in series with the physical gate

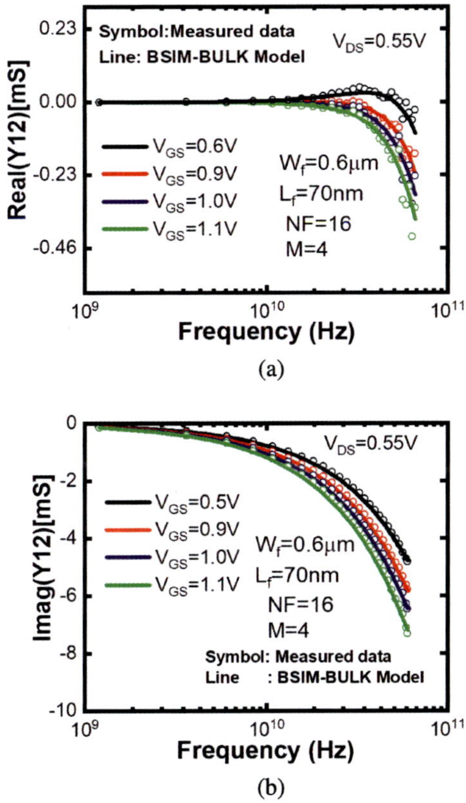


FIGURE 6.17 RF Model validation: (a) $Real(Y_{12})$. (b) $Imag(Y_{12})$. Device used in this RF study has $L = 70\,\text{nm}$, $W = 0.6\,\mu\text{m}$, $NF = 16$ and $M = 4$.

electrode resistance, which is modeled as [31]

$$\frac{1}{R_{ch}} = XRCRG1 \cdot NF \cdot \left( \frac{I_{ds}}{V_{ds}} + XRCRG2 \cdot \frac{W_{eff} \cdot \mu_{eff} \cdot C_{ox} \cdot V_t}{L_{eff}} \right) \tag{6.79}$$

where $I_{ds}$ drain source current, $V_{ds}$ drain source voltage, $\mu_{eff}$ is the effective mobility, $C_{ox}$ is the oxide capacitance per unit area, $V_t$ is thermal voltage, and $XRCRG1$ and $XRCRG2$ are used as fitting parameters.

BSIM-BULK provides four options for modeling gate electrode resistance (bias-independent) and intrinsic-input resistance (IIR, bias-dependent). The IIR model considers the relaxation-time effect due to the distributive RC nature of the channel region and therefore describes the first-order nonquasi-static effect. The model selector RGATEMOD is used to choose different options. There are four model selectors for gate resistance network:

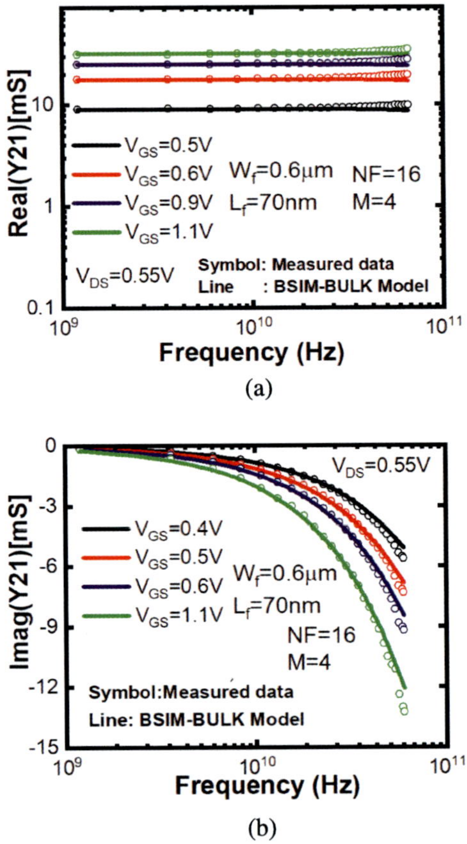


FIGURE 6.18 RF Model validation: (a) $Real(Y_{21})$. (b) $Imag(Y_{21})$. Device used in this RF study has $L = 70\,\text{nm}$, $W = 0.6\,\mu\text{m}$, $NF = 16$ and $M = 4$.

**RGATEMOD** = 0 (zero-resistance): In this case, no gate resistance is generated (see Fig. 6.22(a)).

**RGATEMOD** = 1 (constant-resistance): In this case, only the electrode gate resistance (bias-independent) is generated by adding an internal gate node, as shown in Fig. 6.22(b). The expression of $Rgeltd$ is given in (6.77).

**RGATEMOD** = 2 (IIR model with variable resistance): In this case, the gate resistance is the sum of the electrode gate resistance $Rgeltd$ (6.77) and the intrinsic-input resistance $R_{ii}$ as given by (6.79). An internal gate node will be generated, as shown in Fig. 6.22(c).

**RGATEMOD** = 3 (IIR model with two nodes): In this case, the gate electrode resistance $Rgeltd$ is in series with the intrinsic-input resistance $R_{ii}$ through two internal gate nodes so that the overlap capacitance current will not pass through the intrinsic-input resistance, as illustrated in

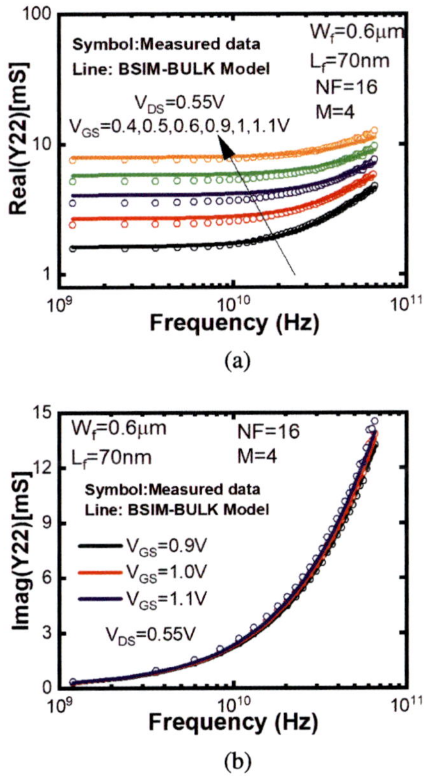


FIGURE 6.19 RF Model validation: (a) $Real(Y_{22})$. (b) $Imag(Y_{22})$. Device used in this RF study has $L = 70\,\text{nm}$, $W = 0.6\,\mu\text{m}$, $NF = 16$ and $M = 4$.

Fig. 6.22(d). Here, $C_{gso}$ and $C_{gdo}$ represent the gate to source and gate to drain overlap capacitances [31].

### 6.8.1 Nonquasi-static effect

The fundamental difference between quasi-static (QS) and nonquasi-static (NQS) model formulations can be briefly outlined as follows. In QS modeling, the charge at any instant of time $t'$ with the time varying voltage $V(t')$ is assumed to be identical to those that would be found, if a DC voltage $V_{DC} = V(t')$ had been frozen permanently at that value [36]. This implied that charge is a function of terminal voltages only, and it does not have any other time explicitly.

$$Q(t) = Q(V_G(t), V_D(t), V_S(t), V_B(t)) \tag{6.80}$$

However, in reality the charge carriers in the channel have nonzero mass, there by it does have some nonnegligible inertia. So, there is a nonzero

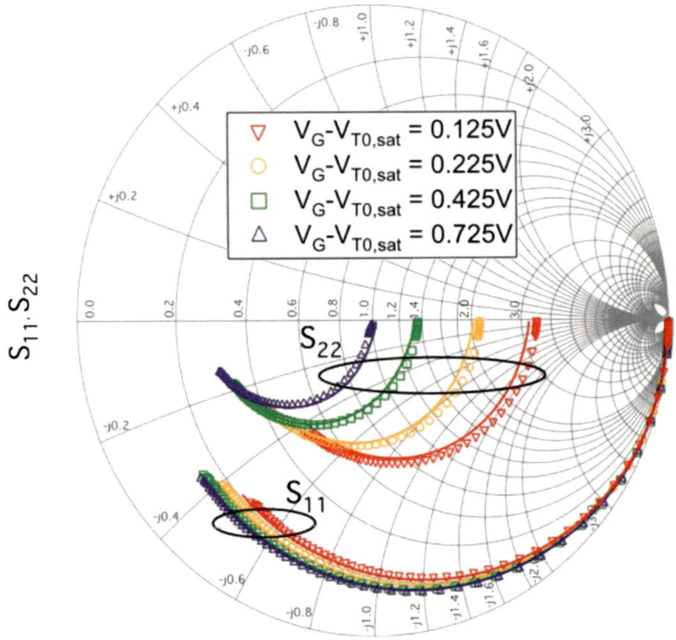


FIGURE 6.20 Smith chart of $S_{11}$ and $S_{22}$ parameters at $V_{GS}$-$V_{THO}$ = [0.125, 0.225, 0.425, 0.725] V and $V_{DS} = 1.1$ V [44].

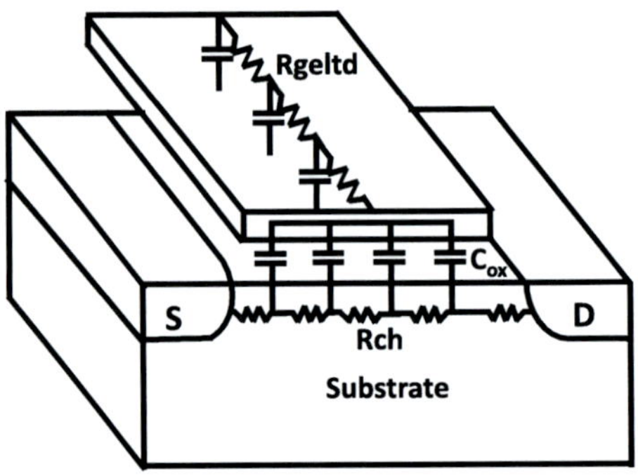


FIGURE 6.21 Illustration of distributed nature of gate electrode resistance $R_{geltd}$, channel resistance $R_{ch}$, and gate capacitance $C_{ox}$.

delay between the cause (any change in terminal voltage) and the effect (corresponding change in current). Therefore the channel charge in this case is not only a function of terminal voltages but also a function of explicit time. This is called the nonquasi-static (NQS) effect [36].

$$Q(t) = Q(t, V_G(t), V_D(t), V_S(t), V_B(t)) \tag{6.81}$$

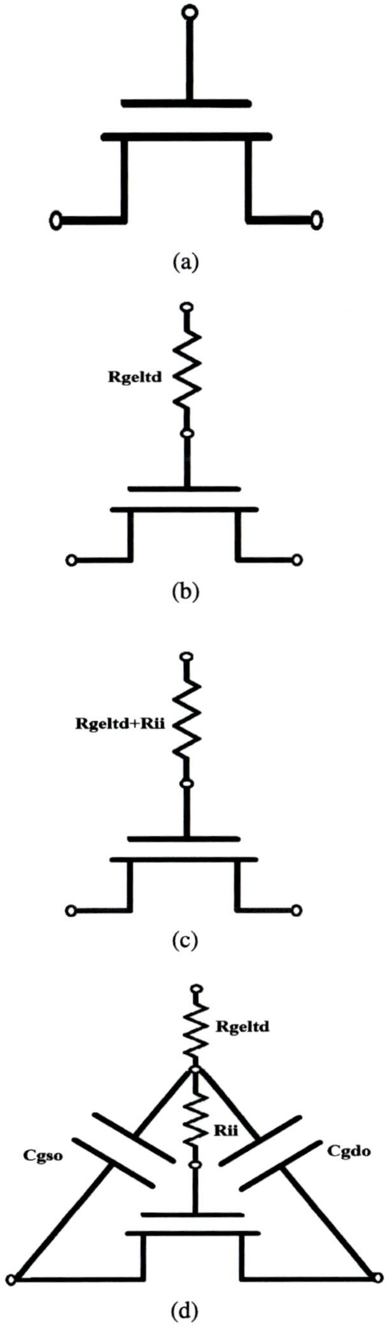


FIGURE 6.22 Gate resistance network for (a) $RGATEMOD = 0$ (b) $RGATEMOD = 1$ (c) $RGATEMOD = 2$ (d) $RGATEMOD = 3$.

If the rise time of the signal is much less than the channel transit time, there will be a nonzero delay between the cause and the effect. This finite time results from the fact that electrons indeed move with a finite speed in the channel. Assuming that the electrons respond with infinite velocity is equivalent to assuming that the resistivity of the channel to be zero. This ignores the distributed channel RC network shown in Fig. 6.21, which is needed to take care of finite channel charging time to reach equilibrium [34].

Similar to BSIM4, BSIM-BULK also offers a first-order channel-induced gate resistance model RGATEMOD=3, as shown in the Fig. 6.22(d) [31] for modeling the NQS effects. In this approach, a virtual gate resistance $R_{ch}$ is in series with the physical gate electrode resistance $R_{geltd}$. This model is accurate up to some frequency range due to the first-order approximations in NQS resistance formulation. In addition, BSIM models also offer a channel segmentation-based nonquasi-static model. In this approach, the channel is broken down into multiple channel segments and current continuity equations at the intermediate internal nodes are solved within the SPICE environment [37]. As the number of channel segments increases, the accuracy increases but at the price of longer simulation time.

## 6.8.2 Improved nonquasi-static effect

Fig. 6.23 shows the measured data (symbols) of $Real(Y_{21})$ versus frequency. In the existing NQS models, which have been used for the RF modeling in the industry standard MOSFET models, either (a) the first-order approximation of channel-reflected NQS resistance (in BSIM-BULK model) or (b) segmentation method is used. But both first-order approximation and segmentation methods (consisting of 10-segments) are insufficient for frequency range above 30 GHz, and, hence, the NQS effect in the millimeter wave band cannot be captured using conventional models. Although segmentation model with segments $\geq 17$ can capture the NQS effect up to $50 f_t$ (see black dotted line), it is computationally expensive. Thus, an accurate and computationally efficient transmission line-based improved physical equivalent circuit model for NQS is derived and implemented in BSIM-BULK to capture the NQS effects [38,39].

#### 6.8.2.0.1 Improved model formulation

**Modeling of Distributed Channel Impedance** As already discussed that in the first-order NQS model used in BSIM4 and BSIM-BULK models, a virtual gate resistance, $R_{ch}$, is used in series with the physical gate electrode resistance $R_{geltd}$, [22,31]. In the improved model [38,39], the input impedance of distributed channel RC network is derived by analyzing the transmission line model at $V_{ds} = 0\,\text{V}$, as shown in Fig. 6.24 [40]. Defining

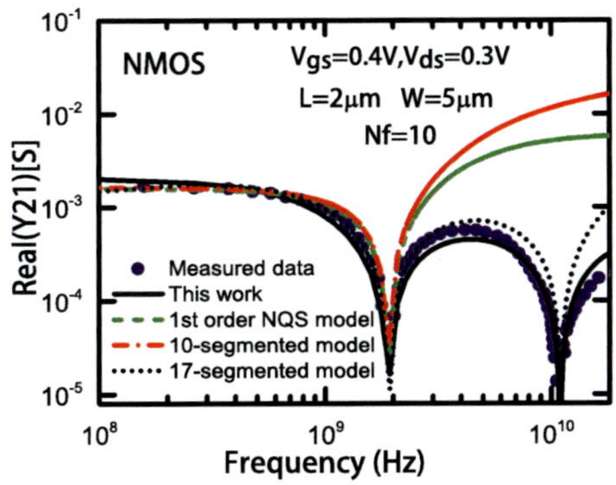


FIGURE 6.23 $Real(Y_{21})$ vs. frequency at $V_{GS} = 0.4$ V and $V_{DS} = 0.3$ V. Symbols indicate the measured data [33], and it shows the NQS effect at very high frequencies. Conventional models in [34] and [35] cannot capture the NQS trend in the millimeter wave range, whereas segmentation model with segments $\geq 17$ can capture the NQS trend, but it significantly increases the computational time. Therefore, to capture the NQS up to $50 f_t$, improved gate resistance network model is required.

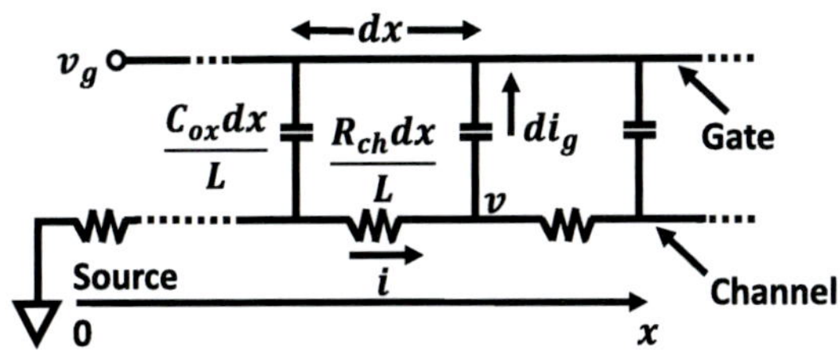


FIGURE 6.24 Transmission line model of MOSFET at $V_{ds} = 0$ V [40].

$C'_{ox} = \frac{C_{ox}}{L}$, $R'_{ch} = \frac{R_{ch}}{L}$ and $u = v - v_g$, from Fig. 6.24, we can write

$$\frac{di_g}{dx} = j\omega C'_{ox}(v - v_g) = \frac{di}{dx} \tag{6.82}$$

$$\frac{du}{dx} = -i R'_{ch} \tag{6.83}$$

Therefore

$$\frac{d^2u}{dx^2} + \gamma^2 u = 0 \tag{6.84}$$

where $\gamma = \sqrt{j\omega R'_{ch} C'_{ox}}$.

Integrating (6.82), we get [40]

$$Z_{in} = \sqrt{\frac{R_{ch}}{C_{ox} \cdot 4 \cdot s}} coth(\sqrt{R_{ch} C_{ox} \cdot s/4}) \tag{6.85}$$

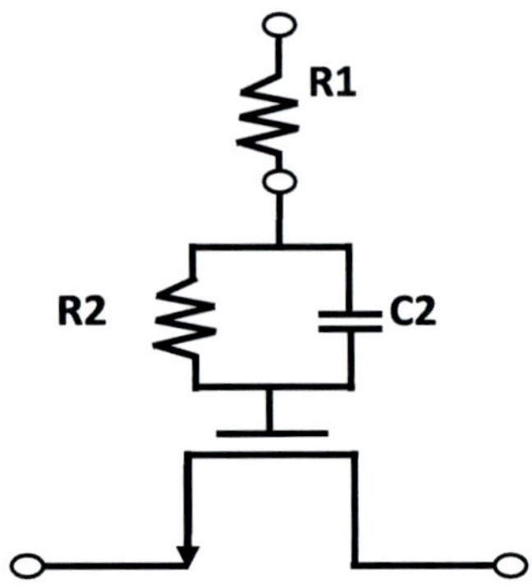


FIGURE 6.25 Proposed channel reflection network, the term $\frac{1}{C_{ox} \cdot s}$ represents the total gate capacitance realized using MOSFET, while the first term in (6.87) is realized using a RC network.

where $s = j\omega$. Using the Taylor series expansion in (6.85), we get

$$Z_{in} = \frac{R_{ch}}{12} + \frac{1}{C_{ox} \cdot s} - \frac{1}{720} \cdot R_{ch}^2 \cdot C_{ox} \cdot s + \frac{1}{30240} \cdot R_{ch}^3 \cdot C_{ox}^2 \cdot s^2 - \frac{1}{1209600} \cdot R_{ch}^4 \cdot C_{ox}^3 \cdot s^3 + ... \tag{6.86}$$

The existing NQS model in BSIM-BULK [31] keeps only the first two terms in (6.86) and ignores the higher-order terms. For $\omega << \frac{1}{R_{ch} C_{ox}}$, by considering first four terms in the expansion, (6.86) can be simplified as,

$$Z_{in} = \frac{R_{ch}}{12} \left( \frac{1 + \frac{1}{140} \cdot R_{ch} \cdot C_{ox} \cdot s}{1 + \frac{1}{42} \cdot R_{ch} \cdot C_{ox} \cdot s} \right) + \frac{1}{C_{ox} \cdot s} \tag{6.87}$$

This $Z_{in}$ in (6.87) can be realized as shown in Fig. 6.25. The impedance of the RC network on top of the MOSFET is given by

$$Z_{in} = \left( R_1 + R_2 \right) \left( \frac{1 + \frac{R_1 R_2}{R_1 + R_2} C_2 s}{1 + R_2 C_2 s} \right) \tag{6.88}$$

By comparing the first term in (6.87) and (6.88), we get

$$\frac{1}{R_1} = XRCRG1 \cdot NF \cdot G_{ch} \tag{6.89}$$

$$\frac{1}{R_2} = XRCRG2 \cdot NF \cdot G_{ch} \tag{6.90}$$

$$C_2 = XRCCG \cdot CGGI \tag{6.91}$$

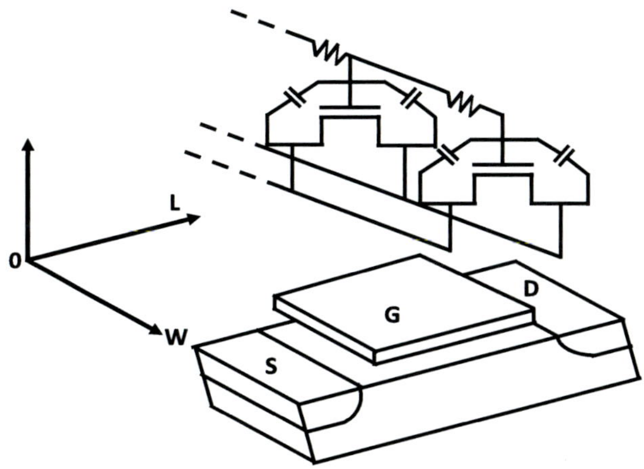


FIGURE 6.26 A visual representation of the gate-distributed RC Network in MOS transistor structure. The gate electrode resistance is distributed along the gates of this individual subtransistor.

where,

$$CGGI = C_{gsi} + C_{gbi} + C_{gdi} \tag{6.92}$$

Here, CGGI is intrinsic gate capacitance and $XRCRG1 = 40$, $XRCRG2 = \frac{120}{7}$ and $XRCCG = \frac{20}{49}$. However, all these coefficients have been kept as fitting parameters. This derivation assumes that channel is a uniformly distributed RC network (which is true for $V_{DS} = 0$), but for nonzero $V_{DS}$, channel is nonuniform from source to drain. Furthermore, there could be some layout effects in the actual fabricated device; in that case, the fitting parameters can provide more flexibility to capture the NQS effects in the millimeter frequency range. Whereas, $G_{ch} = 1/R_{ch}$ is the same as in [31].

**Modeling of Distributed Gate Electrode RC network**

A MOSFET with large gate width (relative to its gate length) can be modeled as an array of transistors distributed along its gate width as shown in Fig. 6.26. The gate electrode resistance is distributed along the gates of this individual subtransistor. For multi-finger device, this distributed effect can be neglected. However, as the device operating frequency approaches the millimeter wave frequency band, this effect cannot be neglected. Here across W, the transmission line can be viewed as in Fig. 6.27

$$\text{where } R'_g = \frac{R_g}{W}, C'_{gs} = \frac{C_{gs}}{W} \text{ and } C'_{gd} = \frac{C_{gd}}{W}. \tag{6.93}$$

Here it is assumed that the DC bias conditions remain the same along the gate width; however, AC small signals applied at the gate region needed

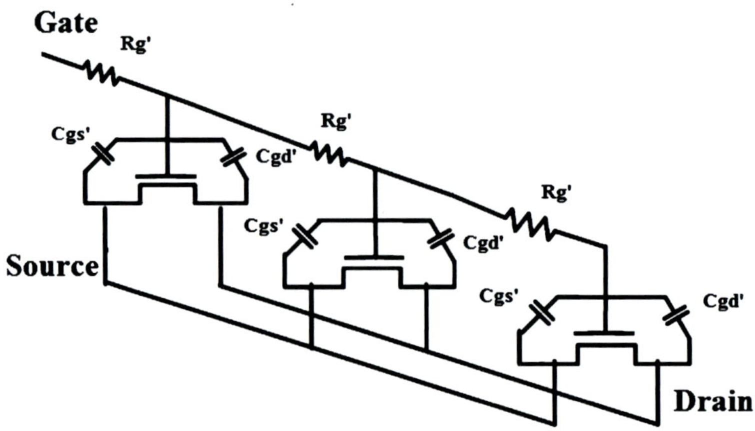


FIGURE 6.27 Transmission line along gate width.

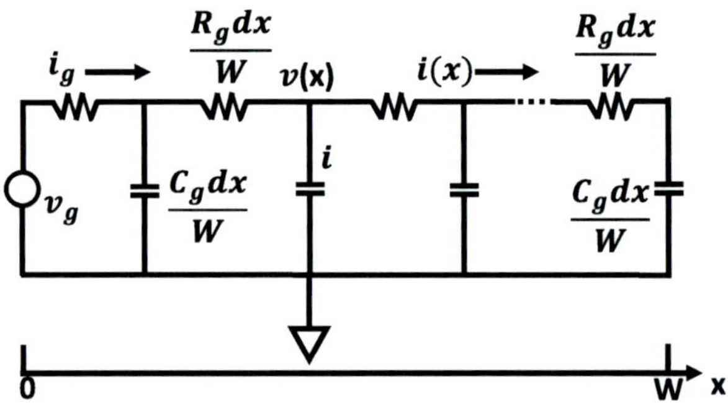


FIGURE 6.28 Small signal equivalent circuit.

to be considered. For calculating the input impedance across gate width, $V_{ds} = 0\,\text{V}$ condition is used. Thus small signal equivalent network at the input side from Fig. 6.28 is reduced to [38,39]. Using the transmission line analysis along gate width [41–43],

$$\frac{dv(x)}{dx} = -i(x)R'_g \tag{6.94}$$

Also,

$$\frac{di(x)}{dx} = j\omega C'_g v(x) \tag{6.95}$$

combining (6.94) and (6.95)

$$\frac{d^2v(x)}{dx^2}+\gamma^2v(x)=0 \tag{6.96}$$

$$\text{where } \gamma=\sqrt{j\omega R'_g C'_g}$$

The general solution for (6.96) is

$$v(x)=K_3sinh(\gamma x)+K_4cosh(\gamma x) \tag{6.97}$$

Using the boundary condition

$$v(0)=v_g, i(W)=0 \tag{6.98}$$

Hence

$$K_3=-v_gtanh(\gamma W) \tag{6.99}$$

and

$$K_4=v_g \tag{6.100}$$

Therefore

$$\frac{di(x)}{dx}=j\omega C'_g[K_3sinh(\gamma x)+K_4cosh(\gamma x)] \tag{6.101}$$

Integrating

$$i_g=\int_0^W j\omega C'_g[K_3sinh(\gamma x)+K_4cosh(\gamma x)]dx \tag{6.102}$$

$$i_g=\frac{j\omega C'_g}{\gamma}[K_3(cosh(\gamma W)-1)+K_4sinh(\gamma W)] \tag{6.103}$$

substituting $K_3$ and $K_4$ and simplifying above expression

$$i_g=\frac{j\omega C'_g}{\gamma}v_g[tanh(\gamma W)] \tag{6.104}$$

the input impedance is given by

$$Z_{in}=R_g\frac{coth(\gamma W)}{\gamma W} \tag{6.105}$$

$$C_g=C_{gsi}+C_{gbi}+C_{gdi}+C_{gso}+C_{gdo}+C_{fring} \tag{6.106}$$

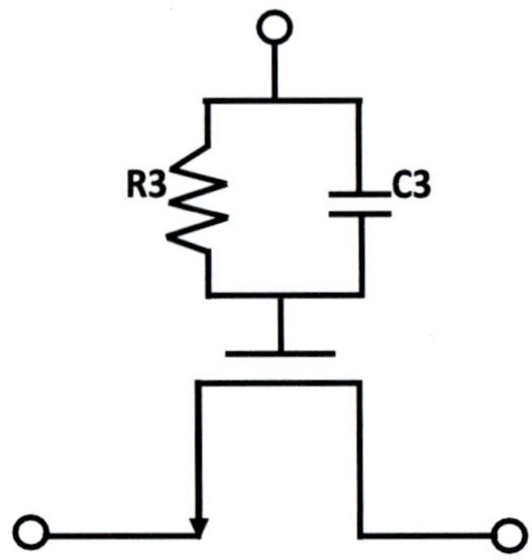


FIGURE 6.29 Realization gate RC network, the term $\frac{1}{C_g \cdot s}$ represents a MOSFET, while the second term in (6.109) is realized using parallel RC network. Here we do not take into account the channel-distributed RC network for the simplicity of analysis.

where $C_{gsi}$ and $C_{gdi}$ are the intrinsic gate-source and gate-drain capacitances, $C_{gso}$ and $C_{gdo}$ are the gate-source and gate-drain overlap capacitances, and $C_{fring}$ is the fringing capacitance. Using the Taylor series expansion,

$$Z_{in} = \frac{R_g}{3} + \frac{1}{C_g \cdot s} - \frac{2}{15} \cdot R_g^2 \cdot C_g \cdot s + ... \tag{6.107}$$

Hence,

$$Z_{in} = \frac{1}{C_g \cdot s} + \frac{R_g}{3}\left(1 - \frac{6}{15} \cdot R_g \cdot C_g \cdot s + ...\right) \tag{6.108}$$

Then, for $\omega << \frac{1}{R_g C_g}$, (6.108) can be simplified as

$$Z_{in} = \frac{1}{C_g \cdot s} + \frac{\frac{R_g}{3}}{1 + \frac{6}{15} \cdot R_g \cdot C_g \cdot s} \tag{6.109}$$

Eq. (6.109) can be realized as shown in Fig. 6.29, where

$$R_3 = \frac{R_g}{3}, C_3 = \frac{6C_{ox}}{5} \tag{6.110}$$

The resistance, $R_3$, uses the same gate electrode resistance model $R_{geltd}$ as in the BSIM-BULK NQS model, and

$$C_3 = XGCCG \cdot CGG \tag{6.111}$$

where

$$CGG = CGGI + C_{gso} + C_{gdo} + C_{fringe}$$

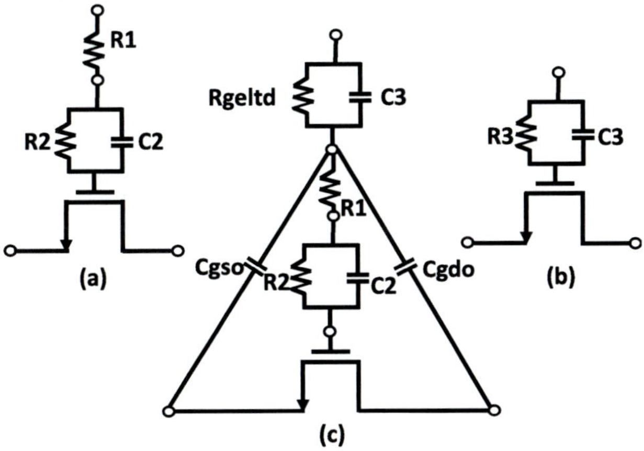


FIGURE 6.30 Proposed model, the equivalent network realized in Fig. 6.25 (Channel-distributed RC network for NQS) and equivalent circuit realization in Fig. 6.29 (Gate electrode-distributed RC network) are merged to produce overall equivalent high-frequency network.

$XGCCG = \frac{6}{5}$ is used as a fitting parameter for accounting the distributed nature of gate electrode RC network, CGG is the total gate capacitance, including both intrinsic gate capacitance CGGI, overlap capacitances at the gate electrode, and its fringing capacitances.

### Complete Model

The complete improved model of NQS after the formulation of channel-distributed RC network (NQS effect), and longitudinal gate electrode distributed RC network is shown in Fig. 6.30 [38,39].

In this new model, one must observe that intrinsic gate capacitance is taken into account twice. The reason for this formulation is that the same intrinsic capacitance makes two distributed RC networks in orthogonal directions, that is, along the channel length with $R_{ch}$ and along the channel width with $R_{geltd}$. Since we are using fitting parameters for accounting this distribution effect, that is $XRCCG$ and $XGCCG$ in capacitances $C_2$ and $C_3$, this double counting will not create any problem in the model accuracy. And the comparison between the old first-order NQS model and the new improved model is illustrated in Fig. 6.31.

Using default values of RC distributed network parameters (coming from derivation), i.e., $XRCRG1 = 40$, $XRCRG2 = 120/7$, $XRCCG = 20/49$, $XGCCG = 6/5$, it is shown that with the new model the NQS fitting for $Real(Y_{21})$ and $Real(Y_{12})$ at $V_{GS} = 0.9\,\text{V}$ and for $Real(Y_{21})$ at $V_{GS} = 0.4\,\text{V}$ in Fig. 6.32(a) to Fig. 6.32(c). Although the new NQS model gives the required NQS trends with the default parameter set, it does not give a good fitting at very high frequency. This is due to the fact that this derivation

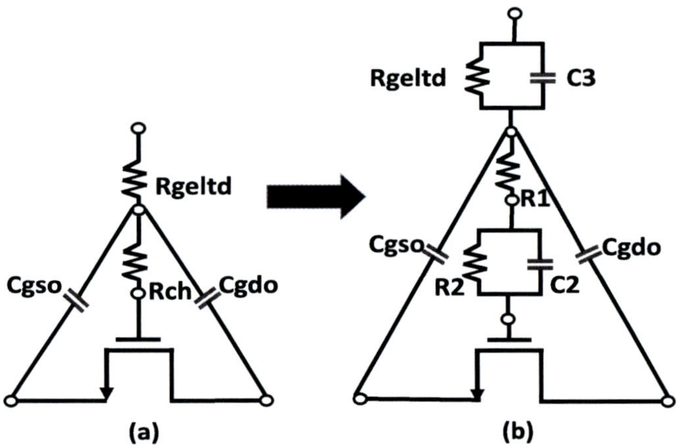


FIGURE 6.31 The transformation of the BSIM-BULK first-order gate network to modified complete proposed model.

is based on the assumption that channel is a uniformly distributed RC network, even for $V_{DS} \neq 0$. But, for nonzero $V_{DS}$, channel is tapered from source to drain, and, hence, it is a nonuniform distributed RC network. Therefore all these coefficients are considered as fitting parameters. Additionally, there could be some layout effects in the real fabricated device; in this case, the fitting parameters can provide more flexibility to capture the NQS effects in the millimeter frequency range. Hence, the complete model (by considering all RC network coefficients as fitting parameters) shows an excellent match with the measured data.

Here, we have used the "gate electrode distributed network" parameters, while validating the model with the measured and 2D TCAD data. Although, 2D TCAD does not consider the impact of gate electrode distributed network on NQS, but the presence of this network in our model increases the fitting flexibility and accuracy of our model at high frequency. The developed model also involves some assumptions to capture distributed effects through lumped circuit. Hence, there ought to be some discrepancy, but this discrepancy is compensated by the fitting parameters and the complete network.

Fig. 6.33(a) - Fig. 6.33(c) show the comparison of our model with 2D TCAD data. It can be easily observed from these figures that when the effect of "gate electrode distributed network" is not taken into account, the model does not match well in high-frequency range (50 GHz - 100 GHz). Fig. 6.33(a) shows the $Real(H_{11})$, which represents the gate impedance. Since model (without gate distributed network) overestimates the gate impedance or $Real(H_{11})$ at high frequency, it does not predict correct trends for $Y_{12}$ and $Y_{21}$ for this frequency range, as shown in Fig. 6.33. The complete model, on the other hand, shows an excellent match with the

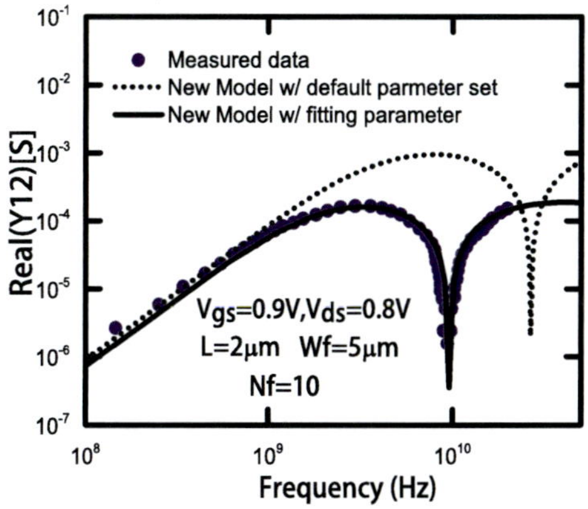


(a)

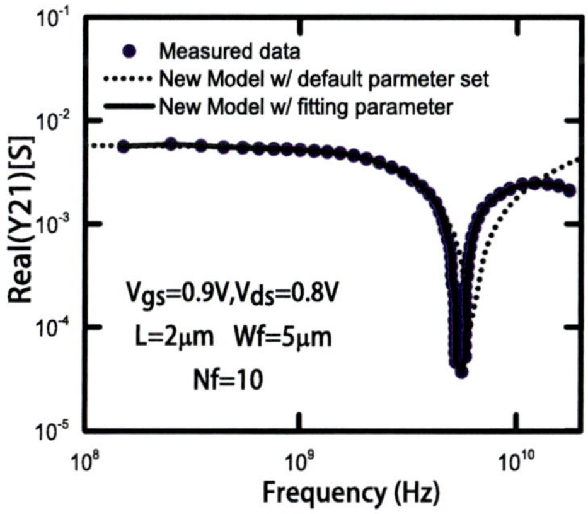


(b)

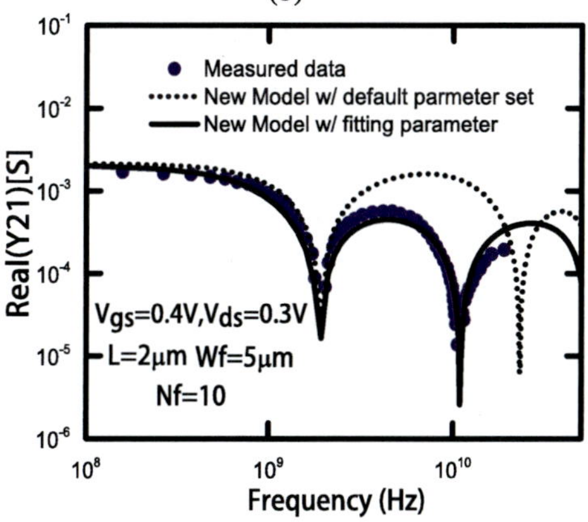


(c)

FIGURE 6.32 (a) shows $Real(Y_{12})$ for $V_{GS} = 0.9\,\text{V}$ and $V_{DS} = 0.8\,\text{V}$, (b) shows $Real(Y_{21})$ for $V_{GS} = 0.9\,\text{V}$ and $V_{DS} = 0.8\,\text{V}$, and (c) shows $Real(Y_{21})$ for $V_{GS} = 0.4\,\text{V}$ and $V_{DS} = 0.3\,\text{V}$. Device used in this RF study has $L = 2\,\mu\text{m}$, $W = 5\,\mu\text{m}$, and $N_f = 10$. Symbols: Measured data, Lines: Model.

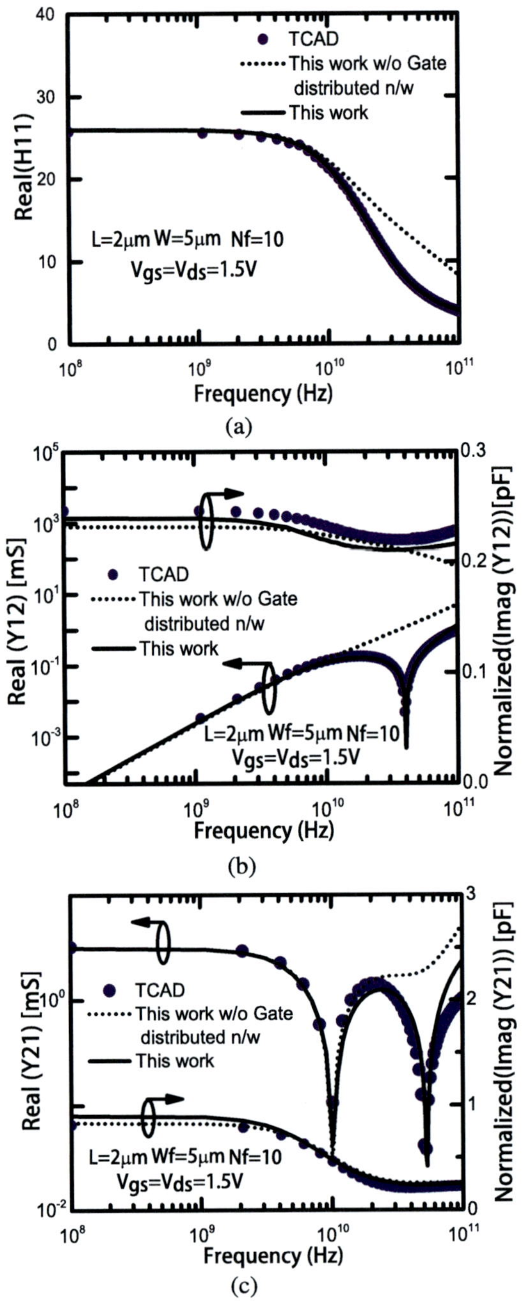


**FIGURE 6.33** RF Model comparison: (a) shows $Real(H_{11})$, (b) shows $Real(Y_{12})$ (Left Y-axis), and Right Y-axis shows the trend of normalized $Imag(Y_{12})$. (c) In $Real(Y_{21})$ (Left Y-axis). Right Y-axis of the same plot shows normalized $Imag(Y_{21})$. Device used in this RF study has $L = 2\,\mu m$, $W = 50\,\mu m$ and $T_{ox} = 3\,nm$. Symbols: TCAD, Lines: Model.

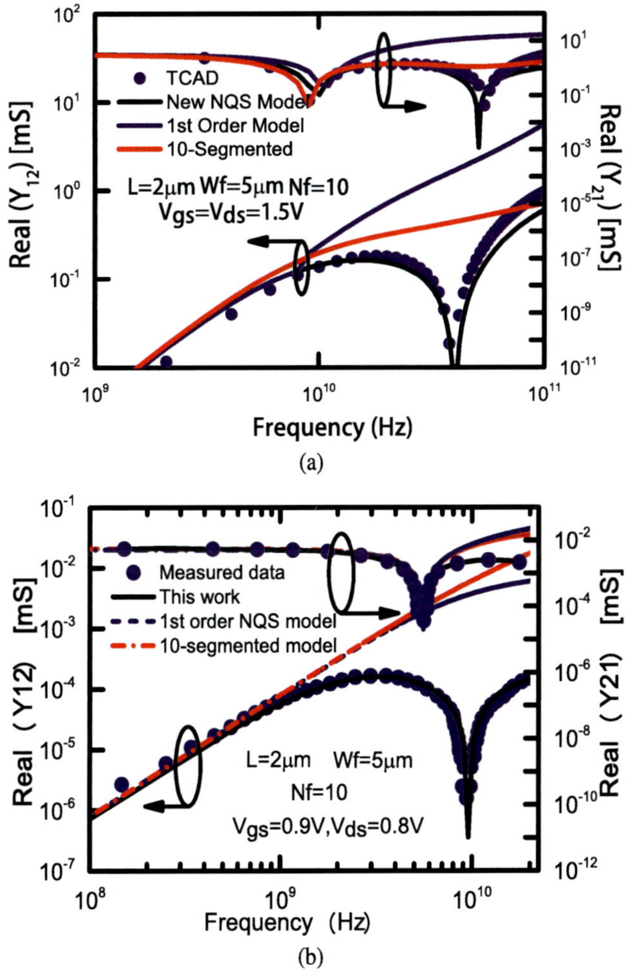


**FIGURE 6.34** Measurement verification of the proposed model for $Real(Y_{12})$ (Left Y-axis) and $Real(Y_{21})$ (Right Y-axis): (a) with TCAD data. (b) with measured data. The BSIM-BULK NQS model and 10-segmented model fail at high-frequency region [38,39]. Solid: Model, Symbols: data.

TCAD data due to an additional parameter "XGCCG", which shows its significant presence at very high frequency.

The model is also applicable for high-frequency small signal and large signal transient analyses. Fig. 6.34(a) and Fig. 6.34(b) prove the accuracy of our NQS model by capturing the significant NQS effects in $Real(Y_{12})$ and $Real(Y_{21})$ up to the frequency range of 100 GHz with both TCAD and measured data. It may be noted here that the existing NQS models cannot fit the TCAD data for high-frequency range. At high frequency, the

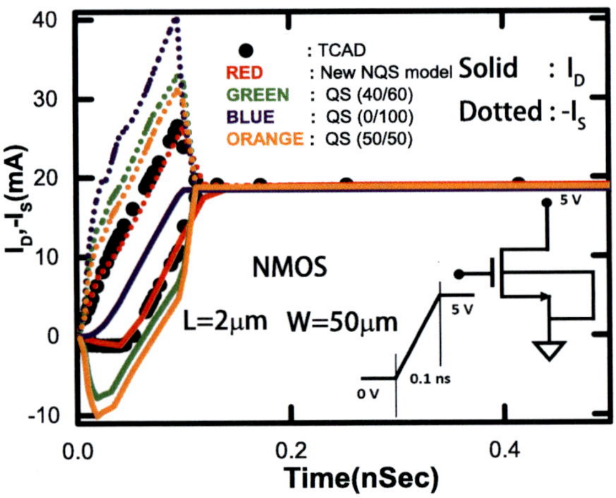


FIGURE 6.35 Drain and source terminal currents for gate switching from 0 to 5 V at a rate of $5X10^{10}$ V/s [38].

substrate network dominates over the gate network, and this leads to a dip in $Real(Y_{12})$ and right dip in $Real(Y_{21})$, as shown in Fig. 6.34(a) and Fig. 6.34(b). The left dip in $Real(Y_{21})$ is due to $g_m$ term in $Real(Y_{21})$ expression [38]. Fig. 6.35, on the other hand, shows the model validation of terminal currents $I_D$ and $I_S$ of MOSFET with the TCAD data. It is evident from Fig. 6.35 that the QS model of BSIM-BULK with charge partitioning drain/source ratios of 0/100; 50/50 and 40/60 cannot capture the transient behavior. However, the new NQS model captures this behavior even for $dV_G/dt = 5X10^{10}$ V/s.

## 6.9 Modeling of the substrate network

The source-body or drain-body junctions form a p-n junction and are playing an integral component of a high-frequency analysis. At very high frequencies, this junction capacitance behaves as a short, and the output conductance becomes influenced by the substrate network.

For CMOS RF circuit simulation, it is essential to consider the high-frequency coupling through the substrate. BSIM-BULK offers a flexible built-in substrate resistance network. This network is constructed such that little simulation efficiency penalty will result. Note that the substrate resistance parameters should be extracted for the total device, not on a per-finger basis.

#### 6.9.0.1 Model selector and topology

The model selector **RBODYMOD** can be used to turn on or turn off the resistance network.

**RBODYMOD** = 0 (Off): No substrate resistance network is generated at all.

**RBODYMOD** = 1 (On): All five resistances $RBPS$, $RBPD$, $RBPB$, $RBSB$, and $RBDB$ in the substrate network as shown schematically below are present simultaneously. A minimum conductance, GBMIN, is introduced in parallel with each resistance, preventing infinite resistance values, which would otherwise cause poor convergence. GBMIN is merged into each resistance to simplify the representation of the model topology. Note that the intrinsic model substrate reference point, in this case, is the internal body node **bNodePrime**, into which the impact ionization current $I_{ii}$ and the GIDL current $I_{GIDL}$ flow.

**RBODYMOD** = 2 (On: Scalable Substrate Network): The schematic is similar to **RBODYMOD** = 1, but all the five resistors in the substrate network are now scalable with a possibility of choosing either five resistors, three resistors or one resistor as the substrate network. The resistors of the substrate network are scalable with respect to channel length (L), channel width (W), and number of fingers (NF). The scalable model allows to account for both horizontal and vertical contacts. The scalable resistors RBPS and RBPD are evaluated through:

$$RBPS = RBPS0 \cdot \left(\frac{L}{10^{-6}}\right)^{RBPSL} \cdot \left(\frac{W}{10^{-6}}\right)^{RBPSW} \cdot NF^{RBPSNF} \tag{6.112}$$

$$RBPD = RBPD0 \cdot \left(\frac{L}{10^{-6}}\right)^{RBPDL} \cdot \left(\frac{W}{10^{-6}}\right)^{RBPDW} \cdot NF^{RBPDNF} \tag{6.113}$$

The resistor RBPB consists of two parallel resistor paths, one to the horizontal contacts and the other to the vertical contacts. These two resistances are scalable and RBPB is given by a parallel combination of these two resistances.

$$RBPBX = RBPBX0 \cdot \left(\frac{L}{10^{-6}}\right)^{RBPBXL} \cdot \left(\frac{W}{10^{-6}}\right)^{RBPBXW} \cdot NF^{RBPDNF} \tag{6.114}$$

$$RBPBY = RBPBY0 \cdot \left(\frac{L}{10^{-6}}\right)^{RBPBYL} \cdot \left(\frac{W}{10^{-6}}\right)^{RBPBYW} \cdot NF^{RBPDNF} \tag{6.115}$$

$$RBPB = \frac{RBPBX \cdot RBPBY}{RBPBX + RBPBY} \tag{6.116}$$

The resistors RBSB and RBDB share the same scaling parameters but have different scaling prefactors. These resistors are modeled in the same way as RBPB. The equations for RBSB are shown below. The calculation for RBDB follows RBSB.

$$RBSBX = RBSBX0 \cdot \left(\frac{L}{10^{-6}}\right)^{RBSBXL} \cdot \left(\frac{W}{10^{-6}}\right)^{RBSBXW} \cdot NF^{RBSDNF} \tag{6.117}$$

$$RBSBY = RBSBY0 \cdot \left(\frac{L}{10^{-6}}\right)^{RBSBYL} \cdot \left(\frac{W}{10^{-6}}\right)^{RBSBYW} \cdot NF^{RBSDNF} \tag{6.118}$$

$$RBSB = \frac{RBSBX \cdot RBSBY}{RBSBX + RBSBY} \tag{6.119}$$

Similarly, the equations for $RBDB$ are as follows

$$RBDBX = RBDBX0 \cdot \left(\frac{L}{10^{-6}}\right)^{RBDBXL} \cdot \left(\frac{L}{10^{-6}}\right)^{RBDBXW} \cdot (NF)^{RBDBXNF} \tag{6.120}$$

$$RBDBY = RBDBY0 \cdot \left(\frac{L}{10^{-6}}\right)^{RBDBYL} \cdot \left(\frac{L}{10^{-6}}\right)^{RBDBYW} \cdot (NF)^{RBDBYNF} \tag{6.121}$$

$$RBDB = \frac{RBDBX \times RBDBY}{RBDBX + RBDBY} \tag{6.122}$$

The implementation of **RBODYMOD** = 2 allows the user to chose between the 5-R network (with all five resistors), 3-R network (with RBPS, RBPD and RBPB), and 1-R network (with only RBPB). If the user does not provide both the scaling parameters RBSBX0 and RBSBY0 for RBSB or both the scaling parameters RBDBX0 and RBDBY0 for RBDB, then the conductances for both RBSB and RBDB are set to GBMIN. This converts the 5-R schematic to 3-R schematic where the substrate network consists of the resistors RBPS, RBPD and RBPB. RBPS, RBPD and RBPB are then calculated using (6.112), (6.113), and (6.116). In BSIM-BULK, the substrate region is modeled as a five resistor network as shown in Fig. 6.36.

If the user chooses not to provide either of RBPS0 or RBPD0, then the 5-R schematic is converted to 1-R network with only one resistor RBPB. The conductances for RBSB and RBDB are set to GBMIN. The resistances RBPS and RBPD are set to 1e-3 Ohm. The resistor RBPB is then calculated

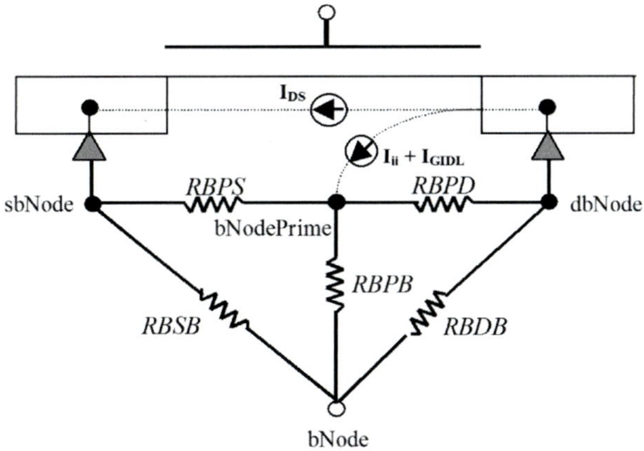


FIGURE 6.36 Topology with the substrate resistance network turned on.

using (6.116). In all other situations, 5-R network is used with the resistor values calculated from the equations aforementioned.

## References

[1] P. Langevin, Comptes Rendus 146 (1908) 530.
[2] D. Lu, Compact Models for Future Generation CMOS, Ph.D. Thesis, UC Berkeley, 2011.
[3] A. Scholten, R. Van Langevelde, L. Tiemeijer, D. Klaassen, Compact modeling of noise in CMOS, in: IEEE Custom Integrated Circuits Conference 2006, 2006, pp. 711–716.
[4] F.N. Hooge, 1/f noise is no surface effect, Physics Letters A 29 (3) (1969) 139–140.
[5] A.L. McWhorter, 1/f noise and germanium surface properties, in: Semiconductor Surface Physics, 1957.
[6] K. Hung, P.K. Ko, C. Hu, Y.C. Cheng, A physics-based MOSFET noise model for circuit simulators, IEEE Transactions on Electron Devices 37 (1990) 1323–1333.
[7] K. Hung, P.-K. Ko, C. Hu, Y. Cheng, A unified model for the flicker noise in metal-oxide-semiconductor field-effect transistors, IEEE Transactions on Electron Devices 37 (3) (Mar 1990) 654–665.
[8] K. Benaissa, et al., New cost-effective integration schemes enabling analog and high-voltage design in advanced CMOS SOC technologies, in: Symposium on VLSI Technology (VLSIT), June 2010, pp. 221–222.
[9] J.-W. Wu, et al., Pocket implantation effect on drain current flicker noise in analog nMOSFET devices, IEEE Transactions on Electron Devices 51 (8) (Aug 2004) 1262–1266.
[10] H. Agarwal, et al., Analysis and modeling of low frequency noise in presence of doping non-uniformity in MOSFETs, in: 2016 3rd International Conference on Emerging Electronics (ICEE), Mumbai, India, 2016, pp. 1–4, https://doi.org/10.1109/ICEmElec.2016.8074572.
[11] C. Blanc, K. Rochereau, S. Bordez, Impact of pocket implantation on low frequency noise performances in MOSFET devices, AIP Conference Proceedings 922 (2007) 137–140.
[12] A.S. Roy, et al., Mechanism of long-channel drain-induced barrier lowering in halo MOSFETs, IEEE Transactions on Electron Devices 58 (4) (April 2011) 979–984.
[13] N. Paydavosi, et al., Flicker noise in advanced CMOS technology: effects of halo implant, in: Solid-State Device Research Conference (ESSDERC), Proceedings of the European, 2013.

[14] A.S. Roy, C.C. Enz, J.M. Sallese, Noise modeling methodologies in the presence of mobility degradation and their equivalence, IEEE Transactions on Electron Devices 53 (2) (Feb 2006) 348–355.
[15] A. Cappy, W. Heinrich, High-frequency FET noise performance: a new approach, IEEE Transactions on Electron Devices 36 (2) (Feb 1989) 403–409.
[16] F.M. Klaassen, Characterization of low 1/f noise in MOS transistors, IEEE Transactions on Electron Devices 18 (10) (Oct 1971) 887–891.
[17] H. Agarwal, S. Khandelwal, S. Dey, C. Hu, Y.S. Chauhan, Analytical modeling of flicker noise in halo implanted MOSFETs, IEEE Journal of the Electron Devices Society 3 (4) (2015) 355–360, https://doi.org/10.1109/JEDS.2015.2424686.
[18] H. Agarwal, P. Kushwaha, C. Gupta, S. Khandelwal, C. Hu, Y.S. Chauhan, Analysis and modeling of flicker noise in lateral asymmetric channel MOSFETs, Solid-State Electronics 115 (2016) 33–38. [Online]. Available: https://www.sciencedirect.com/science/article/pii/S0038110115002750.
[19] H. Agarwal, S. Venugopalan, M.-A. Chalkiadaki, N. Paydavosi, J.P. Duarte, S. Agnihotri, C. Yadav, P. Kushwaha, Y.S. Chauhan, C.C. Enz, A. Niknejad, C. Hu, Recent enhancements in BSIM6 bulk MOSFET model, in: Simulation of Semiconductor Processes and Devices (SISPAD), International Conference on, Sept 2013, https://doi.org/10.1109/SISPAD.2013.6650572.
[20] R. Jindal, Compact noise models for MOSFETs, IEEE Transactions on Electron Devices 53 (9) (2006) 2051 2061.
[21] R. Jayaraman, C. Sodini, A 1/f noise technique to extract the oxide trap density near the conduction band edge of silicon, IEEE Transactions on Electron Devices 36 (9) (1989) 1773–1782.
[22] BSIM4 Technical Manual. [Online]. Available: http://www-device.eecs.berkeley.edu/bsim/?page=BSIM4.
[23] G. Gildenblat, Compact Modeling: Principles, Techniques and Applications, Springer, 2010.
[24] H. Agarwal, S. Khandelwal, Y.S. Chauhan, C. Hu, Noise modeling in BSIM6 compact model, in: NSTI Nanotech, June 2014.
[25] BSIM6 Technical Manual. [Online]. Available: http://www-device.eecs.berkeley.edu/bsim/?page=BSIM6_LR, 2014.
[26] H. Agarwal, Modeling of Drain Current, Transconductance and Flicker Noise in Presence of Doping Non-Uniformity, PhD Thesis, IIT Kanpur, 2017.
[27] A.A. Abidi, RF-CMOS comes of age, IEICE Transactions on Electronics 87 (6) (2004) 840–853.
[28] S. Voinigescu, S. Tarasewicz, T. MacElwee, J. Ilowski, An assessment of the state-of-the-art 0.5/spl mu/m bulk CMOS technology for RF applications, in: Proceedings of International Electron Devices Meeting, IEEE, 1995, pp. 721–724.
[29] C. Enz, An MOS transistor model for RF IC design valid in all regions of operation, IEEE Transactions on Microwave Theory and Techniques 50 (1) (2002) 342–359.
[30] M.-A. Chalkiadaki, C.C. Enz, RF small-signal and noise modeling including parameter extraction of nanoscale MOSFET from weak to strong inversion, IEEE Transactions on Microwave Theory and Techniques 63 (7) (2015) 2173–2184.
[31] A. Sharma, R. Goel, C. Dabhi, G. Pahwa, Y. Chauhan, S. Salahuddin, C. Hu, BSIM-BULK107.1.0 MOSFET compact model, 2022.
[32] S. Venugopalan, From Poisson to Silicon - Advancing Compact Spice Models for IC Design, University of California, Berkeley, 2013.
[33] M. Bucher, A. Bazigos, S. Yoshitomi, N. Itoh, A scalable advanced RF IC design-oriented MOSFET model, International Journal of RF and Microwave Computer-Aided Engineering: Co-sponsored by the Center for Advanced Manufacturing and Packaging of Microwave, Optical, and Digital Electronics (CAMPmode) at the University of Colorado at Boulder 18 (4) (2008) 314–325.

[34] X. Jin, J.-J. Ou, C.-H. Chen, W. Liu, M.J. Deen, P.R. Gray, C. Hu, An effective gate resistance model for CMOS RF and noise modeling, in: International Electron Devices Meeting 1998. Technical Digest (Cat. No. 98CH36217), IEEE, 1998, pp. 961–964.
[35] A.C. Aarts, G.D. Smit, A.J. Scholten, D.B. Klaassen, A PSP-based small-signal MOSFET model for both quasi-static and nonquasi-static operations, IEEE Transactions on Electron Devices 55 (6) (2008) 1424–1432.
[36] Y. Tsividis, Operation and Modeling of the MOS Transistor, McGraw-Hill, Inc., 1987.
[37] S. Venugopalan, K. Dandu, S. Martin, R. Taylor, C. Cirba, X. Zhang, A.M. Niknejad, C. Hu, A non-iterative physical procedure for RF CMOS compact model extraction using BSIM6, in: Proceedings of the IEEE 2012 Custom Integrated Circuits Conference, IEEE, 2012, pp. 1–4.
[38] C. Gupta, N. Mohamed, H. Agarwal, R. Goel, C. Hu, Y.S. Chauhan, Accurate and computationally efficient modeling of nonquasi static effects in MOSFETs for millimeter-wave applications, IEEE Transactions on Electron Devices 66 (1) (2019) 44–51.
[39] N. Mohamed, H. Agarwal, C. Gupta, Y.S. Chauhan, Compact modeling of nonquasi-static effect in bulk MOSFETs for RF circuit design in sub-THz regime, in: 2016 3rd International Conference on Emerging Electronics (ICEE), 2016, pp. 1–4.
[40] G. Gildenblat, X. Li, W. Wu, H. Wang, A. Jha, R. Van Langevelde, G.D. Smit, A.J. Scholten, D.B. Klaassen, PSP: an advanced surface-potential-based MOSFET model for circuit simulation, IEEE Transactions on Electron Devices 53 (9) (2006) 1979–1993.
[41] S.F. Tin, A.A. Osman, K. Mayaram, Comments on "A small-signal MOSFET model for radio frequency IC applications", IEEE Transactions on Computer-Aided Design of Integrated Circuits and Systems 17 (4) (1998) 372–374.
[42] E. Abou-Allam, T. Manku, A small-signal MOSFET model for radio frequency IC applications, IEEE Transactions on Computer-Aided Design of Integrated Circuits and Systems 16 (5) (1997) 437–447.
[43] E. Abou-Allam, T. Manku, An improved transmission-line model for MOS transistors, IEEE Transactions on Circuits and Systems. 2, Analog and Digital Signal Processing 46 (11) (1999) 1380–1387.
[44] Y.S. Chauhan, et al., BSIM6: analog and RF compact model for bulk MOSFET, IEEE Transactions on Electron Devices 61 (2) (Feb. 2014) 234–244, https://doi.org/10.1109/TED.2013.2283084.

CHAPTER

# 7

# Junction Diode and Layout Dependent Parasitic Model

## 7.1 Basic formulation

An analytic expression for the ideal diode current from the Boltzmann transport model is as follows [1]:

$$I_j = I_{j0}.\left(e^{\frac{qV_j}{KT}} - 1\right) \tag{7.1}$$

where $V_j$ is the voltage bias across the diode, and $I_{j0}$ is the reverse-bias saturation current of the junction. While modeling, the source/drain-substrate junction diode is divided into three distinct regions – the junction bottom or the area component, the shallow trench isolation (STI) side junction or the perimeter component, and the MOSFET gate-edge component. These are modeled using various parameters related to the source and drain sides.

The reverse-bias saturation current attributed from the source/drain side diode region is given by:

$$I_{sbs(jd0)} = A_{seff(deff)} \cdot JSS(JSD) + P_{seff(deff)} \cdot JSSWS(JSSWD) + W_{eff} \cdot NF \cdot JSSWGS(JSSWGD) \tag{7.2}$$

where $A_{seff(deff)}$ is the area of the source junction, $P_{seff(deff)}$ is the perimeter length of the source junction, and $W_{eff}$ is the effective width of the MOSFET. Further, **JSS(JSD)**, **JSSWS(JSSWD)**, and **JSSWGS(JSSWGD)** are the saturation current densities of the source (drain)/bulk area junction, source (drain)/bulk sidewall junction, and gate-edge [2]. For certain values of reverse-bias voltage, it is also possible that the diode current increases exponentially with the applied voltage due to the breakdown phenomenon. As the reverse-bias voltage increases, the strength of the electric field across the depletion region of the p-n junction increases. As a

BSIM-Bulk MOSFET Model for IC Design - Digital, Analog, RF and High-Voltage
https://doi.org/10.1016/B978-0-32-385677-5.00011-1

result, the mobile electrons moving through the depletion layer are accelerated and gain high kinetic energy. When these mobile electrons collide with atoms comprising a crystal lattice, their kinetic energy excites their valence electrons, creating more electron-hole pairs and contributing to the current avalanche. Under similar conditions of high field, some highly doped junctions attribute to a quantum mechanical effect known as tunneling, due to which the electrons can jump from one side of the junction.

## 7.2 Asymmetric MOS junction diode models

### 7.2.1 Junction diode IV model

In BSIM-BULK, there is only one diode model (DIOMOD=2 from BSIM4), which includes resistance and breakdown. BSIM-BULK models the diode breakdown with current limiting in both forward **IJTHSFWD** or **IJTHDFWD** and reverse operations **XJBVS**, **XJBVD**, **BVS**, and **BVD** [3–6].

#### 7.2.1.0.1 Source/body junction diode

The equations for the source-side diode are as follows:

$$I_{bs} = I_{sbs}\left[exp\left(\frac{V_{bs}}{NJS \cdot V_t}\right) - 1\right] \cdot f_{breakdown} \tag{7.3}$$

where $I_{sbs}$ is the total saturation current consisting of the components through the gate-edge ($J_{sswgs}$) and isolation-edge sidewalls ($J_{ssws}$) and the bottom junction ($J_{ss}$),

$$I_{sbs} = A_{seff} J_{ss}(T) + P_{seff} J_{ssws}(T) + W_{eff} \cdot NF \cdot J_{sswgs}(T) \tag{7.4}$$

where the breakdown factor, $f_{breakdown}$, is empirically modeled as

$$f_{breakdown} = 1 + XJBVS \cdot exp\left(\frac{-(V_{bs} + BVS)}{NJS \cdot V_t}\right) \tag{7.5}$$

In the above equations, $V_{bs}$ is the substrate to source terminal bias voltage. **XJBVS** represents the breakdown coefficient, **BVS** is the source-to-substrate breakdown voltage, and **NJS** is the nonideality factor used to capture any second-order effects in the junction diode current model with respect to a real diode, for example, an additional leakage currents due to traps in the junction region.

In BSIM-BULK, the effect of series resistance offered by the quasi-static regions of the diodes is modeled by considering the diode junction current to switch smoothly from the exponential to linear behavior, depending on the applied voltage across the diode. In the forward-bias region, the value of the current after which the switching occurs is indicated by the

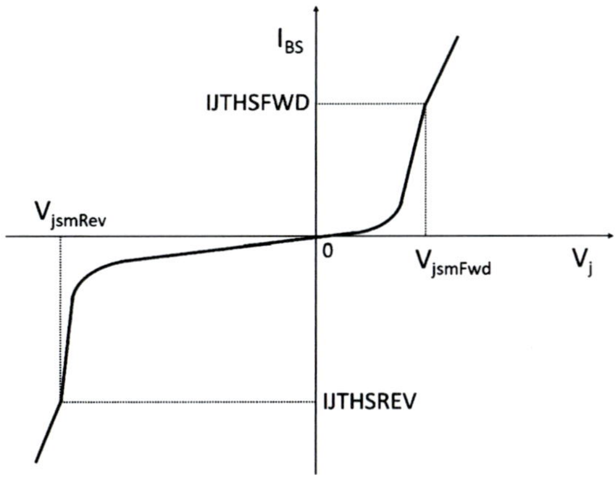


FIGURE 7.1 Junction diode current for the three regions as described in the BSIM-BULK model.

parameter **IJTHSFWD** in the model, and by **IJTHSREV** in the reverse bias condition. As a result, the diode current has been modeled into three regions: less than **IJTHSREV**, in between **IJTHSREV** and **IJTHSFWD**, and greater than **IJTHSFWD** (Fig. 7.1). Let us consider the first case when the diode junction current is greater than **IJTHSFWD**. Thus the bias voltage, $V_{jsmFwd}$, at which this occurs can be found by substituting **IJTHSFWD** and $V_{jsmFwd}$ into Eq. (7.3). After some mathematical manipulations, one can obtain the quadratic equation

$$X^2 - C_1 \cdot X - C_2 = 0 \tag{7.6}$$

$$\textbf{where, } X = exp\left(\frac{VjsmFwd}{NJS \cdot V_t}\right)$$

$$C_1 = 1 + \frac{IJTHSFWD}{I_{sbs}} - XJBVS \cdot exp\left(\frac{-BVS}{NJS \cdot V_t}\right)$$

$$C_2 = XJBVS \cdot exp\left(\frac{-BVS}{NJS \cdot V_t}\right)$$

The derived value of $V_{jsmFwd}$ is given below:

$$VjsmFwd = NJS \cdot V_t \cdot log\left(\frac{C_1 + \sqrt{C_1^2 + 4 \cdot C_2}}{2}\right) \tag{7.7}$$

For the junction current greater than **IJTHSFWD**, the diode current is obtained using a first-order Taylor expansion of Eq. (7.3) at $V_{bs} = VjsmFwd$ and is given as follows:

$$I_{js} = IJTHSFWD + k_{SlopeFwd} \cdot (V_{bs} - VjsmFwd) \tag{7.8}$$

If one would take out the first-order derivative of Eq. (7.3) with respect to $V_{bs}$ at $V_{bs} = VjsmFwd$, the slope factor $k_{SlopeFwd}$ can be obtained:

$$k_{SlopeFwd} = \frac{q.I_{sbs}}{NJS.KT} \cdot \left( exp\left(\frac{q.VjsmFwd}{NJS.KT}\right) + XJBVS.exp\left(\frac{-BVS + V_{bs}}{NJS.KT}\right)\right) \quad (7.9)$$

Under reverse breakdown, the diode junction current can be approximated as follows:

$$I_{bs,rev} = I_{sbs} \cdot f_{breakdown} \quad (7.10)$$

The bias voltage $V_{jsmRev}$ at which we transition to a linear behavior can be found by substituting $I_{bs,rev} = IJTHSREV$ and $V_{bs} = V_{jsmRev}$ into Eq. (7.10)

$$VjsmRev = -BVS - NJS \cdot V_t \cdot log\left(\frac{IJTHSREV - I_{sbs}}{XJBVS \cdot I_{sbs}}\right) \quad (7.11)$$

The junction diode current for junction current greater than **IJTHSREV** can be obtained using a first-order Taylor expansion of Eq. (7.10) at $V_{bs} = VjsmRev$ and is given as follows:

$$I_{js} = \left[exp\left(\frac{V_{bs}}{NJS \cdot V_t}\right) - 1\right] \cdot [IJTHSREV + k_{SlopeRev}.(V_{bs} - VjsmRev)] \quad (7.12)$$

Finally, the overall source-side diode junction current in the BSIM-BULK model is given as [3,4]

$$I_{bs} = \begin{cases} \left[exp\left(\frac{V_{bs}}{NJS \cdot V_t}\right) - 1\right] \cdot [IJTHSREV + k_{SlopeRev}.(V_{bs} - VjsmRev)] \\ \quad \text{if } V_{bs} < VjsmRev \\ I_{sbs}\left[exp\left(\frac{V_{bs}}{NJS \cdot V_t}\right) - 1\right] \cdot f_{breakdown} \quad \text{if } VjsmRev < V_{bs} < VjsmFwd \\ IJTHSFWD + k_{SlopeFwd}.(V_{bs} - VjsmFwd) \quad \text{otherwise} \end{cases} \quad (7.13)$$

**Temperature Dependence of Source Diode Current Characteristics:** The p-n junction exhibits the temperature dependence in the current, which comes from the saturation current term, which, in turn, gets the temperature dependence from the intrinsic carrier density ($n_i$) and the energy band gap ($E_g$) [1,7,8]. If we talk about the saturation current, it contains two different components: 1) Generation Current due to the thermal generation inside the space-charge area and 2) Diffusion current attributed to minority carriers diffusing across the depletion region. Both these components of the generation current are the strong function of temperature:

$$I_{generation} \propto n_i \propto T^{3/2}.exp\left(\frac{-E_g}{2KT}\right), \tag{7.14}$$

$$I_{generation} \propto n_i^2 \propto T^{3}.exp\left(\frac{-E_g}{KT}\right). \tag{7.15}$$

Both (7.14) and (7.15) can be combined into a single equation:

$$I_{Saturation} \propto T^{XTIS}.exp\left(\frac{-E_g}{NJS.KT}\right), \tag{7.16}$$

$$I_{Saturation} \propto exp(ln(T^{XTIS})).exp\left(\frac{-E_g}{NJS.KT}\right), \tag{7.17}$$

$$\text{or } I_{Saturation} \propto exp\left(\frac{\frac{-E_g}{KT} + XTIS.lnT}{NJS}\right). \tag{7.18}$$

The parameter **XTIS** is a junction current temperature exponent, and **NJS** is a junctions emission coefficient.

Therefore the temperature dependence of the source-side current densities is as follows [3–6]:

$$J_{SS}(T) = JSS.exp\left(\frac{\frac{E_{g,TNOM}}{K.TNOM} - \frac{E_g}{K.T} + XTIS.ln\left(\frac{T}{TNOM}\right)}{NJS}\right), \tag{7.19}$$

$$J_{SWS}(T) = JSSWS.exp\left(\frac{\frac{E_{g,TNOM}}{K.TNOM} - \frac{E_g}{K.T} + XTIS.ln\left(\frac{T}{TNOM}\right)}{NJS}\right), \tag{7.20}$$

$$J_{SWGS}(T) = JSWGS.exp\left(\frac{\frac{E_{g,TNOM}}{K.TNOM} - \frac{E_g}{K.T} + XTIS.ln\left(\frac{T}{TNOM}\right)}{NJS}\right). \tag{7.21}$$

#### 7.2.1.0.2 Drain/body junction diode

The equations for the drain-side diode are as follows [3–6]:

$$I_{bd} = I_{sbd}\left[\exp\left(\frac{V_{bd}}{NJD \cdot V_t}\right) - 1\right] \cdot f_{breakdown} \tag{7.22}$$

where $I_{sbd}$ is the total saturation current consisting of the components through the gate-edge ($J_{sswgd}$) and isolation-edge sidewalls ($J_{sswd}$) and the bottom junction ($J_{ss}$),

$$I_{sbd} = A_{deff} J_{SD}(T) + P_{deff} J_{SSWD}(T) + W_{eff} \cdot NF \cdot J_{SSWGD}(T) \tag{7.23}$$

The exponential term in (7.22) is linearized at both the limiting current **IJTHDFWD** in the forward-bias mode and the limiting current **IJTH-DREV** in the reverse-bias mode. In (7.22), $f_{breakdown}$ is given by

$$f_{breakdown} = 1 + XJBVD \cdot exp\left( - \frac{\cdot(BVD + V_{bd})}{NJD \cdot V_t} \right) \tag{7.24}$$

Also, the temperature dependence of the drain-side current densities is as follows:

$$J_{SD}(T) = JSD.exp\left( \frac{\frac{E_{g,TNOM}}{K.TNOM} - \frac{E_g}{K.T} + XTID.ln\left(\frac{T}{TNOM}\right)}{NJD} \right), \tag{7.25}$$

$$J_{SSWD}(T) = JSSWD.exp\left( \frac{\frac{E_{g,TNOM}}{K.TNOM} - \frac{E_g}{K.T} + XTID.ln\left(\frac{T}{TNOM}\right)}{NJD} \right), \tag{7.26}$$

$$J_{SSWGD}(T) = JSSWGD.exp\left( \frac{\frac{E_{g,TNOM}}{K.TNOM} - \frac{E_g}{K.T} + XTID.ln\left(\frac{T}{TNOM}\right)}{NJD} \right). \tag{7.27}$$

Here, the parameter **XTID** is a junction current temperature exponent, and **NJD** is a junctions emission coefficient.

#### 7.2.1.0.3 Reverse-bias additional leakage model

A "trap" is a defect attributed to impurities and physical lattice defects in the silicon in a crystal lattice, which creates a "deep" energy state where charge carriers (electrons or holes) can become trapped. These traps lead to Shockley-Read-Hall (SRH) current. Also, these traps generate electron and hole tunneling across the depletion region in the junction. This trap-assisted tunneling (TAT) leakage current component is more dominant in the reverse-bias condition in the presence of the high built-in electric field of the diode. Thus TAT is also known as electric-field enhanced SRH current, where the electrons from the valence band of the p-type region get trapped into a trap in the depletion region and move out into the conduction band of the n-type region of the diode. The TAT leakage current shows its presence in both forward and reverse bias regimes of the diode junction current. However, in the forward bias, their impact is seen only for low bias well before the diode turn on, which in the model will be taken care of by the nonideality factor (NJS/NJD). On the other hand, under reverse-bias operation, this leakage component is significant.

The TAT current is given by [9,10]

$$I_{jn,TAT} = K_{SRH}.(1 + \Gamma_{TAT}).W_{dep}.\left( e^{\frac{V_{jn}}{2KT}} - 1 \right) \tag{7.28}$$

where $K_{SRH}$ accounts for the SRH electron/hole trap capture cross-section and trap density in the depletion region, $\Gamma_{TAT}$ is a field enhancement factor, and $V_{jn}$ is the bias voltage across the junction. Further, $W_{dep}$ is the depletion width. In the BSIM-BULK model, the reverse junction leakage current for the source-side junction is given by [3,4]:

$$\begin{aligned} I_{bs\_total} &= I_{bs} \\ &- W_{eff} \cdot NF \cdot J_{tsswgs}(T) \\ &\cdot \left[exp\left(\frac{-V_{bs}}{NJTSSWG(T) \cdot Vtm0} \cdot \frac{VTSSWGS}{VTSSWGS - V_{bs}}\right)\right] \\ &- P_{seff} J_{tssws}(T)\left[exp\left(\frac{-V_{bs}}{NJTSSW(T) \cdot Vtm0} \cdot \frac{VTSSWS}{VTSSWS - V_{bs}}\right) - 1\right] \\ &- A_{seff} J_{tss}(T)\left[exp\left(\frac{-V_{bs}}{NJTS(T) \cdot Vtm0} \cdot \frac{VTSS}{VTSS - V_{bs}}\right) - 1\right] \end{aligned} \tag{7.29}$$

where $J_{tss}(T)$, $J_{tssws}(T)$, and $J_{tsswgs}(T)$ are the temperature-dependent pre-exponential tuning factors for the bottom area, STI side perimeter, and gate side perimeter junctions. **NJTS(T)**, **NJTSSW(T)**, and **NJTSSWG(T)** are the corresponding temperature-dependent nonideality factors. $V_{bs}$ is the junction substrate to source terminal bias voltage. The term within the exponential has been added of the form $VTX/(VTX\text{-}V_{bs})$ so that, under the forward-bias condition, the effect of the exponential terms rapidly decreases to zero. This is due to the fact that the influence of the SRH leakage component has already been taken into account using the term **NJS** in Eq. (7.29).

$$\begin{aligned} I_{bd\_total} &= I_{bd} \\ &- W_{eff} \cdot NF \cdot J_{tsswgd}(T) \\ &\cdot \left[exp\left(\frac{-V_{bd}}{NJTSSWGD(T) \cdot Vtm0} \cdot \frac{VTSSWGD}{VTSSWGD - V_{bd}}\right)\right] \\ &- P_{deff} J_{tsswd}(T)\left[exp\left(\frac{-V_{bd}}{NJTSSWD(T) \cdot Vtm0} \cdot \frac{VTSSWD}{VTSSWD - V_{bd}}\right) - 1\right] \\ &- A_{deff} J_{tsd}(T)\left[exp\left(\frac{-V_{bd}}{NJTD(T) \cdot Vtm0} \frac{VTSD}{VTSD - V_{bd}}\right) - 1\right] \end{aligned} \tag{7.30}$$

where $J_{tsd}(T)$, $J_{tsswd}(T)$, and $J_{tsswgd}(T)$ are the temperature-dependent pre-exponential tuning factors for the bottom area, STI side perimeter, and gate side perimeter junctions. **NJTD(T)**, **NJTSSWD(T)**, and **NJTSSWGD(T)** are the corresponding temperature-dependent nonideality factors. $V_{bs}$ is the junction substrate to source terminal bias voltage. Fig. 7.2 shows the $I_{bs}$ versus $V_{bs}$ plot, including the tunneling component (Eqs. (7.13) and (7.29)).

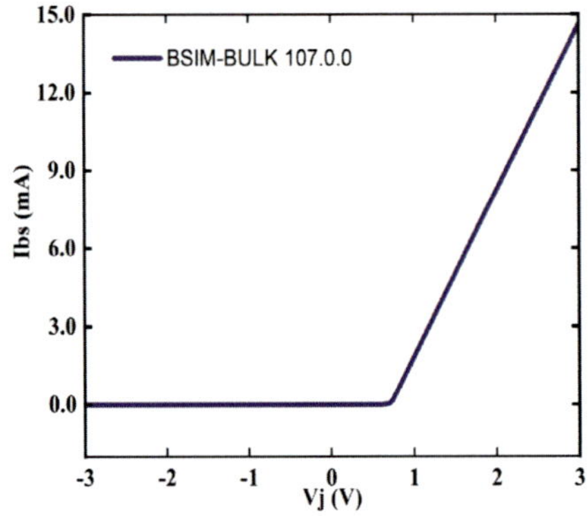


FIGURE 7.2 Total Junction diode current, including tunneling current of BSIM-BULK, is plotted.

**Temperature Dependence of Trap-Assisted Tunneling Current:** The temperature dependence for the trap-assisted tunneling saturation current has been modeled as follows.

On the source side, we have

$$J_{tss}(T) = JTSS.exp\left(\frac{E_{g,TNOM}.XTSS.\left(\frac{T}{TNOM} - 1\right)}{}\right), \tag{7.31}$$

$$J_{tssws}(T) = JTSSWS.exp\left(\frac{E_{g,TNOM}.XTSSWS.\left(\frac{T}{TNOM} - 1\right)}{}\right), \tag{7.32}$$

$$J_{tsswgs}(T) = JTSSWGS.\left(\sqrt{\frac{JTWEFF}{W_{eff0}}} + 1\right) .exp\left(\frac{E_{g,TNOM}.XTSSWGS.\left(\frac{T}{TNOM} - 1\right)}{}\right), \tag{7.33}$$

where **JTSS** is the current density for the junction bottom, **JTSSWS** is the current density for the junction sidewall, and **JTSSWGS** is the current density for the gate sidewall.

$$NJTS(T) = NJTS.\left[1 + TNJTS.\left(\frac{T}{TNOM} - 1\right)\right], \tag{7.34}$$

$$NJTSSW(T) = NJTSSW.\left[1 + TNJTSSW.\left(\frac{T}{TNOM} - 1\right)\right], \tag{7.35}$$

$$NJTSSWG(T) = NJTSSWG.\left[1 + TNJTSSWG.\left(\frac{T}{TNOM} - 1\right)\right], \tag{7.36}$$

where **NJTS** corresponds to the junction bottom, **NJTSSW** corresponds to the sidewall, and **NJTSSWG** corresponds to the gate sidewall.

Similarly, we can also write on the drain side:

$$J_{tsd}(T) = JTSD.exp\left(\frac{E_{g,TNOM}.XTSD.\left(\frac{T}{TNOM} - 1\right)}{}\right), \tag{7.37}$$

$$J_{tsswd}(T) = JTSSWD.exp\left(\frac{E_{g,TNOM}.XTSSWD.\left(\frac{T}{TNOM} - 1\right)}{}\right), \tag{7.38}$$

$$J_{tsswgd}(T) = JTSSWGD.\left(\sqrt{\frac{JTWEFF}{W_{eff0}}} + 1\right)$$
$$.exp\left(\frac{E_{g,TNOM}.XTSSWGD.\left(\frac{T}{TNOM} - 1\right)}{}\right), \tag{7.39}$$

where **JTSD** is the current density for the junction bottom, **JTSSWD** is the current density for the junction sidewall, and **JTSSWGD** is the current density for the gate sidewall.

$$NJTSD(T) = NJTSD.\left[1 + TNJTSD.\left(\frac{T}{TNOM} - 1\right)\right], \tag{7.40}$$

$$NJTSSWD(T) = NJTSSWD.\left[1 + TNJTSSWD.\left(\frac{T}{TNOM} - 1\right)\right], \tag{7.41}$$

$$NJTSSWGD(T) = NJTSSWGD.\left[1 + TNJTSSWGD.\left(\frac{T}{TNOM} - 1\right)\right], \tag{7.42}$$

where **NJTSD** corresponds to the junction bottom, **NJTSSWD** corresponds to the sidewall, and **NJTSSWGD** corresponds to the gate sidewall.

### 7.2.2 Junction diode CV model

Source/Drain junction capacitance can be divided into three components: the bottom junction capacitance $C_{jb}$, the sidewall periphery junction capacitance $C_{jpsw}$ (of the field oxide edge), and the gate-edge periphery junction capacitance $C_{jpg}$.

#### 7.2.2.0.1 Source/body junction diode

The source-side junction capacitance can be calculated by

$$C_{bs} = A_{seff}C_{jbs} + P_{seff}C_{jbssw} + W_{eff} \cdot NF \cdot C_{jbsswg} \tag{7.43}$$

where $C_{jbs}$ is the unit-area bottom S/B junction capacitance, $C_{jbssw}$ is the unit-length S/B junction sidewall capacitance along the isolation edge, and $C_{jbsswg}$ is the unit-length S/B junction sidewall capacitance along the gate edge. The effective area and perimeters in (7.43) are given in Section 7.3.

First, we will derive a junction charge model for the well-known simple case of a single junction p-n diode capacitance model for reverse bias [9]. From the Poisson equation describing the electrostatics of the charge carriers in the p- and n-type regions for a diode, the junction depletion depth is given by

$$W_{dep,jn} = W_{dep,0} \cdot \left(1 - \frac{V_{bs}}{PBS}\right)^{MJS} \tag{7.44}$$

where $V_{bs}$ is the voltage across the source-side junction, $W_{dep,0}$ is the zero-bias junction depletion width, **PBS** represents the source-side junction built-in voltage, and **MJS** is the junction doping grading coefficient. The junction capacitance is then given as

$$C_{jbs} = \frac{\epsilon_0 . \epsilon_r}{W_{dep,jn}} = \frac{C_{jbs0}}{\left(1 - \frac{V_{bs}}{PBS}\right)^{MJS}} \tag{7.45}$$

where $\epsilon_r$ is the relative dielectric constant of the junction material, and $C_{jbs0}$ is the zero-bias junction capacitance.

From (7.44) and (7.45), the total charge per unit area in the depletion region as a function of $C_{jbs0}$ is given below

$$Q'_j = 2.PBS.C_{jbs0} \cdot \sqrt{1 - \frac{V_{bs}}{PBS}} \tag{7.46}$$

There are two important observations related to the junction capacitance: 1) For the reverse-bias and small forward-bias conditions, $Q'_j$ is the dominant store charge, as compared to the injected diffusion charge. 2) For the condition when $V_{bs}$ approaches **PBS**, then the depletion charge ($Q'_j$) becomes zero, as a result $C_{jbs}$ tends to infinity. Thus, for that case equation, (7.45) is not valid. So, to overcome this major issue, the SPICE depletion capacitance model has been used [7]:

$$Q'_j = \begin{cases} C_{jbs0}\int_0^{V_{bs}}\left(1-\frac{V}{PBS}\right)^{-MJS} dV & \text{if } \frac{V_{bs}}{PBS(T)} \le x_0 \\ C_{jbs0}F_1 + \frac{C_{jbs0}}{F_2}\cdot\int_{x_0.PBS}^{V_{bs}}\left(F_3 + \frac{MJS.V}{PBS}\right) dV & \text{otherwise} \end{cases} \tag{7.47}$$

Therefore

$$C'_j = \frac{dQ'_j}{dV_{bs}} = \begin{cases} C_{jbs0}\cdot\left(1-\frac{V_{bs}}{PBS}\right)^{-MJS} & \text{if } \frac{V_{bs}}{PBS(T)} \le x_0 \\ \frac{C_{jbs0}}{F_2}\left(F_3 + \frac{MJS.V_{bs}}{PBS}\right) & \text{otherwise} \end{cases} \tag{7.48}$$

where $F_1$, $F_2$ and $F_3$ are SPICE constants given by:

$$F_1 = \frac{PBS}{1-MJS}[1-(1-x_0)^{1-MJS}]$$
$$F_2 = (1-x_0)^{1+MJS}$$
$$F_3 = 1 - x_0(1+MJS)$$

Fig. 7.3 shows the junction capacitance derivative comparison between BSIM4 and BSIM-BULK models. The second derivative of charge produces a kink around $V_j = 0$ for BSIM4. However, in BSIM-BULK, this point has been shifted away from $V_j = 0$ (to large $V_j$) so that the derivatives remain continuous for all practical applications. Since the updated diode junction capacitance model of BSIM-BULK (1.43) is infinitely differentiable for the complete $V_{bs}$ range and symmetric around $V_{bs} = 0$, it is evident in Fig. 7.4 that it gives smooth junction capacitance. BSIM4, on the other hand, shows asymmetry in the second derivative (around $V_{bs} = 0$) and therefore has issues with the ac symmetry test [11].

The final unified expressions for $C_{jbs}$, $C_{jbssw}$ and $C_{jbsswg}$ are given below [3–5]:

**Cjbs is given by**

$$C_{jbs} = \begin{cases} CJS(T)\cdot\left(1-\frac{V_{bs}}{PBS(T)}\right)^{-MJS} & \text{if } \frac{V_{bs}}{PBS(T)} \le x_0 \\ CJS(T)\cdot\frac{1}{(1-x_0)^{MJS}}\cdot\left[1+MJS\left(1+\frac{\frac{V_{bs}}{PBS}-1}{1-x_0}\right)\right] & \text{otherwise} \end{cases} \tag{7.49}$$

where the value of $x_0$ is taken as 0.9.

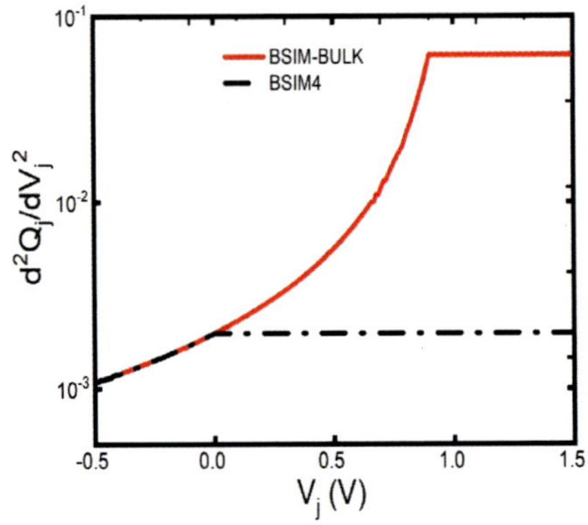


FIGURE 7.3 Junction capacitance model: BSIM4 vs. BSIM-BULK. Symmetry issues around $V_{DS} = 0$ can be seen in BSIM4. This is removed in BSIM-BULK, which is now infinitely differentiable at that point [5].

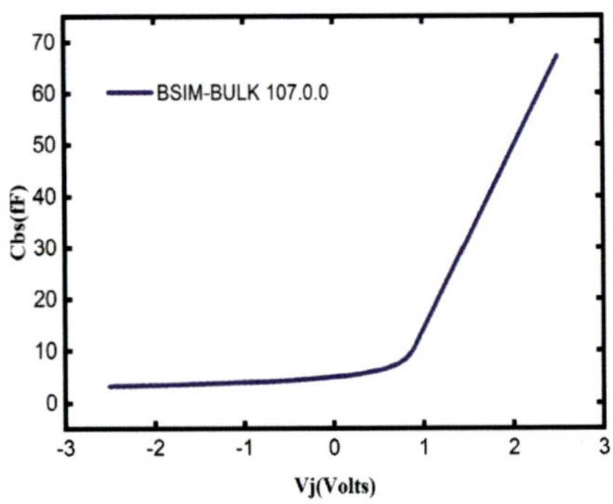


FIGURE 7.4 Junction capacitance ($C_{bs}$) plot of BSIM-BULK.

## Cjbssw is given by

$$C_{jbssw} = \begin{cases} CJSWS(T) \cdot \left(1 - \frac{V_{bs}}{PBSWS(T)}\right)^{-MJSWS} & \text{if } \frac{V_{bs}}{PBSWS(T)} \leq x_0 \\ CJSWS(T) \cdot \frac{1}{(1-x_0)^{MJSWS}} \cdot \left[1 + MJSWS\left(1 + \frac{\frac{V_{bs}}{PBSWS(T)} - 1}{1-x_0}\right)\right] \\ \text{otherwise} \end{cases} \tag{7.50}$$

where the value of $x_0$ is taken as 0.9.

**Cjbsswg is given by**

$$C_{jbsswg} = \begin{cases} CJSWGS(T) \cdot \left(1 - \frac{V_{bs}}{PBSWGS(T)}\right)^{-MJSWGS} & \text{if } \frac{V_{bs}}{PBSWGS(T)} \leq x_0 \\ CJSWGS(T) \cdot \frac{1}{(1-x_0)^{MJSWGS}} & \\ \quad \cdot \left[1 + MJSWGS\left(1 + \frac{\frac{V_{bs}}{PBSWGS(T)} - 1}{1-x_0}\right)\right] & \text{otherwise} \end{cases} \tag{7.51}$$

where the value of $x_0$ is taken as 0.9.

**7.2.2.0.2 Drain/body junction diode**

The drain-side junction capacitance can be calculated by

$$C_{bd} = A_{deff}C_{jbd} + P_{deff}C_{jbdsw} + W_{effcj} \cdot NF \cdot C_{jbdswg} \tag{7.52}$$

where $C_{jbd}$ is the unit-area bottom D/B junction capacitance, $C_{jbdsw}$ is the unit-length D/B junction sidewall capacitance along the isolation edge, and $C_{jbdswg}$ is the unit-length D/B junction sidewall capacitance along the gate edge. The effective area and perimeters in (7.52) are given in Section 7.3.

**Cjbd is calculated by**

$$C_{jbd} = \begin{cases} CJD(T) \cdot \left(1 - \frac{V_{bs}}{PBD(T)}\right)^{-MJD} & \text{if } \frac{V_{bs}}{PBD(T)} \leq x_0 \\ CJD(T) \cdot \frac{1}{(1-x_0)^{MJD}} \cdot \left[1 + MJD\left(1 + \frac{\frac{V_{bs}}{PBD} - 1}{1-x_0}\right)\right] & \text{otherwise} \end{cases} \tag{7.53}$$

where the value of $x_0$ is taken as 0.9.

**Cjbdsw is calculated by**

$$C_{jbdsw} = \begin{cases} CJSWD(T) \cdot \left(1 - \frac{V_{bs}}{PBSWD(T)}\right)^{-MJSWS} & \text{if } \frac{V_{bs}}{PBSWD(T)} \leq x_0 \\ CJSWD(T) \cdot \frac{1}{(1-x_0)^{MJSWD}} & \\ \quad \cdot \left[1 + MJSWD\left(1 + \frac{\frac{V_{bs}}{PBSWD(T)} - 1}{1-x_0}\right)\right] & \text{otherwise} \end{cases} \tag{7.54}$$

where the value of $x_0$ is taken as 0.9.

**Cjbdswg is calculated by**

$$C_{jbdswg} = \begin{cases} CJSWGD(T) \cdot \left(1 - \frac{V_{bs}}{PBSWGD(T)}\right)^{-MJSWGD} \\ \text{if } \frac{V_{bs}}{PBSWGD(T)} \leq x_0 \\ CJSWGD(T) \cdot \frac{1}{(1-x_0)^{MJSWGD}} \\ \cdot \left[1 + MJSWGD\left(1 + \frac{\frac{V_{bs}}{PBSWGD(T)} - 1}{1-x_0}\right)\right] & \text{otherwise} \end{cases} \tag{7.55}$$

where the value of $x_0$ is taken as 0.9.

#### 7.2.2.0.3 Temperature dependence

On the source side, the temperature dependence of the junction capacitances and potentials is defined as

$$\begin{aligned} CJS(T) &= CJS[1 + TCJ(T - TNOM)] \\ CJSWS(T) &= CJSWS[1 + TCJSW(T - TNOM)] \\ CJSWGS(T) &= CJSWGS[1 + TCJSWG(T - TNOM)] \\ PBS(T) &= PBS(TNOM) - TPB(T - TNOM) \\ PBSWS(T) &= PBSWS(TNOM) - TPBSW(T - TNOM) \\ PBSWGS(T) &= PBSWGS(TNOM) - TPBSWG(T - TNOM) \end{aligned}$$

where **CJS** is the junction capacitance at the bottom, **CJSWS** is the junction capacitance at the source sidewall, **CJSWGS** is the junction capacitance at the gate sidewall, **PBS** is the junction potential at the bottom, **PBSWS** is the junction capacitance at the source sidewall, and **PBSWGS** is the junction capacitance at the gate sidewall.

The temperature dependence of the junction capacitances and potentials at the drain side is defined as

$$\begin{aligned} CJD(T) &= CJD[1 + TCJ(T - TNOM)] \\ CJSWD(T) &= CJSWD[1 + TCJSW(T - TNOM)] \\ CJSWGD(T) &= CJSWGD[1 + TCJSWG(T - TNOM)] \\ PBD(T) &= PBD(TNOM) - TPB(T - TNOM) \\ PBSWD(T) &= PBSWD(TNOM) - TPBSW(T - TNOM) \\ PBSWGD(T) &= PBSWGD(TNOM) - TPBSWG(T - TNOM) \end{aligned}$$

where **CJD** is the junction capacitance at the bottom, **CJSWD** is the junction capacitance at the drain sidewall, **CJSWGD** is the junction capacitance at the gate sidewall, **PBD** is the junction potential at the bottom, **PBSWD** is the junction capacitance at the drain sidewall, and **PBSWGD** is the junction capacitance at the gate sidewall.

## 7.3 Layout-dependent parasitics models

BSIM-BULK provides a comprehensive and versatile geometry/layout-dependent parasitics model [11]. It offers the modeling of series and multi-finger device layout or a combination of these two configurations. These models impact every BSIM-BULK submodel except the substrate resistance network model. Note that the narrow-width effect in the per-finger device with multi-finger configuration is accounted for by this model. Since the extension resistance model of BSIM-BULK has the same mathematical form as of BSIM4 model, the former adopts the same parameter naming conventions as of later [3,4].

Like BSIM4, two options for source/drain resistance implementation are also provided in BSIM-BULK. If **RDSMOD = 0**, the extension resistance is implemented as a current degradation factor multiplied by the drain current. On the other hand, if **RDSMOD = 1**, the extension resistance is implemented as two external resistors in series with the transistor [3,4].

To align with the prior BSIM4 model, BSIM-BULK provides its users with the option to set **RGEOMOD = 0** to specify geometry-dependent parasitic source/drain resistances via the sheet resistance parameters **RSHS** and **RSHD**, where

$$R_{S,geo} = NRS.RSHS$$
$$R_{D,geo} = NRD.RSHD$$

In such a case, the spreading and contact resistance models are not used for calculations. On the other hand, if the user sets **RGEOMOD = 1**, the spreading and contact resistance formulations will be used.

### 7.3.1 Model formulation and options

#### *7.3.1.1 Effective junction perimeter and area*

In the following, only the source-side case is illustrated. The same approach is used for the drain side. The effective junction perimeter on the source side is calculated by [3,4]. If the parameter PS is not specified, then $PS_{eff}$ is computed from NF, DWJ, GEOMOD, DMCG, DMCI, DMDG,

DMCGT, RSH, and MIN. Otherwise, it is obtained as follows:

$$PS_{eff} = \begin{cases} PS & \text{if PERMOD=0} \\ PS - W_{eff}.NF & \text{otherwise} \end{cases} \tag{7.56}$$

Similarly, if AS and AD are not specified, their effective values ($AS_{eff}$, $AD_{eff}$) are computed from NF, DWJ, GEOMOD, DMCG, DMCI, DMDG, DMCGT, RSH, and MIN. These effective values are used to calculate junction diode IV and CV. $PS_{eff}$ does not include the gate-edge perimeter.

#### 7.3.1.2 Source/drain diffusion resistance

If the parameter NRS (number of source squares) is not specified and RDGEMOD and RSH set to value greater than 0, $R_{Sdiff}$ is computed from NF, DWJ, GEOMOD, DMCG, DMCI, DMDG, DMCGT, RSH, and MIN. Otherwise, it is obtained as follows [3,4]:

$$R_{Sdiff} = \begin{cases} NRS \cdot RSHS & \text{if NRS is specified} \\ 0 & \text{otherwise} \end{cases} \tag{7.57}$$

Similarly, the drain diffusion resistance is also calculated.

#### 7.3.1.3 Option for source/drain connections

Table 7.1 lists the options for source/drain connections through the model selector **geoMod** [3,4]. For multi-finger devices, all inside S/D diffusions are assumed to be shared.

TABLE 7.1 **geoMod** options.

| geoMod | End source | End drain | Note |
|---|---|---|---|
| 0 | isolated | isolated | NF=Odd |
| 1 | isolated | shared | NF=Odd, Even |
| 2 | shared | shared | NF=Odd, Even |
| 3 | shared | isolated | NF=Odd, Even |
| 4 | isolated | merged | NF=Odd |
| 5 | shared | merged | NF=Odd, Even |
| 6 | merged | isolated | NF=Odd |
| 7 | merged | shared | NF=Odd, Even |
| 8 | merged | merged | NF=Odd |
| 9 | sha/iso | shared | NF=Even |
| 10 | shared | sha/iso | NF=Even |

#### 7.3.1.4 Option for source/drain contacts

Table 7.2 lists the options for source/drain contacts through the model selector **rgeoMod** [3,4].

TABLE 7.2 **rgeoMod** options.

| rgeoMod | End-source contact | End-drain contact |
|---|---|---|
| 0 | No $R_{sdiff}$ | No $R_{ddiff}$ |
| 1 | wide | wide |
| 2 | wide | point |
| 3 | point | wide |
| 4 | point | point |
| 5 | wide | merged |
| 6 | point | merged |
| 7 | merged | wide |
| 8 | merged | point |

# References


[1] S. Sze, Physics of Semiconductor Devices, Wiley, New York, 1981.

[2] Y. Cheng, C. Hu, MOSFET Modeling & BSIM3 User's Guide, Springer Science & Business Media, 1999.

[3] BSIM6 Technical Manual. [Online]. Available: http://www-device.eecs.berkeley.edu/bsim/.

[4] Y.S. Chauhan, S. Venugopalan, M.A. Chalkiadaki, M.A.U. Karim, H. Agarwal, S. Khandelwal, N. Paydavosi, J.P. Duarte, C.C. Enz, A.M. Niknejad, C. Hu, BSIM6: analog and RF compact model for bulk MOSFET, IEEE Transactions on Electron Devices 61 (2) (Feb 2014) 234–244.

[5] H. Agarwal, S. Venugopalan, M.-A. Chalkiadaki, N. Paydavosi, J.P. Duarte, S. Agnihotri, C. Yadav, P. Kushwaha, Y.S. Chauhan, C.C. Enz, A. Niknejad, C. Hu, Recent enhancements in BSIM6 bulk MOSFET model, in: Simulation of Semiconductor Processes and Devices (SISPAD), International Conference on, Sept 2013, pp. 53–56, https://doi.org/10.1109/SISPAD.2013.6650572.

[6] C. Gupta, R. Goel, H. Agarwal, C. Hu, Y.S. Chauhan, BSIM-bulk: accurate compact model for analog and RF circuit design, in: 2019 IEEE Custom Integrated Circuits Conference (CICC), 2019, pp. 1–8.

[7] P. Antognetti, G. Massobrio, Semiconductor Device Modeling with SPICE, McGraw-Hill, Inc., 1993.

[8] N.D. Arora, MOSFET Models for VLSI Circuit Simulation: Theory and Practice, Springer Science & Business Media, 2012.

[9] H. Fan, Temperature dependence of the energy gap in semiconductors, Physical Review 82 (6) (1951) 900.

[10] B. Van Zeghbroeck, Principles of Semiconductor Devices, vol. 34, Colorado University, 2004.

[11] N. Paydavosi, T.H. Morshed, D.D. Lu, W.M. Yang, M.V. Dunga, X.J. Xi, J. He, W. Liu, M.C. Kanyu, X. Jin, et al., BSIM4v4. 8.0 MOSFET Model, University of California, Berkeley (CA), 2013.

CHAPTER

# 8

# Compact Modeling of High Voltage Transistors

## 8.1 Introduction

Fig. 8.1 shows the schematic representation of a high-voltage LDMOS device. The drift region exists between the drain and the channel region where the most potential drop occurs, allowing HV devices to withstand high drain voltage. The drift region generally comprises two sections, one under the thin and the other under the thick gate oxide. Resistance of the former is modulated by the gate voltage, while that of the latter is a weak function of the applied gate voltage [1]. Furthermore, as the drain voltage increases, the carriers in the drift region undergo velocity saturation. Therefore the drift resistance is also a strong function of applied drain voltage. Starting from BSIM-BULK107.0.0, BSIM-BULK includes a compact model of HV devices, which can be activated by setting RDSMOD=1 and HVMOD=1. Fig. 8.2(a) shows the equivalent circuit for the HV device in BSIM-BULK107.0.0.

## 8.2 Drift region resistance modeling

The channel and the drift region under the thin oxide are modeled using BSIM-BULK model, which is based on physical charge-based core [4,5]. Therefore the high-voltage model benefits from the advancements in the BSIM-BULK MOS model [6–12]. In BSIM-BULK107.0.0, drift resistance model is derived using velocity saturation model [2], and resistance is expressed as a function of drain-source current (see Fig. 8.2(a)). This formulation is later improved in BSIM-BULK 107.1.0 as shown in Fig. 8.2(b). We will first discuss the current-based formulation of drift resistance and then the updated model.

BSIM-Bulk MOSFET Model for IC Design - Digital, Analog, RF and High-Voltage
https://doi.org/10.1016/B978-0-32-385677-5.00012-3

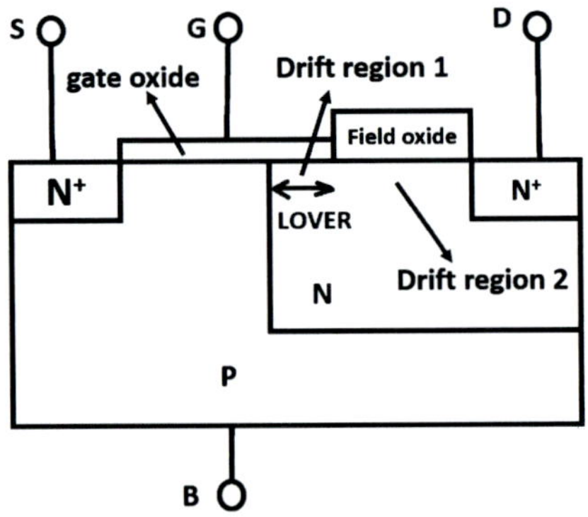


FIGURE 8.1 Schematic representation of LDMOS High-Voltage transistor. Large part of the applied drain voltage drops across the drift region, enabling the device to withstand higher voltage.

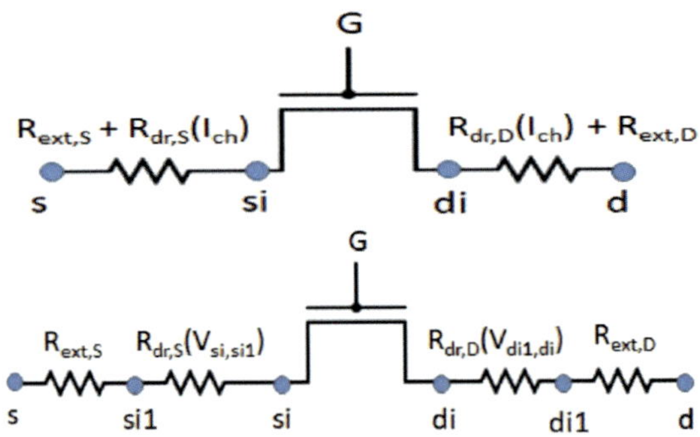


FIGURE 8.2 Modeling HV devices: equivalent circuit representation, showing only the newly modeled drift resistor part. (a) Current-based resistor model implemented in BSIM-BULK107.0.0 [2] (b) Updated model, available from beta 5 and higher [3]. The gated channel portion and the drift region are modeled using the BSIM-BULK MOS model and the newly derived physics-based model, respectively. Drift resistance at the source can be turned off for the asymmetric devices. RDSMOD and HVMOD must be set to 1 to enable HV models.

Current ($I_{dr}$) through an idealized semiconductor bar of thickness $t$, width $W$, length $L_{dr}$, doping $N_{DR}$ can be expressed as:

$$I_{dr} = W \cdot (q.N_{DR}.t) \cdot v_{dr} \text{ where } v_{dr} = \mu \cdot \frac{E}{1 + \frac{E}{E_c}} \tag{8.1}$$

here $E_c$ is the critical field for velocity saturation, and $v_{dr}$ is the carrier velocity. $\mu$ represents low-field carrier mobility, and E is the electric field. For $E >> E_c$, $v_{dr} = \mu E_c$; otherwise, $v_{dr} = \mu E$.

$$I_{dr} = \frac{W}{\rho_s} \frac{\frac{dV}{dx}}{1 + \frac{1}{E_c}\frac{dV}{dx}} \text{ where } \rho_s^{-1} = q.\mu.N_{DR}.t \tag{8.2}$$

$$R_{dr} = \frac{V_{dr}}{I_{dr}} = \frac{R_0}{1 - \frac{I_{dr}}{I_{sat}}} \tag{8.3}$$

with

$$I_{sat} = W.q.N_{DR}.t.VDRIFT \tag{8.4}$$

where $R_0 = \rho_s \frac{L_{dr}}{W}$ and VDRIFT is the carrier saturation velocity. $\rho_s$ represents sheet resistance. In this derivation, it is assumed that $t$ is independent of applied bias [1]. We will return to this assumption later. The above model is derived using the simple velocity-saturation model. While it enables an analytical solution, the real velocity saturation in the devices may be more complex. Therefore we introduce a model parameter MDRIFT to provide more flexibility in modeling the velocity saturation effect. (8.3) is modified to:

$$R_{dr} = R_0 \left[ 1 - \left( \frac{I_{dr}}{I_{sat}} \right)^{MDRIFT} \right]^{\frac{-1}{MDRIFT}} \tag{8.5}$$

The typical range of MDRIFT that we observed during the parameter extraction process is between 0.8 and 2.5. However, one may need to use it beyond this range depending on the process. There is, however, an issue here. It is well known that higher-order derivatives of a model like (8.5) do not exist for all values of MDRIFT [2]. In order to resolve this issue, we further modified the above model by introducing $\delta$:

$$R_{dr} = RDLCW \left[ 1 - \delta * \left( \frac{I_{dr}}{I_{sat}} \right)^{MDRIFT} \right]^{\frac{-1}{MDRIFT}} \tag{8.6}$$

where

$$\delta = \frac{I_{dr}^{4-MDRIFT}}{I_{dr}^{4-MDRIFT} + \alpha * I_{sat}^{4-MDIFT}} \tag{8.7}$$

$\alpha$ is a model parameter with default value of $10^{-4}$. Drift resistance $R_{dr,D}$ and $R_{dr,S}$ in Fig. 8.2 are modeled as in (8.5) with $I_{dr} = I_{ds}$ from current continuity, where $I_{ds}$ represents the channel current. $L_{dr}$, $\rho_s$ and VDRIFT are treated as a model parameters. For asymmetric devices with drift region only on the drain side, $R_{dr,S}$ can be set to 0. This is the final equation that is used in BSIM-BULK107.0.0 ($R_0$ is taken as a model parameter RDLCW). The importance of the term $\delta$ in (8.5) is illustrated by comparing the gummel symmetry test [13] results for two cases: $\delta = 1$, i.e., model in (8.5) and model in (8.9). MDRIFT is set to 0.8. The symmetric HVMOS model is invoked by connecting drift resistance to both the drain and the source side.

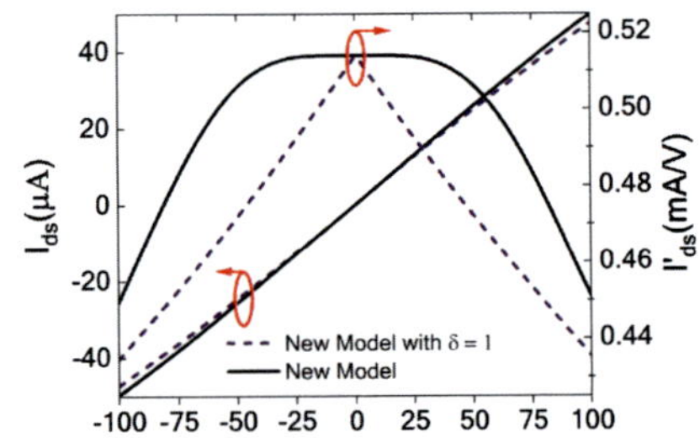


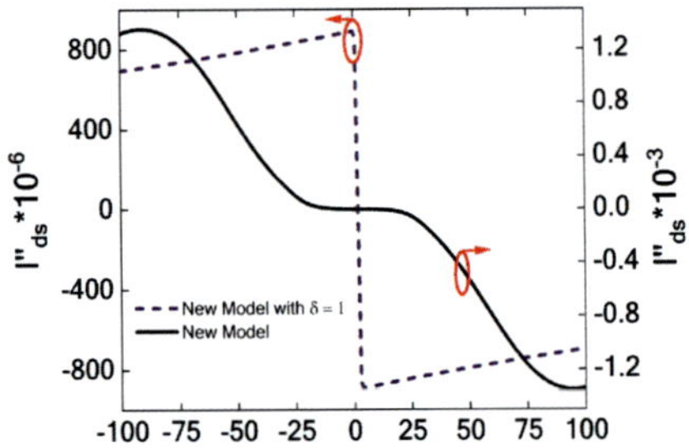


FIGURE 8.3 Testing continuity of the of the HV model: (a) Current and its first derivative (b) the second derivative. Second- and higher-order derivatives do not exist for $\delta = 1$. $\beta = 0.8$, $\alpha$ =1e-3 are used for these simulations [2].

Fig. 8.3(a) compares current and its first derivative. Note that the current is nearly the same in both cases, which indicate that the introduction of $\delta$ has little impact on the DC extraction. The second derivative is shown in Fig. 8.3(b), and, clearly, it does not exist at $V_x = 0\,\mathrm{V}$ for the $\delta = 1$ case.

## 8.2.1 Updated drift resistance model

In the BSIM-BULK 107.1.0, several updates are made to the drift resistance model. We will discuss them in this section.

The first update is regarding the model behavior under transient operative conditions. (8.6) assumed that the drift current ($I_{dr}$) is equal to the channel transport current ($I_{ch}$). This assumption breaks down in the presence of impact ionization current ($I_{ii}$) or transient charging-discharging current. Fig. 8.4(a) and (b) show drain current and drift resistance behavior when a ramp gate voltage is applied. We clearly see a nonphysical hump in current and drift resistance. To fix this, $I_{dr} = I_D$ must be ensured instead of $I_{ch}$, where $I_D$ is given as:

$$I_D = I_{ch} + I_{ii} + \frac{dQ_{d_i}}{dt} + \frac{dQ_{par}}{dt} + \frac{dQ_{bdj}}{dt} \tag{8.8}$$

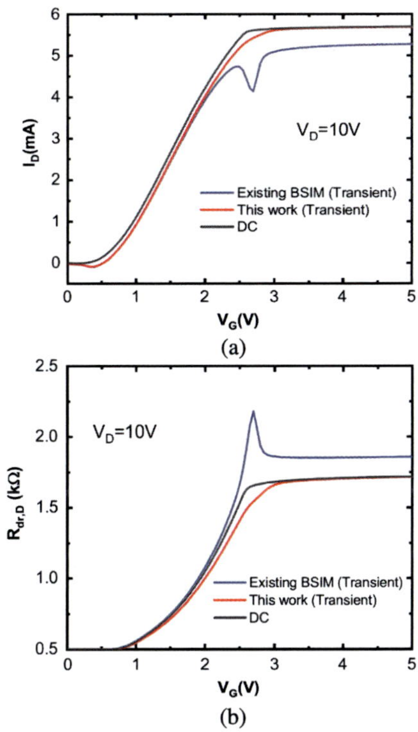


FIGURE 8.4 Updated drift resistance model: drain current and drift resistance under transient operating conditions (a) drain current and (b) drift resistance as functions of gate ramp. Transient results are compared with the dc results. The current-dependent model shows a nonphysical kink, which is rectified in BSIM-BULK 107.1.0 release by ensuring the total current (DC + transient) is used to calculate Rdrift [3].

here $Q_{d_i}$, $Q_{par}$, and $Q_{bdj}$ represent drain terminal charge, fringing capacitance charges, and drain junction charges.

In addition to the above change, the drift resistance model is modified from the current-based formulation to the voltage-based formulation. This is done considering the fact that commercial SPICE simulators use modified nodal analysis (MNA) to solve a circuit, where voltage nodes are the variables to be solved via a matrix formulation. Furthermore, another internal node $V_{d_{i1}}$ is introduced to ensure $I_{dr} = I_D$, as shown in Fig. 8.2(b). The updated drift resistance model is given as [3]:

$$R_{dr} = RDLCW\left[1 + \delta_v * \left(\frac{|V_{d_{i1},d_i}|}{V_c}\right)^{MDRIFT}\right]^{\frac{1}{MDRIFT}} \tag{8.9}$$

where

$$\delta_v = \frac{|V_{d_{i1},d_i}|^{4-MDRIFT}}{|V_{d_{i1},d_i}|^{4-MDRIFT} + \alpha * V_c^{4-MDRIFT}} \tag{8.10}$$

where

$$V_c = I_{sat} * RDLCW \tag{8.11}$$

In addition, BSIM-BULK 107.1.0 was updated to include the following changes in the calculations of $V_c$ (see (8.9)). New $V_c$ is given as:

$$V_c = V_c * f(V_{sb}) * f(V_{gb}) * f(V_{dr}) \tag{8.12}$$

$$f(V_{sb}) = 1 - PDRWB * V_{sb} \tag{8.13}$$

$$f(V_{dr}) = 1 + GADRIFT * (|V_{dr}| - RDVDS) \tag{8.14}$$

$$f(V_{gb}) = 1 + PTWGHV * T2 \tag{8.15}$$

where

$$T2 = (q_s - PTWGH1)||(10 * PSATXHV) \tag{8.16}$$

In the above equations, PDRWB, GADRIFT, RDVDS, PTWGHV, and PTWGHV1 are the model parameters. $f(V_{sb})$, $f(V_{gb})$, and $f(V_{dr})$ are smoothly limited to avoid nonphysical values of $V_c$.

## 8.3 Drift region charge model

The presence of the drift region not only affects the dc behavior but also has a significant impact on the terminal capacitance. The drift region 1 (see Fig. 8.1) is directly under the influence of the gate through the thin oxide. It undergoes from inversion to accumulation as the gate voltage is swept from the large negative value to the positive value (for an n-channel transistor). Asymptotically, it contributes to W * LOVER * $C_{ox}$ at the two extreme bias points (in the accumulation and strong inversion). To model the charge contribution, surface potential in the overlap region is determined using the BSIM-BULK core model:

$$v_{gfb,drift} = \psi_{p,drift} + \gamma * \sqrt{e^{-\psi_{p,drift}} - 1 + \psi_{p,drift}} \tag{8.17}$$

where

$$v_{gfb,drift} = \frac{V_{g,di} - VFBOV}{V_t}; \; \gamma = \frac{2q\epsilon_{si} NDR}{C_{ox} * \sqrt{V_t}} \tag{8.18}$$

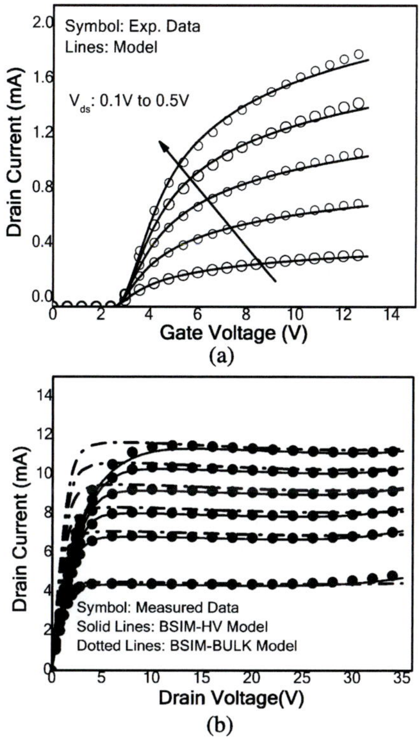


FIGURE 8.5 Model validation: (a) input characteristics, $I_{ds}$-$V_{gs}$ for different $V_{ds}$ (b) output characteristics, $I_{ds}$-$V_{ds}$ for different $V_{gs}$ [2]. Symbols: experimental data, Lines: BSIM-HV model.

$\psi_{p,drift}$ is the normalized pinch-off potential, and $C_{ox}$ is the oxide capacitance per unit area. VFBOV and NDR are the model parameters representing flat-band voltage and doping of the drift region, respectively. (8.17) is solved analytically for $\psi_{p,drift}$, and then the inversion charge in the drift region ($q_{drift}$) is obtained in the way BSIM-BULK does. The inversion and accumulation charges are then expressed as:

$$Q_{inv,drift} = \beta * 2n_{qhv} * q_{drift} \tag{8.19}$$

$$Q_{acc,drift} = \beta * (v_{gfb,drift} - \psi_{p,drift} - 2n_{qhv} * q_{drift}) \tag{8.20}$$

where $\beta = W * LOVER * C_{ox} * V_t$. $Q_{inv,drift}$ and $Q_{acc,drift}$ are assigned to the body and the drain node, respectively. Here, W represents the channel width.

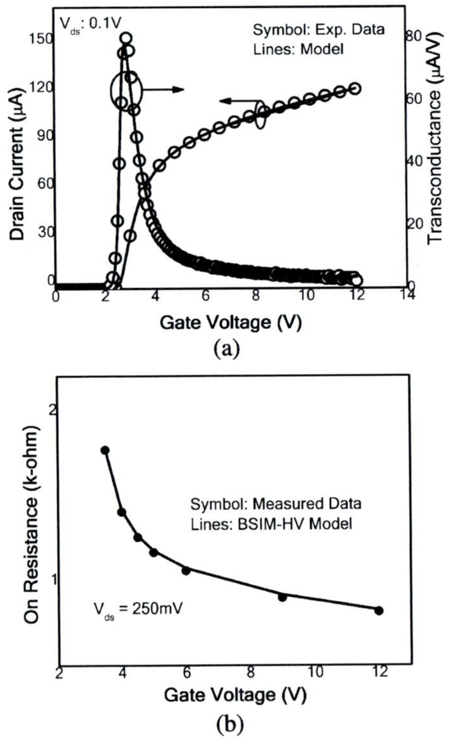


FIGURE 8.6 Model validation: (a) input characteristics, $I_{ds}$-$V_{gs}$ (b) on-resistance vs. $V_{gs}$ [2]. $R_{on} = V_{ds}/I_{ds}$. Symbols: experimental 90 V LDMOS data, Lines: BSIM-HV model.

## 8.4 Model results and validation

We have tested the model with multiple sets of experimental data, ranging from medium- to high-voltage devices. Fig. 8.5 shows the model validation with the experimental 35 V LDMOS device having $L_g = 1.2\,\mu\text{m}$ and the length of the drift region in 4 μm [14]. Fig. 8.5(a) shows drain current versus gate voltage in the linear region for various $V_{ds}$ while Fig. 8.5(b) shows output characteristics $I_{ds}$-$V_{ds}$ at different gate voltages. Fig. 8.5(b) also shows model results without including velocity-saturation effect in the newly derived drift resistance model, i.e., setting $R_{dr} = RDLCW$ in (8.9). The model accurately captures the LDMOS device characteristics.

Next, the model is validated with 90 V LDMOS device (drift region length 10 μm [15]). Fig. 8.6(a) shows input characteristics, $I_{ds}$-$V_{gs}$ and its derivative (transconductance, $g_m$). Current and $g_m$ are both accurately modeled. Fig. 8.6(b) shows model-data overlay of the on-resistance ($R_{on}$) plotted as a function of $V_{gs}$. $R_{on}$ is defined as $R_{on} = V_{ds}/I_{ds}$, and it is calculated in the on-state (i.e., in strong inversion). The reduction of

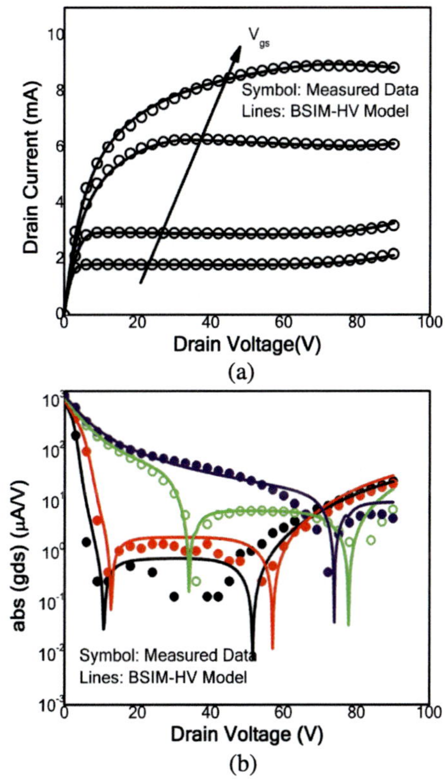


FIGURE 8.7 Model validation with experimental data: (a) $I_{ds}$-$V_{ds}$ at different gate voltages (b) output conductance, $|g_{ds}|$ [2]. L=1.2 µm, $L_{dr}$ = 10 µm. Symbols: experimental 90 V LDMOS data, Lines: BSIM-HV model.

the on-resistance with gate voltage is accurately reproduced. Compact model captures the transfer characteristics very well across the gate bias. Fig. 8.6(a) shows output characteristics, $I_{ds}$-$V_{ds}$ for different gate biases. Quasi-saturation effect is clearly observed, which is well modeled by the BSIM-HV model. Fig. 8.7(b) shows model validation with output conductance, $|g_{ds}|$-$V_{ds}$. Accurate modeling of $g_{ds}$ is considered challenging, as we see that the model is able to accurately reproduce two distinctive humps in $g_{ds}$, arising due to self-heating effect and impact ionization effect.

After the model validation with lateral HV device, it is further validated with 40 V experimental vertical DMOS HV transistor. Fig. 8.8(a) shows $I_{ds}$-$V_{gs}$ for several $V_{ds}$, and Fig. 8.8(b) shows output characteristics $I_{ds}$-$V_{ds}$. As observed from these figures, BSIM-HV is able to successfully model the device behavior. Finally, the model's capability to mimic the experimental data for different $V_{bs}$ is shown in Fig. 8.9.

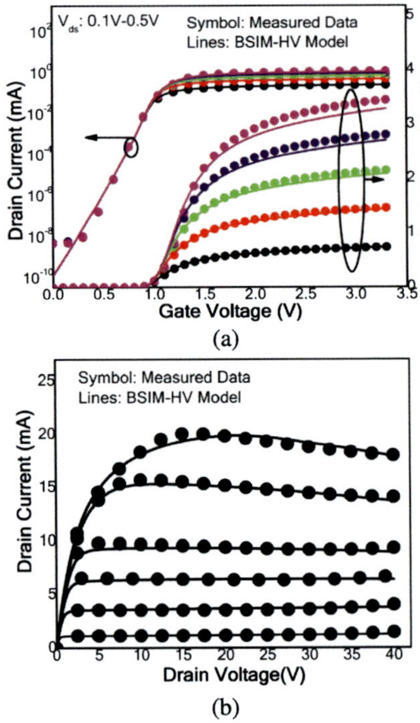


FIGURE 8.8 Model validation with experimental data of vertical DMOS: (a) $I_{ds}$-$V_{gs}$ (b) $I_{ds}$-$V_{ds}$ [2].

**Charge Model Validation:** In BSIM-HV, the newly developed charge model can be invoked by setting HVCAP=1 (along with RDSMOD=1 and HVMOD=1). Fig. 8.10 shows model and TCAD capacitances $C_{gg}$, $C_{sg}$, $C_{dg}$, $C_{bg}$ at $V_{ds} = 0\,\text{V}$. As the simulated device is asymmetric with drift region only at the drain, we see that $C_{sg} \neq C_{dg}$, which the model is able to capture. Further, as the gate voltage increases, the channel starts to form from the drain side, and, eventually, the source side takes control at high $V_{gs}$ causing drain capacitance to fall [16–18]. This peaky behavior of the $C_{dg}$ is also well captured by the model.

The model is also validated with experimental data. Fig. 8.11(a) shows experimental total gate capacitance as a function of $V_{gs}$. BSIM-HV model results with and without the new charge model are also shown, highlighting the importance of modeling charges in the drift region. In Fig. 8.11(b), $C_{gg}$ vs. $V_{gs}$ is shown, for multiple $V_{ds}$. Gate capacitance of the HV devices is very sensitive to the applied drain bias, which the model is able to reproduce. Fig. 8.12 shows the model results with experimental $C_{gs}$ vs. $V_{gs}$ and $C_{gd}$ vs. $V_{ds}$. Again, the model is able to correctly capture the device behavior across gate and drain biases.

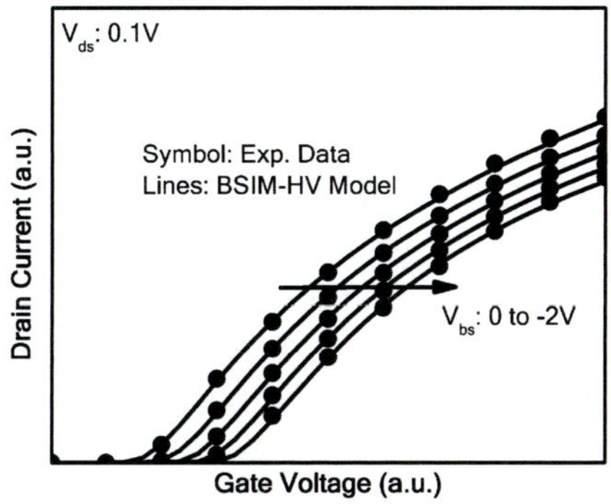

(a)

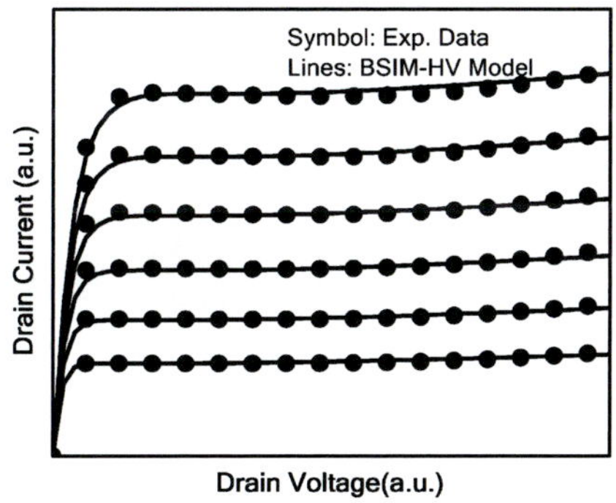

(b)

FIGURE 8.9 Model validation with experimental data at different back biases: (a) $I_{ds}$-$V_{gs}$ and $g_m$-$V_{gs}$ (b) output characteristics, $I_{ds}$-$V_{ds}$ [2].

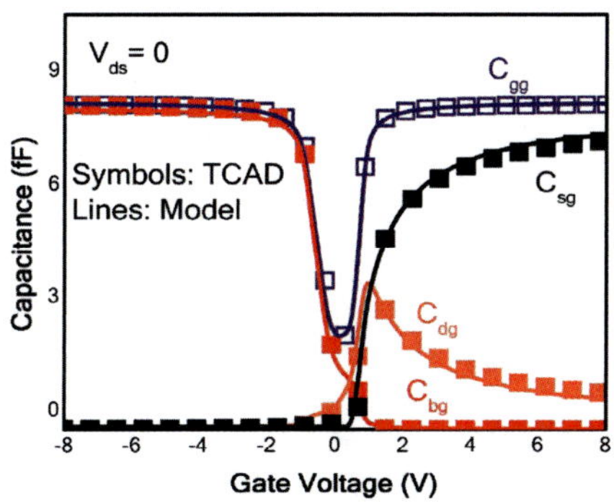


FIGURE 8.10 BSIM-HV capacitance model: $C_{xg}$ vs $V_{gs}$; $x$ =s/d/g/b at $V_{ds}$ = 0. The model accurately captures the capacitance characteristics [2]. LOVER=1 µm and $L_{ch}$ = 2 µm. Symbols represent TCAD data.

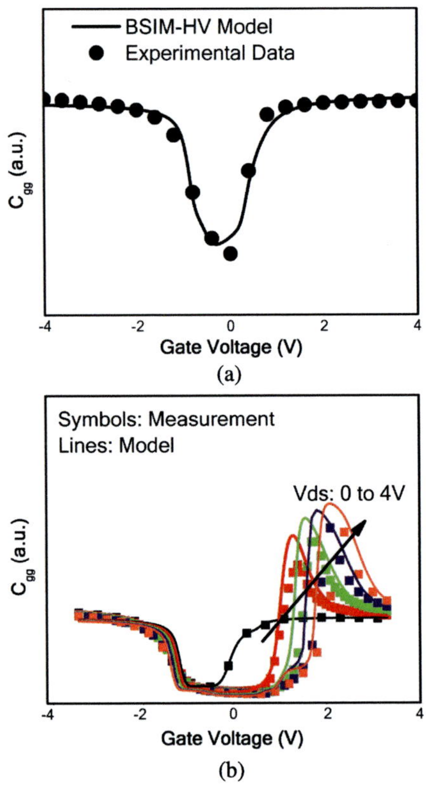


FIGURE 8.11 The BSIM-HV charge model validation with the experimental data: Gate capacitance $C_{gg}$ vs. gate voltage (a) device 1 for $V_{ds} = 0\,\text{V}$ (b) device 2 for different $V_{ds}$ [2].

# References


[1] A.C.T. Aarts, W.J. Kloosterman, Compact modeling of high-voltage LDMOS devices including quasi-saturation, IEEE Transactions on Electron Devices 53 (4) (2006) 897–902, https://doi.org/10.1109/TED.2006.870423.

[2] H. Agarwal, C. Gupta, R. Goel, P. Kushwaha, Y.-K. Lin, M.-Y. Kao, J.-P. Duarte, H.-L. Chang, Y.S. Chauhan, S. Salahuddin, C. Hu, BSIM-HV: high-voltage MOSFET model including quasi-saturation and self-heating effect, IEEE Transactions on Electron Devices 66 (10) (2019) 4258–4263.

[3] G. Pahwa, A. Sharma, R. Goel, G. Gill, H. Agarwal, Y.S. Chauhan, C. Hu, Robust compact model of high voltage MOSFET's drift region, IEEE Transactions on Computer-Aided Design of Integrated Circuits and Systems 42 (1) (2023) 337–340.

[4] Y.S. Chauhan, S. Venugopalan, M.A. Chalkiadaki, M.A.U. Karim, H. Agarwal, S. Khandelwal, N. Paydavosi, J.P. Duarte, C.C. Enz, A.M. Niknejad, C. Hu, BSIM6: analog and RF compact model for bulk MOSFET, IEEE Transactions on Electron Devices 61 (2) (Feb 2014) 234–244, https://doi.org/10.1109/TED.2013.2283084.

[5] BSIM-BULK Technical Manual. [Online]. Available: http://bsim.berkeley.edu/models/bsimbulk/.

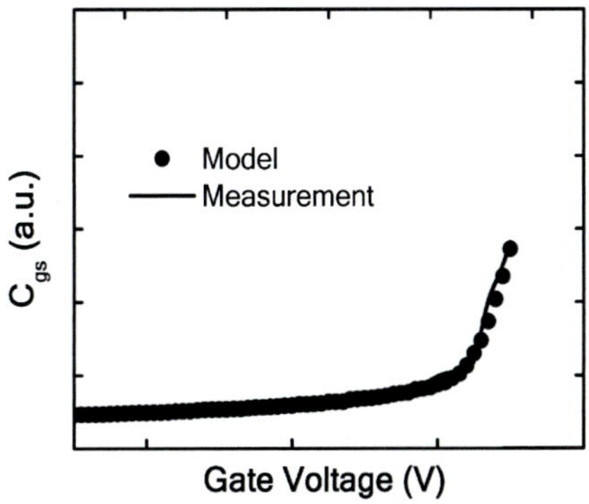


(a)

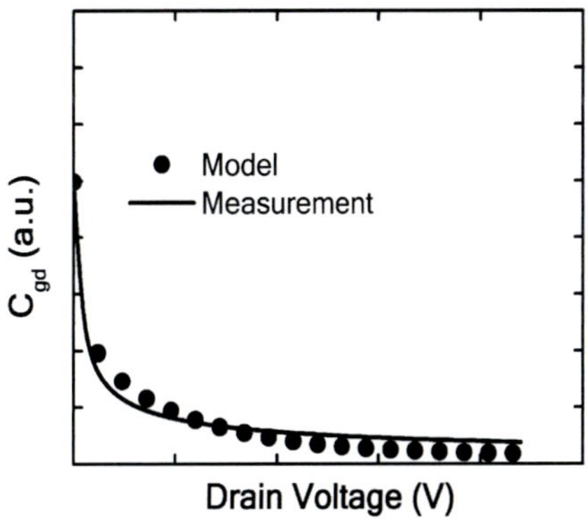


(b)

FIGURE 8.12 The BSIM-HV charge model validation with the experimental data: plot of (a) $C_{gs}$ vs. gate voltage (b) $C_{gd}$ vs. drain voltage.

[6] H. Agarwal, C. Gupta, S. Dey, S. Khandelwal, C. Hu, Y.S. Chauhan, Anomalous transconductance in long channel halo implanted MOSFETs: analysis and modeling, IEEE Transactions on Electron Devices 64 (2) (2017) 376–383, https://doi.org/10.1109/TED.2016.2640279.

[7] S. Khandelwal, H. Agarwal, J.P. Duarte, K. Chan, S. Dey, Y.S. Chauhan, C. Hu, Modeling STI edge parasitic current for accurate circuit simulations, IEEE Transactions on Computer-Aided Design of Integrated Circuits and Systems 34 (8) (April 2015) 1291–1294, https://doi.org/10.1109/TCAD.2015.2419626.

[8] Y.K. Lin, S. Khandelwal, A.S. Medury, H. Agarwal, H.L. Chang, Y.S. Chauhan, C. Hu, Modeling of subsurface leakage current in low VTH short channel MOSFET at accumulation bias, IEEE Transactions on Electron Devices 63 (5) (2016) 1840–1845, https://doi.org/10.1109/TED.2016.2544818.

[9] H. Agarwal, S. Khandelwal, S. Dey, C. Hu, Y.S. Chauhan, Analytical modeling of flicker noise in halo implanted MOSFETs, IEEE Journal of the Electron Devices Society 3 (4) (2015) 355–360, https://doi.org/10.1109/JEDS.2015.2424686.

[10] H. Agarwal, C. Gupta, P. Kushwaha, C. Yadav, J.P. Duarte, S. Khandelwal, C. Hu, Y.S. Chauhan, Analytical modeling and experimental validation of threshold voltage

in BSIM6 MOSFET model, IEEE Journal of the Electron Devices Society 3 (3) (2015) 240–243, https://doi.org/10.1109/JEDS.2015.2415584.

[11] C. Gupta, N. Mohamed, H. Agarwal, R. Goel, C. Hu, Y.S. Chauhan, Accurate and computationally efficient modeling of nonquasi static effects in MOSFETs for millimeter-wave applications, IEEE Transactions on Electron Devices 66 (1) (2019) 44–51, https://doi.org/10.1109/TED.2018.2854671.

[12] C. Gupta, S. Dey, H. Agarwal, R. Goel, C. Hu, Y.S. Chauhan, Analysis and modeling of temperature and bias dependence of current mismatch in halo-implanted MOSFETs, IEEE Transactions on Electron Devices 65 (9) (2018) 3608–3616, https://doi.org/10.1109/TED.2018.2853989.

[13] G. Gildenblat, Compact Modeling: Principles, Techniques and Applications, Springer, 2010.

[14] Y.S. Chauhan, Compact Modeling of High Voltage MOSFETs, Ph.D. Thesis, Swiss Federal Institute of Technology, 2007.

[15] Y.S. Chauhan, C. Anghel, F. Krummenacher, C. Maier, R. Gillon, B. Bakeroot, B. Desoete, S. Frere, A.B. Desormeaux, A. Sharma, M. Declercq, A.M. Ionescu, Scalable general high voltage MOSFET model including quasi-saturation and self-heating effects, Solid-State Electronics 50 (11) (2006) 1801–1813, https://doi.org/10.1016/j.sse.2006.09.002.

[16] Y.S. Chauhan, F. Krummenacher, R. Gillon, B. Bakeroot, M.J. Declercq, A.M. Ionescu, Compact modeling of lateral nonuniform doping in high-voltage MOSFETs, IEEE Transactions on Electron Devices 54 (6) (2007) 1527–1539, https://doi.org/10.1109/TED.2007.896597.

[17] S.F. Frre, J. Rhayem, H.O. Adawe, R. Gillon, M. Tack, A.J. Walton, LDMOS capacitance analysis versus gate and drain biases, based on comparison between TCAD simulations and measurements, in: European Solid-State Device Research Conference, 2001, pp. 219–222, https://doi.org/10.1109/ESSDERC.2001.195240.

[18] Y.S. Chauhan, F. Krummenacher, C. Anghel, R. Gillon, B. Bakeroot, M. Declercq, A.M. Ionescu, Analysis and modeling of lateral non-uniform doping in high-voltage MOSFETs, IEEE Transactions on Electron Devices (2006) 1–4, https://doi.org/10.1109/IEDM.2006.347000.

CHAPTER

# 9

# Self-Heating and Thermal Effects

## 9.1 Basic formulation

For the ease of parameter extraction at different ambient temperatures, BSIM-BULK model equations are developed so that changing temperature-related parameter values do not affect the model results at nominal temperature. For this purpose, a model parameter TNOM is provided, which is set to 27°C by default. Before we discuss the temperature effect models, let us define the following two variables:

$$T_R = \frac{T}{TNOM} \tag{9.1}$$

$$T_{diff} = T - TNOM \tag{9.2}$$

## 9.2 Temperature dependence of band-gap and intrinsic carrier concentration

Energy-band gap of channel ($E_g$): The temperature dependence of $E_g$ is modeled by

$$E_{g0} = BG0SUB - \frac{TBGASUB \cdot TNOM^2}{TNOM + TBGBSUB} \tag{9.3}$$

$$E_g = BG0SUB - \frac{TBGASUB \cdot T^2}{T + TBGBSUB} \tag{9.4}$$

$E_{g0}$ and $E_g$ represents band-gap at nominal temperature (TNOM) and at any operating temperature (T). TBGASUB and TBGBSUB are the band-gap temperature coefficients, and BG0SUB represents band-gap of substrate at 300.15 K with default value of 1.17. Intrinsic carrier concentration of non-silicon channel ($n_i$)

BSIM-Bulk MOSFET Model for IC Design - Digital, Analog, RF and High-Voltage
https://doi.org/10.1016/B978-0-32-385677-5.00013-5

$$n_i = NI0SUB \cdot (T_R)^{3/2} \cdot exp\left(\frac{E_g}{2\frac{k \cdot TNOM}{q}} - \frac{E_g}{2\frac{kT}{q}}\right) \quad (9.5)$$

NI0SUB is the intrinsic carrier concentration of the substrate at 300.15 K.

## 9.3 Temperature dependence of subthreshold characteristics

The temperature dependence of the threshold voltage $V_{th}$ and the subthreshold slope is modeled by

$$\Delta V_{th,temp} = \left(KT1 + KT2 \cdot V_{bsx}\right) \cdot \left(T_R^{KT1EXP} - 1\right) \quad (9.6)$$

$$ETA0(T) = ETA0 + TETA0\left(T_R - 1\right) \quad (9.7)$$

$$NFACTOR(T) = NFACTOR + TFACTOR \cdot \left(T_R - 1\right) \quad (9.8)$$

The term $\Delta V_{th,temp}$ is added to the effective flat-band voltage model and thus provides temperature dependence to the threshold voltage. Parameter TFACTOR provides flexibility to fine-tune the subthreshold slope of the model at different temperatures.

## 9.4 Temperature dependence of the linear-region characteristics

Linear region characteristics of a MOS device are governed by mobility and parasitic series-resistance behavior. Both of these quantities depend on temperature.

### 9.4.1 Mobility

Mobility is very sensitive to temperature. At low temperatures, Coulomb scattering becomes the dominant scattering mechanism while, and at high temperatures, phonon scattering is dominant. Temperature dependence model of mobility model parameters is given below:

$$U0(T) = U0 \cdot T_R^{UTE} \quad (9.9)$$

$$UA(T) = UA[1 + UA1 \cdot T_{diff}] \quad (9.10)$$

$$UC(T) = UC[1 + UC1 \cdot T_{diff}] \quad (9.11)$$

$$UD(T) = UD \cdot T_R^{UD1} \quad (9.12)$$

$$UCS(T) = UCS \cdot T_R^{UCSTE} \quad (9.13)$$

$$EU(T) = EU \cdot [1 + EU1 \cdot (T_R - 1)] \quad (9.14)$$

### 9.4.2 Source-drain resistance

BSIM-BULK offers three different modes for modeling parasitic source-drain resistance (RDSMOD=0, 1 and 2, see Chapter 3). The temperature dependence of the resistance is modeled as:

$$R_{source} = R_{source} \cdot rdstemp \tag{9.15}$$

$$R_{drain} = R_{drain} \cdot rdstemp \tag{9.16}$$

where

$$rdstemp = T_R^{PRT} \tag{9.17}$$

Reader may refer to Chapter 3 for $R_{source}$ and $R_{drain}$ model equations.

## 9.5 Temperature dependence of saturation characteristics

Inversion charge carrier velocity decreases with increasing temperature due to increased lattice vibrations. It is modeled as:

$$VSAT(T) = VSAT \cdot T_R^{-AT} \tag{9.18}$$

$$PTWG(T) = PTWG \cdot (1 - PTWGT \cdot T_{diff}) \tag{9.19}$$

## 9.6 Temperature dependence of junction diode IV

### 9.6.1 Source-side diode

The source-side diode is turned off if both $A_{seff}$ and $P_{seff}$ are zero. Otherwise, the source-side saturation current is given by

$$I_{sbs} = A_{seff} J_{ss}(T) + P_{seff} J_{ssws}(T) + W_{effcj} \cdot NF \cdot J_{sswgs}(T) \tag{9.20}$$

where

$$J_{ss}(T) = JSS \cdot exp\left(\frac{\frac{E_g}{v_t(TNOM)} - \frac{E_g(T)}{v_t(T)} + XTIS \cdot ln(T_R)}{NJS}\right) \tag{9.21}$$

$$J_{ssws}(T) = JSSWS \cdot exp\left(\frac{\frac{E_g}{v_t(TNOM)} - \frac{E_g(T)}{v_t(T)} + XTIS \cdot ln(T_R)}{NJS}\right) \tag{9.22}$$

$$J_{sswgs}(T) = JSSWGS \cdot exp\left(\frac{\frac{E_g}{k_b \cdot TNOM} - \frac{E_g(T)}{k_b \cdot T} + XTIS \cdot ln(T_R)}{NJS}\right) \tag{9.23}$$

where $E_g$ is given in temperature dependence of $E_g$ and $n_i$.

### 9.6.2 Drain-side diode

The drain-side diode is turned off if both $A_{seff}$ and $P_{seff}$ are zero. Otherwise, the drain-side saturation current is given by

$$I_{sbd} = A_{deff} J_{sd}(T) + P_{deff} J_{sswd}(T) + W_{effcj} \cdot NF \cdot J_{sswgd}(T) \tag{9.24}$$

where

$$J_{sd}(T) = JSD \cdot exp\left(\frac{\frac{E_g}{k_b \cdot TNOM} - \frac{E_g(T)}{k_b \cdot T} + XTID \cdot ln(T_R)}{NJD}\right) \tag{9.25}$$

$$J_{sswd}(T) = JSSWD \cdot exp\left(\frac{\frac{E_g}{k_b \cdot TNOM} - \frac{E_g(T)}{k_b \cdot T} + XTID \cdot ln(T_R)}{NJD}\right) \tag{9.26}$$

$$J_{sswgd}(T) = JSSWGD \cdot exp\left(\frac{\frac{E_g}{k_b \cdot TNOM} - \frac{E_g(T)}{k_b \cdot T} + XTID \cdot ln(T_R)}{NJD}\right) \tag{9.27}$$

## 9.7 Temperature dependence of junction diode CV

### 9.7.1 Source-side diode

The temperature dependences of zero-bias unit-length/area junction capacitances on the source side are modeled by

$$CJS(T) = CJS + TCJ \cdot T_{diff} \tag{9.28}$$

$$CJSWS(T) = CJSWS + TCJSW \cdot T_{diff} \tag{9.29}$$

$$CJSWGS(T) = CJSWGS + TCJSWG \cdot T_{diff} \tag{9.30}$$

The temperature dependences of the built-in potentials on the source side are modeled by

$$PBS(T) = PBS - TPB \cdot T_{diff} \tag{9.31}$$

$$PBSWS(T) = PBSWS - TPBSW \cdot T_{diff} \tag{9.32}$$

$$PBSWGS(T) = PBSWGS - TPBSWG \cdot T_{diff} \tag{9.33}$$

### 9.7.2 Drain-side diode

The temperature dependences of zero-bias unit-length/area junction capacitances on the drain side are modeled by

$$CJS(T) = CJS[1 + TCJ \cdot T_{diff}] \tag{9.34}$$

$$CJSWS(T) = CJSWS + TCJSW \cdot T_{diff} \tag{9.35}$$

$$CJSWGS(T) = CJSWGS[1 + TCJSWG \cdot T_{diff}] \tag{9.36}$$

The temperature dependences of the built-in potentials on the drain side are modeled by

$$PBD(T) = PBD - TPB \cdot T_{diff} \tag{9.37}$$

$$PBSWD(T) = PBSWD - TPBSW \cdot T_{diff} \tag{9.38}$$

$$PBSWGD(T) = PBSWGD - TPBSWG \cdot T_{diff} \tag{9.39}$$

### 9.7.3 Trap-assisted tunneling (TAT) and recombination current

$$J_{tsswgs}(T) = J_{tsswgs} \cdot \left(\sqrt{\frac{JTWEFF}{W_{effcj}}} + 1\right) \times exp\left[\frac{-E_g}{k_b T} \cdot X_{tsswgs} \cdot \left(1 - T_R\right)\right] \tag{9.40}$$

$$J_{tssws}(T) = J_{tssws} \cdot exp\left[\frac{-E_g}{k_b T} \cdot X_{tssws} \cdot \left(1 - T_R\right)\right] \tag{9.41}$$

$$J_{tss}(T) = J_{tss} \cdot exp\left[\frac{-E_g}{k_b T} \cdot X_{tss} \cdot \left(1 - T_R\right)\right] \tag{9.42}$$

$$J_{tsswgd}(T) = J_{tsswgd} \cdot \left(\sqrt{\frac{JTWEFF}{W_{effcj}}} + 1\right) \cdot exp\left[\frac{-E_g}{k_b T} \cdot X_{tsswgd} \cdot \left(1 - T_R\right)\right] \tag{9.43}$$

$$J_{tsd}(T) = J_{tsd} \cdot exp\left[\frac{-E_g}{k_b T} \cdot X_{tsd} \cdot \left(1 - T_R\right)\right] \tag{9.44}$$

$$NJTSSWG(T) = NJTSSWG \cdot \left[1 + TNJTSSWG\left(T_R - 1\right)\right] \tag{9.45}$$

$$NJTSSW(T) = NJTSSW \cdot \left[1 + TNJTSSW\left(T_R - 1\right)\right] \tag{9.46}$$

$$NJTS(T) = NJTS \cdot \left[1 + TNJTS\left(T_R - 1\right)\right] \tag{9.47}$$

$$NJTSSWGD(T) = NJTSSWGD \cdot \left[1 + TNJTSSWGD\left(T_R - 1\right)\right] \tag{9.48}$$

$$NJTSSWD(T) = NJTSSWD \cdot \left[1 + TNJTSSWD\left(T_R - 1\right)\right] \tag{9.49}$$

$$NJTSSWD(T) = NJTSSWD \cdot \left[1 + TNJTSSWD\left(T_R - 1\right)\right] \tag{9.50}$$

$$NJTSD(T) = NJTSD \cdot \left[1 + TNJTSD\left(T_R - 1\right)\right] \tag{9.51}$$

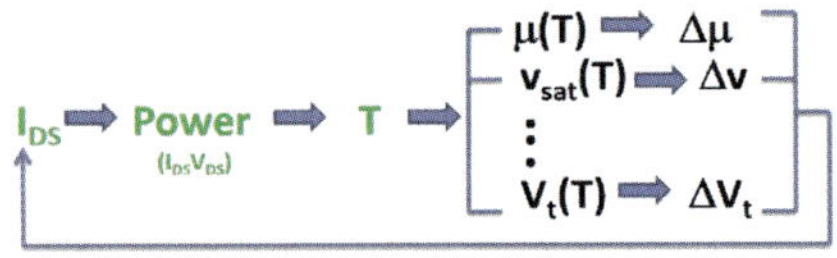


FIGURE 9.1 Illustration of self-heating effect on key device parameters.

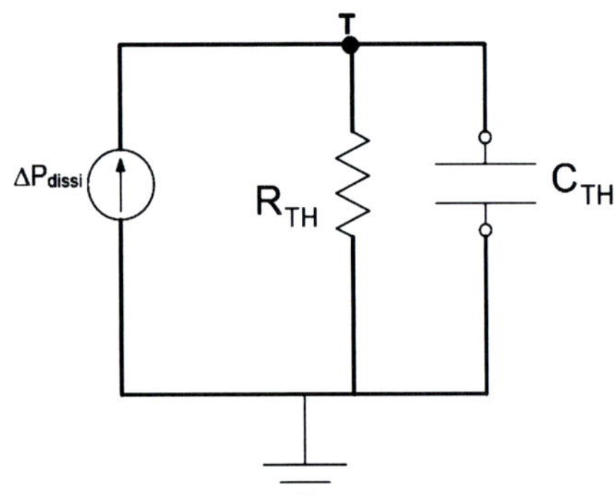


FIGURE 9.2 Subcircuit-based modeling of self-heating effect. Parameters RTH0 and CTH0 determine resistance $R_{TH}$ and capacitance $C_{TH}$ in the circuit. SHE model enabled by setting SHMOD=1.

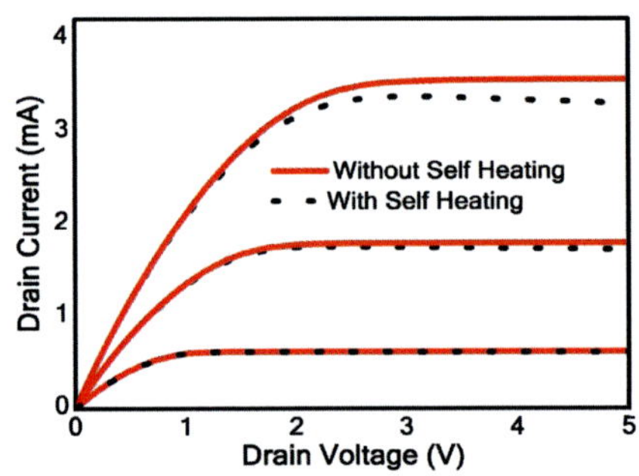


FIGURE 9.3 Illustration of self-heating-effect: Drain current vs. drain voltage with and without the self-heating model.

## 9.8 Self-heating effect

As the current flows through the channel, the power is dissipated in the form of heat. This raises the device temperature, thus changing important properties of the transistor, as shown in Fig. 9.1. Self-heating effect in the BSIM-BULK model is modeled using a subcircuit-based thermal network shown in Fig. 9.2.

Dissipated power ($= I_{ds} \cdot V_{ds}$) drives the network, the more the power dissipation, the higher the voltage at the thermal node 'T'. This voltage represents the temperature rise due to SHE and is added to the ambient temperature in the model. For a given current and voltage, the extent of temperature rise can be controlled by the thermal resistance parameter RTH. It has been shown that the impact of SHE diminishes at higher operating frequency, which is captured by the capacitor in the thermal network. Fig. 9.3 illustrates the impact of the self-heating effect on the output characteristics. At higher gate and drain voltages, SHE is prominent due to higher power dissipation, leading to a current fall.

CHAPTER

# 10

# Parameter Extraction

## 10.1 Parameter extraction methodology

The objective of this section is to provide guidelines for extracting the main model parameters. The procedure is structured so that parameters linked to specific psychical phenomena are extracted from analyses where these effects are prominent. Although parameter extraction is not always a straightforward procedure, the aim is to minimize the effort invested and the number of essential loops performed.

If all the steps of the described procedure are followed, then a global model card is obtained. This means that the model can be used across the entire width/length plane of the technology, although it might not display excellent accuracy for all DUTs. If a local fitting is targeted, then only the parameters of Section 10.2 need to be extracted for each DUT and a more accurate fitting is achieved. However, in that case, binning is necessary if the model card is to be used for the entire geometry range of the technology. Irrespective of the choice between global and local fitting, different model cards should be extracted for NMOS and PMOS devices or different technologies (even for those with the same minimum channel length).

Before proceeding to the extraction of any parameter, it is very important that **TNOM** is set to the value of the temperature at which the measurements were carried out. Also, it is recommended that if they are available, the values of the process parameters are provided. The most common process parameters are shown in Table 10.1 [1].

## 10.2 Extraction of main physical effects & geometry

The first part of the model parameter extraction procedure is to extract the parameters that are related to the main physical phenomena, which define the transistor's behavior, and are also geometry-independent. For that, a wide and long-channel device should be chosen. At this point, **WWIDE** and **LLONG** parameters must be assigned to the values of the

*BSIM-Bulk MOSFET Model for IC Design - Digital, Analog, RF and High-Voltage*
https://doi.org/10.1016/B978-0-32-385677-5.00014-7

TABLE 10.1 Process parameters recommended to be provided before parameter extraction. Parameters followed by * should be considered the most important among them.

| Parameter Name | Physical Description |
|---|---|
| **EPSROX** * | Relative Gate Dielectric Constant |
| **EPSRSUB** | Relative Dielectric Constant of the Channel |
| **TOXE** * | Electrical Gate Equivalent Oxide Thickness |
| **TOXP** or **DTOX** | Physical Gate Equivalent Oxide Thickness |
| **NDEP** * | Channel Doping Concentration |
| **NGATE** | Gate Doping Concentration |
| **NSD** | S/D Doping Concentration |
| **XJ** * | S/D Junction Depth |
| **XW/XL** * | Channel W/L Offset due to Mask/Etch Effect |

width and length of this large chosen DUT. This ensures that once the behavior of the long/wide channel device is fitted, it cannot be further affected by the values scaling parameters that will be extracted in the next steps.

### 10.2.1 $C_{GG}$ vs. $V_G$ analysis @ $V_S = 0\,\mathrm{V}$, $V_D = 0\,\mathrm{V}$ & $V_B = 0\,\mathrm{V}$

At this step process, the parameters and parameters related to the quantum mechanical effect are extracted. Even if values have already been assigned to process parameters, fine-tuning should be made to fit accurately the electrical behavior of the device.

From $C_{GG}$ vs. $V_G$ analysis the following process parameters can be extracted: **NDEP, TOXE, VFB**, and **NGATE**. Each of these parameters affects different regions or the $C_{GG}$ capacitance differently, so they should be extracted accordingly. More specifically:

- **NDEP** is affecting $C_{GG}$ in the depletion region. If possible, **NDEP**, which defines the doping level, is better to be extracted from $C_{GB}$ vs. $V_G$ analysis (S and D terminals are grounded).
- **TOXE** is affecting the deep accumulation and strong-inversion regions.
- **VFB** is defining the flat-band voltage of the device, and it can be extracted by studying the region from depletion till the onset of strong-inversion.
- **NGATE** is related to the poly-silicon depletion effect, so it affects the slope of $C_{GG}$ in the strong-inversion region.

Fig. 10.1, Fig. 10.2, and Fig. 10.3 show the impact of substrate doping, poly-depletion effect and VFB on $C_{gg}$, respectively. Furthermore, the value of $C_{OX}$ is affected by the quantum mechanical effect. So, the parameters: **ADOS, BDOS, QM0** and **ETAQM** are also extracted from $C_{GG}$ vs. $V_G$ analysis, when focusing at the slope of $C_{GG}$ at the onset of the strong-inversion region. Fig. 10.4 shows the impact of ADOS on CV behavior.

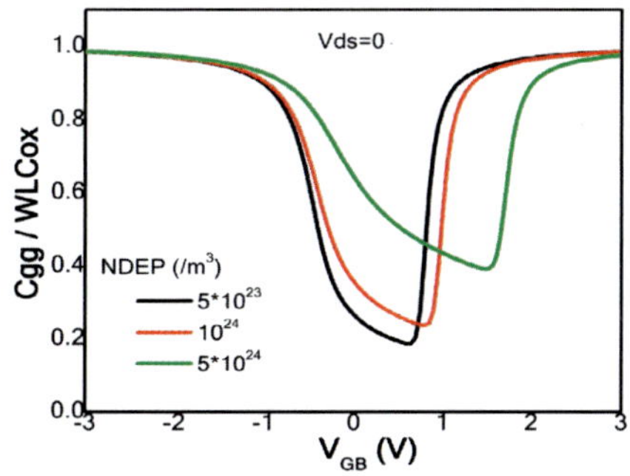


FIGURE 10.1 Gate capacitance model: Impact of substrate doping concentration on capacitance. The minimum capacitance increases as the doping increases. L=W=1 μm and default parameters used for the simulations.

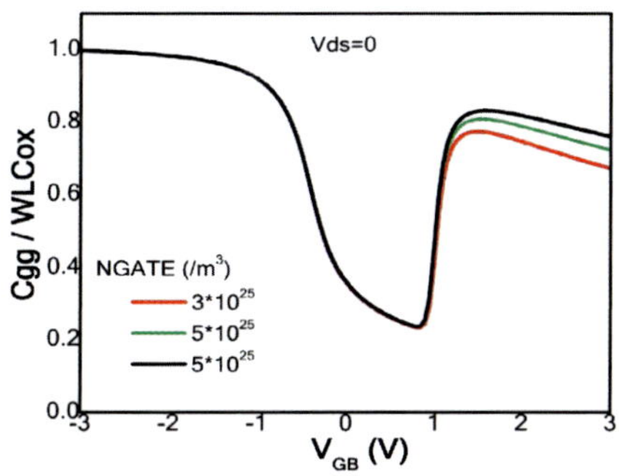


FIGURE 10.2 Gate capacitance model: Impact of poly-depletion effect (PDE): $C_{gg}$ vs. gate voltage for several values of parameter NGATE. The higher the doping of the gate, the lower the impact of poly-gate depletion. Note that the PDE model is only available for CVMOD=1. L=W=1 μm and default parameters used for the simulations.

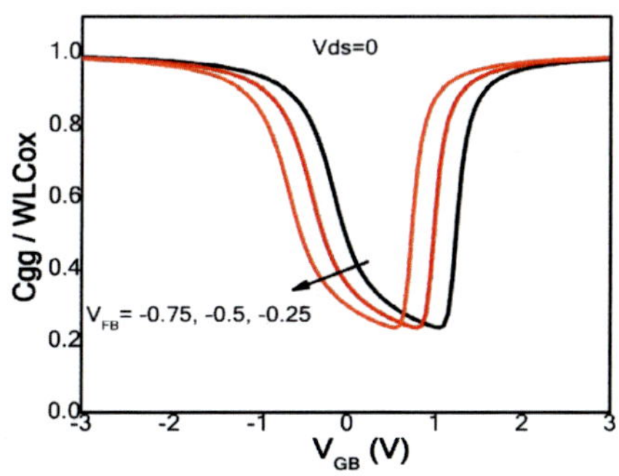


FIGURE 10.3 Gate capacitance model: Impact of flat-band voltage on capacitance. L=W=1 μm and default parameters used for the simulations.

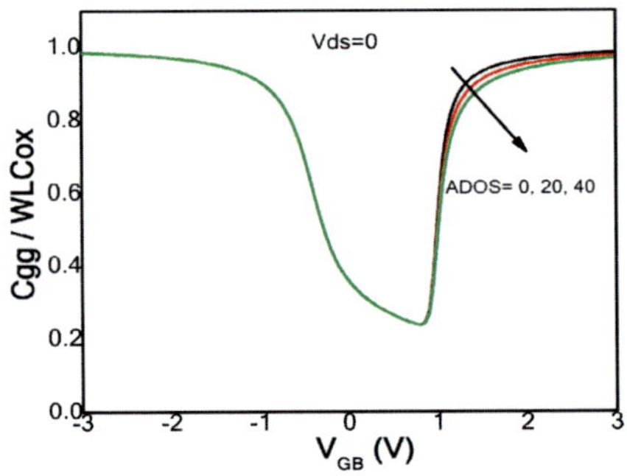


FIGURE 10.4 Gate capacitance model: Impact of quantum capacitance model parameter ADOS on capacitance. L=W=1 μm and default parameters used for the simulations.

### 10.2.2 $I_D$ vs. $V_G$ analysis @ $V_D = [V_{D,lin}, V_{D,sat}]$, $V_S = 0\,\text{V}$ & $V_B = 0\,\text{V}$

In this step, the $V_G$ dependence of the drain current, $I_D$, is studied. Different parameters are extracted in two different regions of operation, namely *linear region* (i.e., $V_D \ll V_G - V_{TH}$) and *saturation* (i.e., $V_D \gg V_G - V_{TH}$). It is very important that during extraction in this step, both $I_D$ and the transconductance, $g_m$, are studied at once.

*Linear Region*

- Focusing in the *weak-inversion* region ($I_D$ vs. $V_G$ characteristic when the y-axis is in a logarithmic scale), **NFACTOR**, which is related to the subthreshold slope of the $I_D$, can be extracted. Furthermore, fine-tuning of the **NDEP** and **VFB** parameters is performed. Fig. 10.5 and Fig. 10.6 show the impact of these two parameters on the IV characteristics. In case the values of **NDEP** and **VFB** obtained during the fitting of $I_D$ vs. $V_G$ characteristic differ much from those obtained during the fitting of $C_{GG}$ vs. $V_G$ characteristic (Section 10.2.1), parameters **NDEPCV** and **VFBCV** can be used for dynamic operation (CV) and **NDEP** and **VFB** for static operation (IV). In general, using different values for **NDEP** and **NDEPCV** for IV and CV operation is not recommended.
- From the *strong-inversion* region, the mobility **U0**, the parameter for the effective field **ETAMOB**, the parameters related to the effect of mobility reduction due to vertical field **UA** and **EU** and the parameters for the Coulomb scattering effect **UD** and **UCS** are extracted. Furthermore, the parameters for S/D series resistances are also extracted under the same bias conditions. If **RDSMOD** = **0** (internal S/D series resistances), **RDSW** is extracted. Otherwise, **RSW** and **RDW** are extracted.

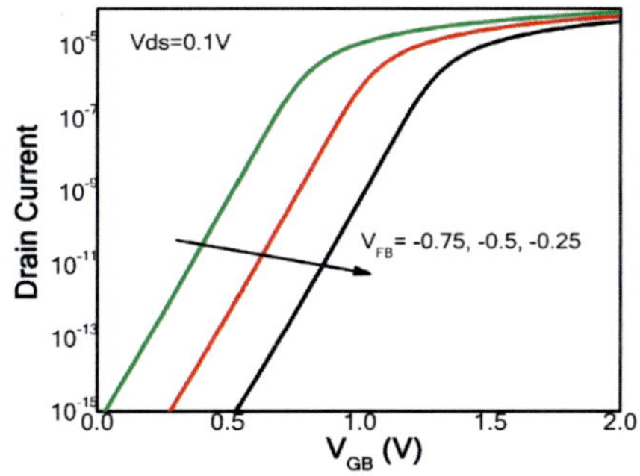


FIGURE 10.5 Impact of parameter VFB on IV characteristics: Drain current vs. gate voltage in the linear region. VFB shifts the threshold voltage but has no impact on subthreshold swing. L=W=1 µm and default parameters used for simulation.

*Saturation*

- From the *weak-inversion* region ($I_D$ vs. $V_G$ characteristic when the y-axis is in a logarithmic scale), **CDSCD** parameter, which is linked to the dependence of the subthreshold slope on drain bias, is extracted.
- Focusing in the *strong-inversion* region, the parameters that are connected to the velocity saturation effect, namely **VSAT**, **PSAT**, **PTWG** and **PSATX**, can be extracted.

Finally, from *accumulation* to *depletion* region, in both *linear region* and *saturation*, the parameters related to GIDL effect are extracted. First, the selector **GIDLMOD** should be set to 1 to activate GIDL/GILS currents and then the parameters **AGIDL**, **BGIDL**, **CGIDL** and **EGIDL** are extracted. Ideally GIDL and GILS currents should be equal, so it is sufficient to extract **AGIDL**, **BGIDL**, **CGIDL** and **EGIDL** parameters. But in case GIDL and GISL currents differ, then parameters **AGISL**, **BGISL**, **CGISL** and **EGISL** can also be used.

### 10.2.3 $I_G$ vs. $V_G$ analysis @ various $V_D$, $V_S = 0\,\mathrm{V}$ & $V_B = 0\,\mathrm{V}$

From $I_G$ vs. $V_G$ analysis, parameters related to the gate current can be extracted. First, the tunneling components should be activated by setting to 1 the selectors **IGCMOD** and **IGBMOD**. Different parameters are extracted in different regions of operation and more specifically:

*Accumulation to weak-inversion Region*

- **AIGBACC**, **BIGBACC**, **CIGBACC** and **NIGBACC**, which are linked to the gate-to-substrate current component determined by ECB.
- **AIGS**, **BIGS** and **CIGS**, which are linked to the tunneling current between the gate and the source diffusion region and **AIGD**, **BIGD** and

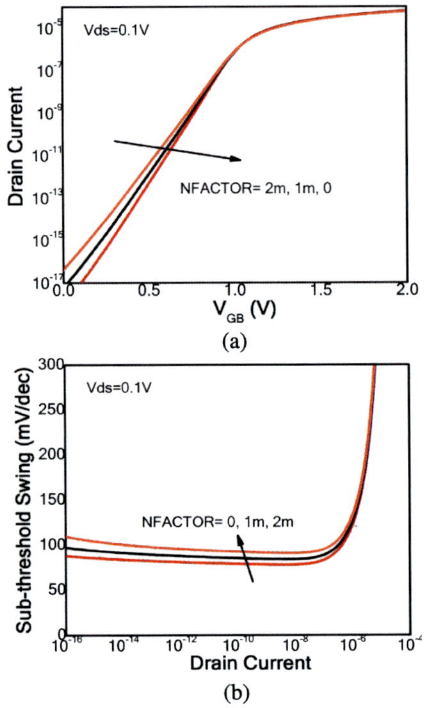


FIGURE 10.6 Impact of parameter NFACTOR: (a) Drain current vs. gate voltage (b) SS -$I_d$ for multiple values of NFACTOR. This parameter should be used to fine-tune the subthreshold swing for small drain biases. L=W=1 μm and default parameters used for the simulations.

**CIGD**, which are linked to the tunneling current between the gate and the drain diffusion region.

- **DLCIG** and **DLCIGD**, which are linked to the S/D overlap length for $I_{GS}$ and $I_{GD}$, respectively.

*Weak to strong-inversion Region*

- **AIGBINV**, **BIGBINV**, **CIGBINV**, **EIGBINV** and **NIGBINV**, which are linked to the gate-to-substrate current component determined by EVB.
- **AIGC**, **BIGC**, **CIGC**, **NIGC** and **PIGCD**, which are linked to the gate-to-channel current. **PIGCD** is expressing the $V_D$ dependence of gate-to-channel current.

### 10.2.4 $I_D$ vs. $V_D$ analysis @ various $V_G$, $V_S = 0\,V$ & $V_B = 0\,V$

In this step, both $I_D$ vs. $V_D$ and output conductance - $g_{ds}$ vs. $V_D$ characteristics are studied at once. Different effects impact both the characteristics, so the parameters related to those effects are extracted. In detail,

- **DELTA**, which is a smoothing factor for the transition between $V_{DS}$ and $V_{DS,sat}$.
- **PDITS** and **PDITSD** linked to DITS effect.
- **PCLM, PCLMG** and **FPROUT** linked to the CLM effect.
- **PDIBLC** linked to the impact of DIBL effect on $R_{out}$.
- **PVAG** linked to the $V_G$ dependence on Early voltage.

### 10.2.5 $C_{GG}$ vs. $V_G$ analysis @ $V_{DS} \neq 0\,\text{V}$ & $V_B = 0\,\text{V}$

Velocity saturation (VS) and channel length modulation (CLM) effects not only affect the static behavior of the transistor but also the dynamic. The extraction of **VSAT** and **PCLM** from $I_D$ vs. $V_G$ and $I_D$ vs. $V_D$ characteristics should be sufficient to capture these effects for CV operation. To verify that, $C_{GG}$ vs. $V_G$ characteristic for different $V_{DS} \neq 0\,\text{V}$ from linear to saturation must be studied. If different values for **VSAT** and **PCLM** are necessary for accurate fitting of CV behavior at different $V_D$ biases, then **VSATCV** and **PCLMCV** can be used.

### 10.2.6 $I_D$ vs. $V_G$ analysis @ $V_D = [V_{D,lin}, V_{D,sat}]$ & various $V_B$ (or various $V_S$)

In this step, almost the same procedure as in Section 10.2.2 will be repeated to extract the parameters that are linked to the body effect. Similar to Section 10.2.2, it is also very important that during the extraction in this step, both $I_D$ and $g_m$ are studied at once.

Linear Mode

- Focusing in the *weak-inversion* region, **CDSCB**, which is linked to the $V_B$ (or $V_S$) dependence of the subthreshold slope, is extracted. Also **K2**, which is linked to the $V_{TH}$ shift due to vertical nonuniform doping, is extracted in the same region.
- In the *strong-inversion* region, **UC**, which is linked to the $V_B$ (or $V_S$) dependence of mobility, is extracted. The parameter for $V_B$ (or $V_S$) dependence of S/D series resistances, **PRWB**, is also extracted under the same bias conditions.

Saturation

- In the *strong-inversion* region, the parameter connected to $V_B$ (or $V_S$) dependence of the velocity saturation effect, i.e., **PSATB**, is extracted.

To validate that the values of the parameters, which are linked to $V_B$ (or $V_S$) dependencies, are correctly extracted, it is useful to check $I_D$ vs. $V_D$ and $g_{ds}$ vs. $V_D$ characteristics @ various $V_G$ & $V_B \neq 0\,\text{V}$ (or $V_S \neq 0\,\text{V}$) and, if needed, to fine-tune the values of the parameters.

### 10.2.7 Fitting verification

When all the extraction steps of this part have been performed, the fitting of the model should be checked for all the analyses carried out up to this point. In case the fitting is not very accurate for all the analyses, all the steps (from 10.2.1 to 10.2.6) must be repeated for fine-tuning the parameters.

## 10.3 Extraction of short channel effects & length scaling parameters

Once the behavior of the wide/long channel device has been accurately modeled, the next step is the extraction of the parameters that are either related to short channel effects or express the different length dependencies. So it this part, devices across the entire length range of the technology, from the shortest to the longest one, are studied simultaneously. To avoid the impact of narrow-channel effects or the width dependencies, these devices should have the same **wide** channel. The extraction that is carried out follows the same flow as in Section 10.2, but now a set of devices with constant **wide** channel but different channel lengths is used.

### 10.3.1 $C_{GG}$ vs. $V_G$ analysis @ $V_S = 0\,\text{V}$, $V_D = 0\,\text{V}$ & $V_B = 0\,\text{V}$

In this step, parameters related to overlap and fringing capacitances, as well as those that are linked to the length dependence of doping concentration and flat-band voltage, are extracted. More specifically:

- **NDEPL1**, **NDEPLEXP1**, **NDEPL1** and **NDEPLEXP1**, which are the length scaling parameters for the doping concentration, are extracted from $C_{GG}$ in the depletion region. If possible, it is recommended that those parameters are extracted from $C_{GB}$ vs. $V_G$ analysis (S and D terminals are grounded).
- Extraction of parameters related to overlap and fringing capacitances is carried out by studying the entire range of $V_G$ bias of $C_{GG}$ vs. $V_G$ characteristic. These parameters are: **CGSO**, **CGDO**, **CGBO**, **CGSL**, **CGDL**, **CKAPPAS**, **CKAPPAD** and **CF**. If possible, it is recommended that **CGSO**, **CGDO**, **CGBO** and **CF** are extracted from $C_{GD}$ vs. $V_G$ at low $V_B$ (when S and D terminals are connected together, and B terminal is grounded), while **CGSL**, **CGDL**, **CKAPPAS** and **CKAPPAD** are extracted from $C_{GD}$ vs. $V_G$ at high $V_B$ (when S, D and B terminals are connected together).
- **DLC**, which is the channel-length offset parameter for the CV model, is extracted in the strong-inversion region of $C_{GG}$.
- **VFBCVL** and **VFBCVLEXP**, which express the length dependence of flat-band voltage at CV, are extracted from the depletion region till the

onset of strong-inversion. In order to be able to use **VFBCVL** and **VFBCVLEXP** parameters, **VFBCV** must be $\neq \mathbf{0}$.

### 10.3.2 $I_D$ vs. $V_G$ analysis @ $V_D = [V_{D,lin}, V_{D,sat}]$, $V_S = 0\,\text{V}$ & $V_B = 0\,\text{V}$

In this step, parameters related to short-channel effects or to length dependencies of $I_D$ vs. $V_G$ are extracted. Similar to the procedure described in Section 10.2.2, the parameters are divided into two groups, those which are extracted in the *linear mode* (i.e., $V_D \ll V_G - V_{TH}$) and those which are extracted in *saturation* (i.e., $V_D \gg V_G - V_{TH}$). It is very important that during the extraction, both $I_D$ and $g_m$ of all the devices are studied at once.

*Linear Mode*

- Focusing in the *weak-inversion* region ($I_D$ vs. $V_G$ characteristic when the y-axis is in a logarithmic scale), **NFACTORL** and **NFACTORLEXP**, which are related to the length dependence of the subthreshold slope of $I_D$ vs. $V_G$, can be extracted. Furthermore, **LINT**, which is the channel length offset parameter, is used to fit both the subthreshold slope and the $V_{TH}$. For fitting the $V_{TH}$ of the devices, also **DVTP0** and **UD** can prove to be useful. **UD** should be used only for fine-tuning because it mainly affects the region above threshold. It is recommended that the parameters **NDEPL1**, **NDEPLEXP1**, **NDEPL1** and **NDEPLEXP1** keep the values extracted from the $C_{GG}$ vs. $V_G$ analysis (Section 10.3.1). But, if the fitting of the $V_{TH}$ across the entire length range cannot be achieved without changing the values of **NDEPL1**, **NDEPLEXP1**, **NDEPL1** and **NDEPLEXP1**, then these parameters are used for static operation (IV) and **NDEPCVL1**, **NDEPCVLEXP1**, **NDEPCVL1** and **NDEPCVLEXP1** parameters are used for dynamic operation (CV).
- In the *strong-inversion* region, the parameters related to the length dependence of: i) the mobility; **U0L** and **U0LEXP**, ii) the effect of mobility reduction due to vertical field; **UAL**, **UALEXP**, **EUL** and **EULEXP** and iii) the coulomb scattering effect; **UDL** and **UDLEXP**, are extracted. Furthermore, parameters for the length dependence of S/D series resistances, namely **RDSWL** and **RDSWLEXP** (when **RDSMOD** $=$ **0**) or **RSWL**, **RSWLEXP**, **RDWL** and **RDWLEXP** (when **RDSMOD** $=$ **1**), are also extracted under the same bias conditions.

*Saturation*

- In the *weak-inversion* region ($I_D$ vs. $V_G$ characteristic when the y-axis is in a logarithmic scale), **CDSCDL** and **CDSCDLEXP** paramerers, which are linked to the length dependence of the subthreshold slope dependence on drain bias, are extracted. Moreover, parameters for DIBL ef-

fect, which control $V_{TH}$ when $V_{DS} \neq 0$, namely **ETA0** and **DSUB**, are also extracted in the same region.
- Focusing in the *strong-inversion* region, the length scaling parameters linked to the velocity saturation effect, i.e., **VSATL**, **VSATLEXP**, **PSATL**, **PSATLEXP**, **PTWGL** and **PTWGLEXP**, can be extracted.

Finally, from *accumulation* to *depletion* region, in both *linear mode* and *saturation*, the parameters **AGIDLL/AGISLL**, which are related the length dependence of GIDL effect (GIDL/GISL currents), are extracted.

### 10.3.3 $I_G$ vs. $V_G$ analysis @ various $V_D$, $V_S = 0\,\text{V}$ & $V_B = 0\,\text{V}$

From $I_G$ vs. $V_G$ analysis, parameters related to the length dependence of gate current are extracted. These parameters are: **AIGCL**, **AIGSL**, **AIGDL** and **PIGCDL**.

### 10.3.4 $I_D$ vs. $V_D$ analysis @ various $V_G$, $V_S = 0\,\text{V}$ & $V_B = 0\,\text{V}$

In this step, both $I_D$ vs. $V_D$ and $g_{ds}$ vs. $V_D$ characteristics should be studied at once. Similar to the procedure described in Section 10.3.4, the parameters extracted are:

- **DELTAL** and **DELTALEXP**, which are related to the length dependence of the smoothing factor for the transition between $V_{DS}$ and $V_{DS,sat}$.
- **PDITSL** linked to the length dependence of DITS effect.
- **PCLML**, **PCLMLEXP**, **FPROUTL** and **FPROUTLEXP** linked to the length dependence of CLM effect.
- **PDIBLCL** and **PDIBLCLEXP** linked to the length dependence of the impact of DIBL effect on $R_{out}$.

It is very important to be mentioned here, that if the slope of $g_{ds}$ vs. $V_D$ characteristic at low levels of inversion is steeper than the measurements, then **ETA0** should be decreased, and **DVTP1** can be used to achieve an accurate fit for the $V_{TH}$ in *saturation*.

### 10.3.5 $C_{GG}$ vs. $V_G$ analysis @ $V_{DS} \neq 0\,\text{V}$ & $V_B = 0\,\text{V}$

The extraction of the length scaling parameters of **VSAT** and **PCLM** from $I_D$ vs. $V_G$ and $I_D$ vs. $V_D$ characteristics (Steps 10.3.2 and 10.3.4) should be sufficient in order to capture VS and CLM effects for CV operation. To verify that, $C_{GG}$ vs. $V_G$ characteristic of all devices, for different $V_{DS} \neq 0\,\text{V}$, from linear mode to saturation, must be studied. If different values for **VSATL**, **VSATLEXP**, **PCLML** and **PCLMLEXP** are necessary for accurate fitting of the CV behavior across L, then **VSATCVL**, **VSATCVLEXP**, **PCLMCVL** and **PCLMCVLEXP** can be used.

### 10.3.6 $I_D$ vs. $V_G$ analysis @ $V_D = [V_{D,lin}, V_{D,sat}]$ & various $V_B$ (or various $V_S$)

In this step, almost the same procedure as in Section 10.2.6 will be repeated to extract the length scaling parameters that are linked to the body effect. Similar to Section 10.2.6, it is also very important that during the extraction in this step, both $I_D$ and $g_m$ of all devices are studied at once.

*Linear Mode*

- Focusing in the *weak-inversion* region, **K2L** and **K2LEXP**, which are linked to the length dependence $V_{TH}$ shift due to vertical nonuniform doping, are extracted.
- In the *strong-inversion* region, **UCL** and **UCLEXP**, which are linked to the length dependence of mobility reduction with $V_B$ (or $V_S$) bias, are extracted. The parameters for the length dependence of S/D series resistances with $V_B$ (or $V_S$) bias, namely **PRWBL** and **PRWBLEXP**, are also extracted under the same bias conditions.

*Saturation*

- In the *weak-inversion* region ($I_D$ vs. $V_G$ characteristic when the y-axis is in a logarithmic scale), the parameters related to length dependence of DIBL effect dependence on $V_B$ (or $V_S$) bias, namely **ETAB** and **ETABEXP**, are extracted.

To validate that the values of the length scaling parameters, which are linked to $V_B$ (or $V_S$) dependencies, are correctly extracted, it is useful to check $I_D$ vs. $V_D$ and $g_{ds}$ vs. $V_D$ characteristics @ various $V_G$ & $V_B \neq 0\,\text{V}$ (or $V_S \neq 0\,\text{V}$) and, if needed, to fine-tune the values of the parameters.

### 10.3.7 Fitting verification

When all the steps for extracting short channel effects and length scaling parameters have been performed, the fitting of the model should be checked for all the analyses carried out in Section 10.3. In case the fitting is not accurate enough, all the steps (from 10.3.1 to 10.3.6) must be repeated for fine-tuning the parameters.

## 10.4 Extraction of narrow channel effects & width scaling parameters

The next step in the parameter extraction procedure is the extraction of the parameters that are either related to narrow channel effects or express the different width dependencies. So it this part, devices across the entire width range of the technology, from the narrowest to the widest, are studied simultaneously. To avoid the impact of short channel effects or the length dependencies, these devices should have the same **long** channel. The extraction that is carried out follows the same flow as in Section 10.3, but now a set of devices with constant **long** channel but different channel widths is used.

### 10.4.1 $C_{GG}$ vs. $V_G$ analysis @ $V_S = 0\,V$, $V_D = 0\,V$ & $V_B = 0\,V$

In this step, parameters related to the width dependencies of the CV behavior of the device, e.g., width dependence of the doping concentration and flat-band voltage, are extracted. More specifically:

- **NDEPW** and **NDEPWEXP**, which are the width scaling parameters for the doping concentration, are extracted from $C_{GG}$ in the depletion region. If possible, it is recommended that those parameters are extracted from $C_{GB}$ vs. $V_G$ analysis (S and D terminals are grounded).
- **DWC**, which is the channel-width offset parameter for the CV model, is extracted in the strong-inversion region of $C_{GG}$.
- **VFBCVW** and **VFBCVWEXP**, which express the width dependence of flat-band voltage at CV, are extracted along the entire $V_G$ bias range of $C_{GG}$ characteristic. In order to be able to use **VFBCVW** and **VFBCVW-EXP** parameters, **VFBCV** must be $\neq$ **0**.

### 10.4.2 $I_D$ vs. $V_G$ analysis @ $V_D = [V_{D,lin}, V_{D,sat}]$, $V_S = 0\,V$ & $V_B = 0\,V$

In this step, parameters related to width dependencies of $I_D$ vs. $V_G$ are extracted. Similar to the procedure described in Section 10.2.2, the parameters are divided into two groups, those which are extracted in the *linear mode* (i.e., $V_D \ll V_G - V_{TH}$) and those which are extracted in *saturation* (i.e., $V_D \gg V_G - V_{TH}$). It is very important that during the extraction, both $I_D$ and $g_m$ of all the devices are studied at once.

Linear Mode

- Focusing in the *weak-inversion* region ($I_D$ vs. $V_G$ characteristic when the y-axis is in a logarithmic scale), **NFACTORW** and **NFACTORWEXP**, which are related to the width dependence of the subthreshold slope of $I_D$ vs. $V_G$, can be extracted. Furthermore, **WINT**, which is the channel width offset parameter, is used to fit both the subthreshold slope and the $V_{TH}$ across W. It is recommended that the parameters **NDEPW** and **NDEPWEXP** keep the values extracted from the $C_{GG}$ vs. $V_G$ analysis (Section 10.4.1). But, if the fitting of the $V_{TH}$ across the entire width range cannot be achieved without changing the values of **NDEPW** and **NDEPWEXP**, then these parameters are used for static operation (IV), and **NDEPCVW** and **NDEPCVWEXP** parameters are used for dynamic operation (CV).

- In the *strong-inversion* region, the parameters related to the width dependence of mobility reduction due to vertical field effect, namely **UAW**, **UAWEXP**, **EUW** and **EUWEXP**, are extracted.

Saturation

- Focusing the in *strong-inversion* region, the width scaling parameters linked to the velocity saturation effect, i.e., **VSATW** and **VSATWEXP**, can be extracted.

Finally, from *accumulation* to *depletion* region, in both *linear mode* and *saturation*, the parameters **AGIDLW/AGISLW**, which are related the width dependence of GIDL effect (GIDL/GISL currents), are extracted.

To validate that the values of the width scaling parameters are correctly extracted, it is useful to check $I_D$ vs. $V_D$ and $g_{ds}$ vs. $V_D$ characteristics @ various $V_G$, $V_S = 0\,\text{V}$ & $V_B = 0\,\text{V}$ (or $V_S \neq 0\,\text{V}$) and, if needed, to fine-tune the values of the parameters.

### 10.4.3 $I_G$ vs. $V_G$ analysis @ various $V_D$, $V_S = 0\,\text{V}$ & $V_B = 0\,\text{V}$

From $I_G$ vs. $V_G$ analysis, parameters related to the width dependence of gate current are extracted. These parameters are: **AIGCW**, **AIGSW** and **AIGDW**.

### 10.4.4 $C_{GG}$ vs. $V_G$ analysis @ $V_{DS} \neq 0\,\text{V}$ & $V_B = 0\,\text{V}$

The extraction of the width scaling parameters of **VSATW** and **VSATWEXP** from $I_D$ vs. $V_G$ and $I_D$ vs. $V_D$ characteristics (Step 10.4.2) should be sufficient to capture VS for CV operation. To verify that, $C_{GG}$ vs. $V_G$ characteristic of all devices, for different $V_{DS} \neq 0\,\text{V}$, from linear mode to saturation, must be studied. If different values for **VSATW** and **VSATWEXP** are necessary for accurate fitting of the CV behavior across W, then **VSATCVW** and **VSATCVWEXP** can be used.

### 10.4.5 $I_D$ vs. $V_G$ analysis @ $V_D = [V_{D,lin}, V_{D,sat}]$ & various $V_B$ (or various $V_S$)

In this step, from the *weak-inversion* region of *linear mode*, **K2W** and **K2WEXP**, which are linked to the width dependence $V_{TH}$ shift due to vertical nonuniform doping, can be extracted. For validation purposes, it is useful to check: i) $I_D$ vs. $V_G$ and $g_m$ vs. $V_G$ characteristics in *weak* and *strong-inversion* and for both *linear mode* and *saturation*, and ii) $I_D$ vs. $V_D$ and $g_{ds}$ vs. $V_D$ characteristics @ various $V_G$ & $V_B \neq 0\,\text{V}$ (or $V_S \neq 0\,\text{V}$) and, if needed, extract **K2W** and **K2WEXP** to fit both (i) and (ii).

### 10.4.6 Fitting verification

When all the extraction steps for the width scaling have been performed, the fitting of the model should be checked for all the analyses carried out in Section 10.4. In case the fitting is not accurate enough, all the steps (from 10.4.1 to 10.4.5) must be repeated for fine-tuning the parameters.

## 10.5 Extraction of parameters for narrow/short channel devices

The final part in the parameter extraction procedure from a geometrical point of view is the extraction of the parameters for narrow/short-channel devices. These devices have the minimum dimensions, so it is more difficult to capture their behavior. Since the narrow/short-channel device parameters can affect the already performed fitting across length and width, it is recommended that two different sets of devices are studied simultaneously, i.e., one set with a constant **short** channel but different channel widths (from the narrowest to the widest) and one set with a constant **narrow** channel but different channel lengths (from the shortest to the longest).

### 10.5.1 $C_{GG}$ vs. $V_G$ analysis @ $V_S = 0\,\mathrm{V}$, $V_D = 0\,\mathrm{V}$ & $V_B = 0\,\mathrm{V}$

In this step, geometry-dependent parameters for modeling the CV behavior of the narrow/short-channel devices, are extracted. More specifically:

- **NDEPWL** and **NDEPWLEXP**, which are used to fit the doping concentration of small-channel devices, are extracted from $C_{GG}$ in the depletion region. If possible, it is recommended that those parameters are extracted from $C_{GB}$ vs. $V_G$ analysis (S and D terminals are grounded).
- **LWLC** and **WWLC**, which are coefficients of length/width dependencies for CV model, are extracted in the strong-inversion region of $C_{GG}$.
- **VFBCVWL** and **VFBCVWLEXP**, which are used to fit the flat-band voltage at CV, are extracted from depletion till the onset of strong-inversion region of $C_{GG}$ characteristic. In order to be able to use **VFBCVWL** and **VFBCVWLEXP** parameters, **VFBCV** must be $\neq \mathbf{0}$.

### 10.5.2 $I_D$ vs. $V_G$ analysis @ $V_D = [V_{D,lin}, V_{D,sat}]$, $V_S = 0\,\mathrm{V}$ & $V_B = 0\,\mathrm{V}$

In this step, geometry-dependent parameters for modeling $I_D$ of the narrow/short-channel devices, are extracted. Similar to the procedure described in Section 10.2.2, the parameters are divided into two groups, those which are extracted in the *linear mode* (i.e., $V_D \ll V_G - V_{TH}$) and those which are extracted in *saturation* (i.e., $V_D \gg V_G - V_{TH}$). It is very important that during the extraction, both $I_D$ and $g_m$ of all the devices are studied at once.

*Linear Mode*

- Focusing in the *weak-inversion* region ($I_D$ vs. $V_G$ characteristic when the y-axis is in a logarithmic scale), **NFACTORWL** and **NFACTORWLEXP**, which are used to fit the subthreshold slope of $I_D$ vs. $V_G$ for small-channel devices, can be extracted. It is recommended that the parameters **NDEPWL** and **NDEPWLEXP** keep the values extracted from the $C_{GG}$ vs. $V_G$ analysis (Section 10.5.1). But, if the fitting of the $V_{TH}$ for both sets of devices cannot be achieved without changing the values of **NDEPWL** and **NDEPWLEXP**, then these parameters are used for static operation (IV), and **NDEPCVWL** and **NDEPCVWLEXP** parameters are used for dynamic operation (CV).
- In the *strong-inversion* region, the parameters which are used to model the effect of mobility reduction due to vertical field in small-channel devices, namely **UAWL**, **UAWLEXP**, **EUWL** and **EUWLEXP**, are extracted.

*Saturation*

- Focusing in the *strong-inversion* region, the parameters, which are used to model to the velocity saturation effect in small-channel devices, i.e., **VSATWL** and **VSATWLEXP**, can be extracted.

To validate that the values of the parameters, modeling the behavior of narrow/short-channel devices, are correctly extracted, it is useful to check $I_D$ vs. $V_D$ and $g_{ds}$ vs. $V_D$ characteristics @ various $V_G$, $V_S = 0\,\text{V}$ & $V_B = 0\,\text{V}$ and, if needed, to fine-tune the values of the parameters.

### 10.5.3 $C_{GG}$ vs. $V_G$ analysis @ $V_{DS} \neq 0\,\text{V}$ & $V_B = 0\,\text{V}$

The extraction of the parameters, which are used to model to the velocity saturation effect in small-channel devices, **VSATWL** and **VSATWEXP**, from $I_D$ vs. $V_G$ and $I_D$ vs. $V_D$ characteristics (Step 10.5.2), should be sufficient to capture VS for CV operation. To verify that, $C_{GG}$ vs. $V_G$ characteristic of all devices, for different $V_{DS} \neq 0\,\text{V}$, from linear mode to saturation, must be studied. If different values for **VSATWL** and **VSATWLEXP** are necessary for accurate fitting of the CV behavior for both sets of devices, then **VSATCVWL** and **VSATCVWLEXP** can be used.

### 10.5.4 $I_D$ vs. $V_G$ analysis @ $V_D = [V_{D,lin}, V_{D,sat}]$ & various $V_B$ (or various $V_S$)

In this step, from the *weak-inversion* region of *linear mode*, **K2WL** and **K2WLEXP**, which are linked to the $V_{TH}$ shift due to vertical nonuniform doping in small-channel devices, can be extracted. For validation purposes, it is useful to check: i) $I_D$ vs. $V_G$ and $g_m$ vs. $V_G$ characteristics in *weak* and *strong-inversion* and for both *linear mode* and *saturation*, and ii) $I_D$ vs. $V_D$ and $g_{ds}$ vs. $V_D$ characteristics @ various $V_G$ & $V_B \neq 0\,V$ (or $V_S \neq 0\,V$) and, if needed, extract **K2WL** and **K2WLEXP** to fit both (i) and (ii).

### 10.5.5 Fitting verification

When all the steps for narrow/short-channel devices have been performed, the fitting of the model should be checked for all the analyses carried out in Section 10.5. In case the fitting is not accurate enough, all the steps (from 10.5.1 to 10.5.4) must be repeated for fine-tuning the parameters.

## 10.6 Extraction of temperature dependence parameters

Up to this point of the parameter extraction procedure, the temperature dependence of the parameters has been ignored since all the parameters were extracted at **TNOM**. In this part, the parameters that are related to the impact of temperature on the behavior of devices are extracted, and for that, data across the temperature range of the technology are necessary. The behavior of devices is studied with the same geometrical sequence as the previous steps, while the temperature dependence parameters are extracted in the same regions of operation as the parameters of the corresponding physical effects.

### 10.6.1 Wide and long channel devices

The first step, in the extraction of temperature dependence parameters, is to study the behavior of a long- and wide-channel devices @ **different T** and for different analyses. It is recommended that the same device as the one in Section 10.2 is used. In detail:

*$I_D$ vs. $V_G$ analysis @ $V_D = V_{D,lin}$, $V_S = 0\,V$ & $V_B = 0\,V$*

- From the *weak-inversion* region ($I_D$ vs. $V_G$ characteristic when the y-axis is in a logarithmic scale), the parameters **TBGASUB** and **TBGBSUB**, which control the temperature dependence of $E_g$, are extracted. Also, **TNFACTOR** is extracted to fit the subthreshold slope of $I_D$ in different T, while **KT1** and **KT1EXP** are extracted for fitting the $V_{TH}$ across T.
- From the *strong-inversion* region, the mobility temperature exponent, **UTE** and the temperature coefficients: i) for mobility reduction due to vertical field effect, namely **UA1** and **UD1**, ii) for Coulomb scattering effect, **UCSTE** and iii) for S/D series resistances, **PRT**, are extracted.

*$I_D$ vs. $V_G$ analysis @ $V_D = V_{D,sat}$, $V_S = 0\,V$ & $V_B = 0\,V$*

- From the *strong-inversion* region, the parameters that are used to model the temperature dependence of velocity saturation effect, i.e., **AT** and **PTWGT**, are extracted.

It is very important that in the above analyses, both $I_D$ and $g_m$ of all the devices are studied at once. Furthermore, from *accumulation* to *depletion* region, in both *linear mode* and *saturation* of $I_D$ vs. $V_G$ analysis, the parameter **TGIDL**, which controls the temperature dependence of GIDL effect, is extracted.

*$I_D$ vs. $V_D$ Analysis @ various $V_G$, $V_S = 0\,V$ & $V_B = 0\,V$*

From $I_D$ vs. $V_D$ analysis in different temperatures, **TDELTA**, which is related to the temperature dependence of the smoothing factor for the transition between $V_{DS}$ and $V_{DS,sat}$, is extracted.

*$I_D$ vs. $V_G$ Analysis @ $V_D = V_{D,lin}$ & various $V_B$ (or various $V_S$)*

- From the *weak-inversion* region ($I_D$ vs. $V_G$ characteristic when the y-axis is in a logarithmic scale) **KT2**, which is linked to the temperature dependence of $V_{TH}$ shift due to vertical nonuniform doping with $V_B$ (or $V_S$) bias, is extracted.
- From the *strong-inversion* region, the temperature coefficient for mobility reduction with $V_B$ (or $V_S$) bias, namely **UC1**, is extracted.

For validation purposes, it is useful to check: i) $I_D$ vs. $V_G$ and $g_m$ vs. $V_G$ characteristics in *weak* and *strong-inversion* and for both *linear mode* and *saturation*, and ii) $I_D$ vs. $V_D$ and $g_{ds}$ vs. $V_D$ characteristics @ various $V_G$ & $V_B \neq 0\,\text{V}$ (or $V_S \neq 0\,\text{V}$) and, if needed, extract **KT2** and **UC1** to fit both (i) and (ii).

## 10.6.2 Length scaling of wide channel devices

The following step in the extraction of temperature dependence parameters is to study the temperatures dependences across L. For this, data **@ different T** of a set of devices with constant **wide** channel but different channel lengths are used.

*$I_D$ vs. $V_G$ analysis @ $V_D = V_{D,lin}$, $V_S = 0\,V$ & $V_B = 0\,V$*

- From the *weak-inversion* region ($I_D$ vs. $V_G$ characteristic when the y-axis is a in logarithmic scale), the parameter **KT1L** is extracted for fitting the $V_{TH}$ across L, at different T.

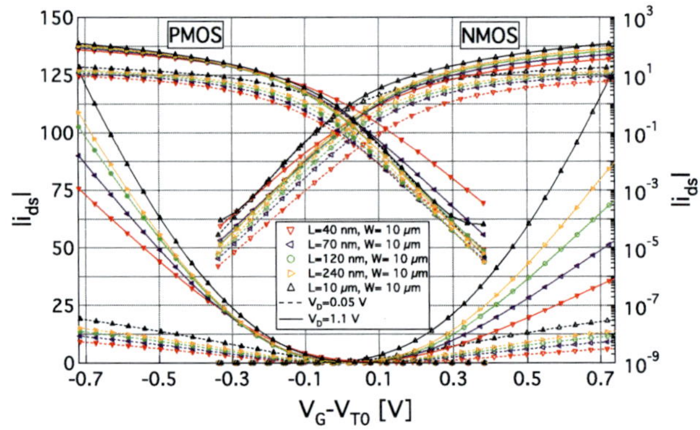


FIGURE 10.7 DC analyses for NMOS and PMOS DC DUTs, with $W = 10\,\mu\text{m}$ and $L = [40n, 70n, 120n, 240n, 10\mu]$ $m$: $|i_{ds}|$ vs. $V_G - V_{T0}$ in linear and logarithmic scales, for $V_D = 0.05\,\text{V}$ (linear mode) and $V_D = 1.1\,\text{V}$ (saturation) [2].

- From the *strong-inversion* region, the length scaling parameters for: i) mobility temperature exponent, **UTEL** and for the temperature coefficients or mobility reduction due to vertical field effect, namely **UA1L** and **UD1L**, are extracted.

*$I_D$ vs. $V_G$ analysis @ $V_D = V_{D,sat}$, $V_S = 0\,V$ & $V_B = 0\,V$*

- From the *weak-inversion* region ($I_D$ vs. $V_G$ characteristic when the y-axis is in a logarithmic scale), the parameter **TETA0**, which is related to the temperature dependence of DIBL effect and thus is controlling the $V_{TH}$ in saturation, is extracted.
- From the *strong-inversion* region, the parameters that are used to model the temperature dependence of velocity saturation effect across L, i.e., **ATL** and **PTWGTL**, are extracted.

It is very important that in the above analyses, both $I_D$ and $g_m$ of all the devices are studied at once. For validating that the values of length scaling parameters for temperature dependence parameters are extracted correctly, it is useful to check also $I_D$ vs. $V_D$ and $g_{ds}$ vs. $V_D$ characteristics and, if needed, to fine-tune their value by repeating Step 10.6.2.

## 10.7 Model validation results

The parameter extraction procedure is tested with experimental data from 40 nm bulk CMOS technology over a wide range of temperature, channel length, and width and biases for both n- and p-channel devices [2]. As shown in Fig. 10.7, Fig. 10.8, Fig. 10.9, and Fig. 10.10, the model has

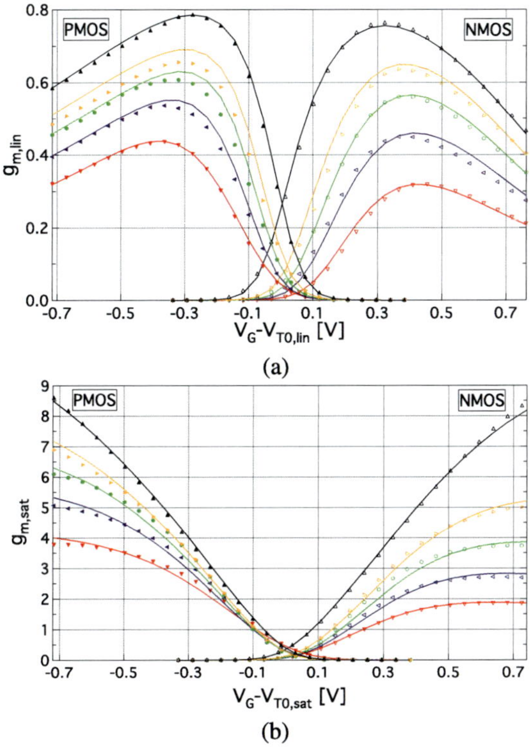


FIGURE 10.8 DC analyses for NMOS and PMOS DC DUTs, with $W = 10\,\mu\text{m}$ and $L = [40n, 70n, 120n, 240n, 10\mu]\ m$: (a) $g_m$ vs. $V_G - V_{T0,lin}$ for $V_D = 0.05\,\text{V}$ (linear mode), (b) $g_m$ vs. $V_G - V_{T0,sat}$ for $V_D = 1.1\,\text{V}$ (saturation). Experimental data is taken from [2].

excellent capabilities to accurately capture the behavior of the experimental devices.

## 10.8 Corner modeling

Modern integrated circuits are mainly fabricated on semiconductor wafers. Tens, hundreds, or even thousands of identical circuits can be built on a single wafer, and usually, dozens of wafers can be manufactured at one time. During the manufacturing process, certain processing steps can cause within-die process variability as well as across-the-chip process variability, leading to the shading of certain characteristics of devices within a wafer. The variation or shading can compromise the performance of the circuits and cause the device to fail a test or specification. This failure reduces the manufacturing yield. Process variability in scaled CMOS technologies critically affects the delay and power variability in VLSI devices,

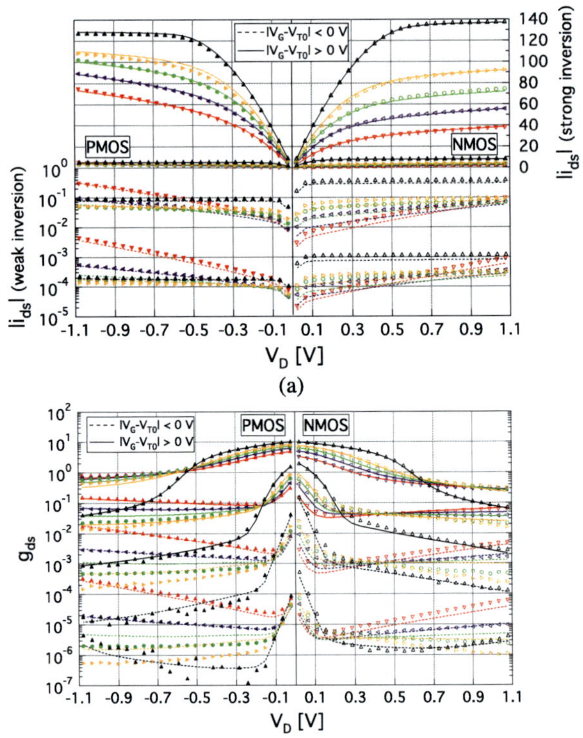


**FIGURE 10.9** DC analyses for NMOS and PMOS DC DUTs, with $W = 10\,\mu\text{m}$ and $L = [40n, 70n, 120n, 240n, 10\mu]\ m$: (a) $i_{ds}$ vs. $V_D$ for $V_G - V_{T0} = [-0.26, -0.07]\,\text{V}$ (weak inversion) and $V_G - V_{T0} = [0.12, 0.72]\,\text{V}$ (strong inversion), (b) $g_{ds}$ vs. $V_D$ for $V_G - V_{T0} = [-0.26, -0.07]\,\text{V}$ (weak inversion) and $V_G - V_{T0} = [0.12, 0.72]\,\text{V}$ (strong inversion). Experimental data is taken from [2].

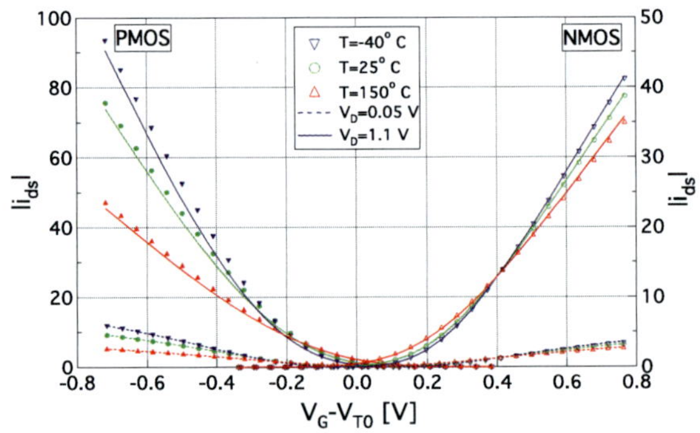


**FIGURE 10.10** DC analyses for NMOS and PMOS DC DUTs, with $W = 10\,\mu\text{m}$ and $L = [40n, 70n, 120n, 240n, 10\mu]\ m$: $|i_{ds}|$ vs. $V_G - V_{T0}$, for $V_D = 0.05\,\text{V}$ (linear mode) and $V_D = 1.1\,\text{V}$ (saturation), for $T = [-40, 25, 250]°\text{C}$ for the shortest DC DUT with $W = 10\,\mu\text{m}$ and $L = 40\,\text{nm}$. Experimental data is taken from [2].

circuits, and chips, and this impact shoot-up as MOSFET devices and CMOS technologies keep shrinking. Thus a compact model addressing the impact of process variability in scaled MOSFET devices is important to implement in the process design kit to support process variability-aware circuit designing. To set the lower and upper limits of process variation, corner modeling is typically used to assess process variation in the circuit performance.

For circuit designing, process variability is modeled by four worst-case corners: two for analog applications and two for digital [1,9]. The first corner for analog applications will come from slow NMOS and slow PMOS to model the worst-case speed and another corner from fast NMOS and fast PMOS (FF) to model the worst-case power. The corners for digital applications, on the other hand, are coming from fast NMOS and slow PMOS (FS) to model the worst-case "1" and from slow NMOS and fast PMOS (SF) to model the worst-case "0". A standard set of model parameters (e.g., $V_{TH}$) is used to account for process variability and model worst-case corner performance of the devices and circuits for the target CMOS technology [1,9].

To account for the effect of process variability, the worst-case corner models are generated by adding an offset of $\pm dP = n.\sigma$ into the particular parameter, $P$, of the typical (TT) compact model, where $n$ is the number of $\sigma$ for $P$. Typically, $3 \leq n \leq 6$ is selected to set the fixed lower limit (LL) and upper limit (UL) of the worst-case models; and TT is the typical compact model extracted against the silicon by characterizing the golden die of golden wafer. However, at the early stages of technology development, measured target characteristics are not available for all electrical targets (ETs) or typical targets (TT). In such cases, one must use the TT targets provided by the process engineers to extract the TT model. Further to this, any early silicon characteristic could be inferior (too far away) to the final targets for a mature technology. As a result, the creation of the compact models involves the magic of 're-targeting' – shifting early silicon performance to the targeted mature technology levels, i.e., one would do the re-targeting on the model card extracted from Section 10.1.

Let us assume that TT target for VTCCSAT ($V_{DS}$=1 V, $I_{constant} = 80n.W/L$) $@L = 60$ nm is 0.4 V and sigma = 30 mV. Therefore the UL and LL targets of VTCCSAT $@L = 60$ nm will be given as $0.4 - 4 * 0.030 = 0.28$ V and $0.4 + 4 * 0.030 = 0.52$ V, respectively. To get the complete trends for TT, UL, and LL across a wide range of length, we will use $VFB = VFB \pm dVFB$, where $dVFB$ is used to set the target LL and UL of the worst-case models, as illustrated in Fig. 10.11.

Further, let us assume that TT target for IDSAT ($V_{GS}$=1 V, $V_{DS}$=1 V) $@L = 60$ nm is 360 µA/µm and sigma = 12.5. Therefore the UL and LL targets of IDSAT $@L = 60$ nm will be given as $360 + 4 * 12.5 = 410$ µA/µm and $360 - 4 * 12.5 = 310$ µA/µm, respectively. To get the complete trends for

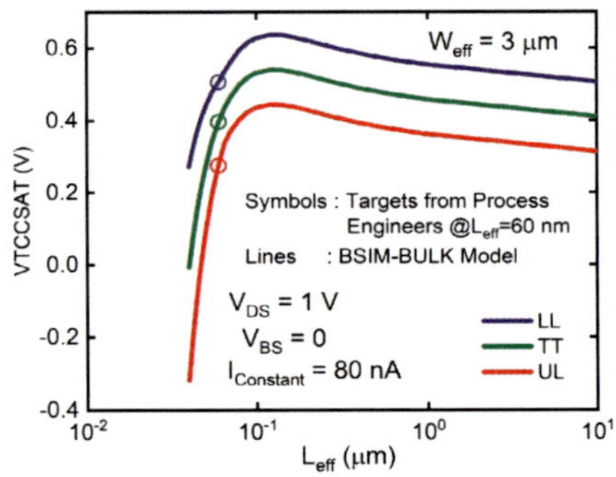


FIGURE 10.11 Length scaling of the BSIM-BULK model: Effective value of channel doping ($NDEP$) as a function of channel length.

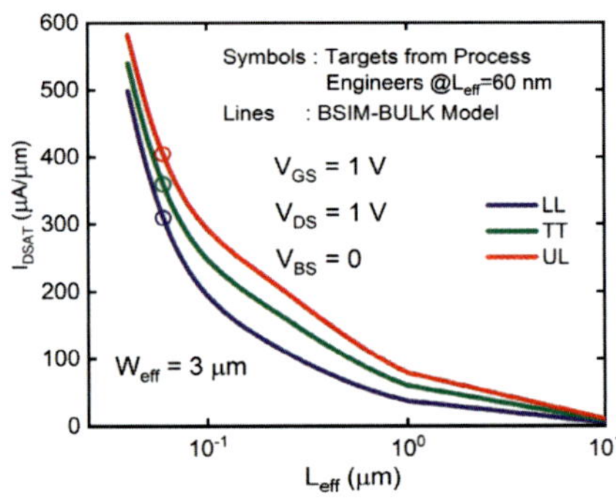


FIGURE 10.12 Length scaling of the BSIM-BULK model: Effective value of channel doping ($NDEP$) as a function of channel length.

TT, UL, and LL across a wide range of length, we will use $U0 = U0 \pm dU0$, where $dU0$ is used to set the target LL and UL of the worst-case models, as illustrated in Fig. 10.12.

Finally, let us assume that TT target for IDOFF ($V_{GS}$=0 V, $V_{DS}$=1 V) @$L$ = 60 nm is 100 pA/μm and sigma = 0.25. Therefore the UL and LL targets of IDOFF @$L$ = 60 nm will be given as $10^{\log(100)+4*0.25}$ = 1000 pA/μm and $10^{\log(100)-4*0.25}$ = 10 pA/um, respectively. To get the complete trends for TT, UL, and LL across a wide range of length, we will use $NFACTOR = NFACTOR \pm dNFACTOR$, where $dNFACTOR$ is used to set the target LL and UL of the worst-case models, as illustrated in Fig. 10.13.

The FF corner is obtained using the UL values of the selected model parameters for both NMOS and PMOS devices whereas SS corner is obtained considering the LL values of the selected model parameters for both NMOS and PMOS devices.

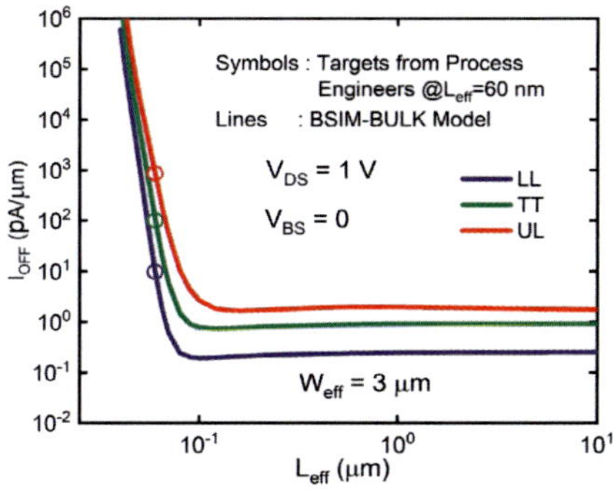


FIGURE 10.13 Length scaling of the BSIM-BULK model: Effective value of channel doping ($NDEP$) as a function of channel length.

## References


[1] BSIM-BULK107.1.0 Technical Manual. [Online]. Available: http://www-device.eecs.berkeley.edu/bsim/.

[2] Y.S. Chauhan, S. Venugopalan, M.A. Chalkiadaki, M.A.U. Karim, H. Agarwal, S. Khandelwal, N. Paydavosi, J.P. Duarte, C.C. Enz, A.M. Niknejad, C. Hu, BSIM6: analog and RF compact model for bulk MOSFET, IEEE Transactions on Electron Devices 61 (2) (Feb 2014) 234–244.

CHAPTER

# 11

# BSIM-BULK Benchmark Tests

## 11.1 Introduction

Compact models need to accurately reproduce the transistor's behavior for a wide range of temperature, bias and geometry configurations. At the same time, the model should be numerically robust and fast. To achieve the same, several assumptions are made during the derivation of the compact model, sometimes to simplify the complex integration and sometimes to impart flexibility to the model. Under such a scenario, how does one ensure that a compact model does not produce any unphysical results when used under varying operating conditions? One way to check the model's physical correctness and asymptotic behavior is through benchmark tests. These tests have been developed over the years by the compact modeling community, and a compact model must pass them. In the subsequent sections, we discuss the BSIM-BULK model's benchmark test results.

## 11.2 DC tests

### 11.2.1 Tree-top test

The tree-top test is used to validate MOSFET compact model's physical behavior from weak to strong inversion transition. It involves observing transconductance ($g_m$) to drain to source current ($I_{DS}$) ratio as the gate voltage is varied from subthreshold to strong inversion. Fig. 11.1 shows the tree-top test result with the BSIM-BULK model. Ratio $g_m/I_{DS}$ varies smoothly from weak inversion to strong inversion [1]. The theoretical maximum value of the ratio is $1/V_t$ ($V_t$ is the thermal voltage), which is practically never achieved. It is important to note that, unlike the BSIM-BULK model, older generation threshold voltage-based models show incorrect constant $g_m/I_{DS}$ in the weak-inversion region. This is due to the semiempirical treatment of the subthreshold region [2].

*BSIM-Bulk MOSFET Model for IC Design - Digital, Analog, RF and High-Voltage*
https://doi.org/10.1016/B978-0-32-385677-5.00015-9

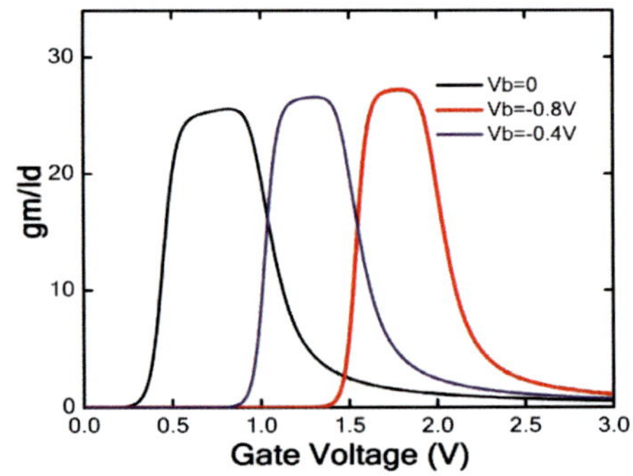


FIGURE 11.1 Tree-top test: plot of transconductance to drain to source current ratio vs. gate bias for different body voltages. L=W=1 µm, default parameters used. $V_{DS} = 0.1\,\text{V}$.

### 11.2.2 Weak inversion conduction test

In this test, output conductance ($g_{ds}$) normalized to $exp(-V_{DS}/V_t)$ is observed as gate voltage is swept from subthreshold to strong inversion. Physically, the ratio $g_{ds} * exp(V_{DS}/V_t)$ should be independent of $V_{DS}$ in the subthreshold region. For the BSIM-BULK compact model, this ratio in the subthreshold region is expressed as follows [3]:

$$i_{ds} = (1 + q_s + q_d) * (q_s - q_d) \approx q_s - q_d \tag{11.1}$$

$$q_s = e^{v_p - v_s}; \quad q_d = e^{v_p - v_d}; \quad i_{ds} = e^{v_p}(1 - e^{-v_{ds}}) \tag{11.2}$$

$$g_{ds} = \frac{di_{ds}}{dv_{ds}} = e^{v_p} e^{-v_{ds}}; \quad g_{ds} \cdot e^{v_{ds}} = e^{v_p} \tag{11.3}$$

Quantities in the above equations are normalized (refer to Chapter 2 for the normalization factor). Note that we have used the approximation $q_{s(d)} << 1$ and $q_{s(d)} \approx exp(v_p - v_{s(d)})$ as the device is biased in the subthreshold region. RHS of (11.3) does not depend on the drain voltage and is a strong function of gate voltage through $v_p$. The weak-inversion conduction test result for the BSIM-BULK model is shown in Fig. 11.2. We see that the model correctly captures the drain bias dependence of normalized output conductance.

### 11.2.3 Slope ratio test

The slope ratio test is used to test MOSFET compact model's transition behavior from weak to strong inversion region [2]. Drain current is obtained as a function of gate voltage, for two drain biases ($V_{DS,1}$, $V_{DS,2}$). Slope ratio is then obtained using the following expression:

$$SR = \frac{(I_{DS,2} + I_{DS,1})(V_{DS,1} - V_{DS,2})}{(I_{DS,2} - I_{DS,1})(V_{DS,2} + V_{DS,1})} \tag{11.4}$$

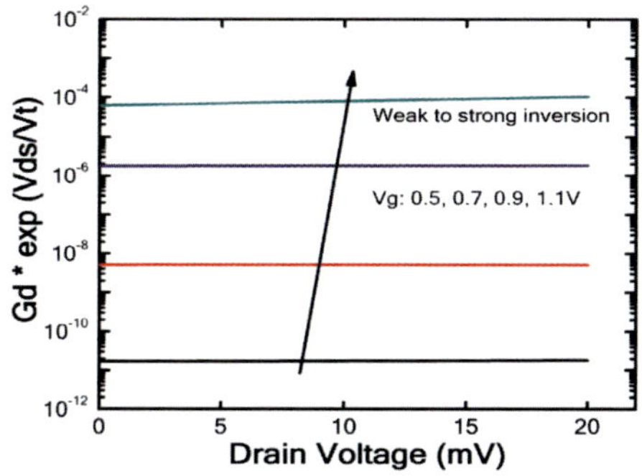


FIGURE 11.2 Weak-inversion conduction test: plot of normalized output conductance vs. drain bias for different gate voltages. **L**=**W**=1 µm, **RDSMOD**=2 default parameters used. $V_{DS} = 0.1\,\text{V}$.

here $I_{DS,1}$ and $I_{DS,2}$ are the drain to source current at drain voltage $V_{DS,1}$, $V_{DS,2}$, respectively. SR should smoothly decrease from 1.3 in the weak inversion to 1.0 in strong inversion. This is indeed true for the BSIM-BULK compact model, as shown in Fig. 11.3(a). Nominal value of $V_{DS,1}$=10 mV and $V_{DS,2}$=20 mV are chosen for SR test. The figure also shows test results for different temperatures, and we see that the limiting value of the slope ratio changes significantly with temperature. To rectify that, the temperature invariant slope ratio test was developed, where $V_{DS,1}$, $V_{DS,2}$ are expressed as:

$$V_{DS,1} = \frac{V_t}{2} - V'; \quad V_{DS,2} = \frac{V_t}{2} + V' \tag{11.5}$$

where $V' << V_t$. Fig. 11.3(b) shows the temperature invariant slope ratio test results for three temperatures: -25, 25, 50°C.

### 11.2.4 Self-heating test

Self-heating effect modeling (SHE) is a crucial component of a compact model. Joule heating due to the power dissipation in the device leads to the rise of the internal temperature of the device. Transistor properties are sensitive to temperature. Current and hence power dissipation, in the device, is a function of the applied bias, making SHE also dependent on the applied bias. SHE is modeled in the BSIM-BULK compact model using a subcircuit network (refer to Chapter 4 for details) [4]. To test the SHE model implementation, $I_{DS}$ from two different simulations are compared [5]: (a) For fixed biases, $I_{DS}$ obtained by enabling the SHE model and sweeping thermal resistance model parameter **RTH0**. Internal temperature is also noted down (b) $I_{DS}$ obtained by turning off the SHE model and sweeping the ambient temperature to values obtained in step (a). If

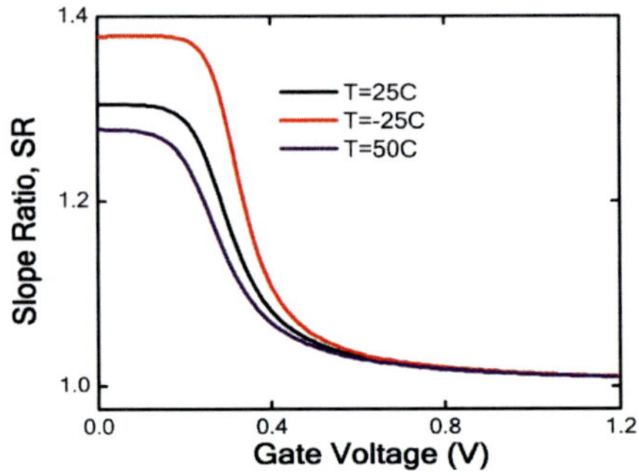


(a)

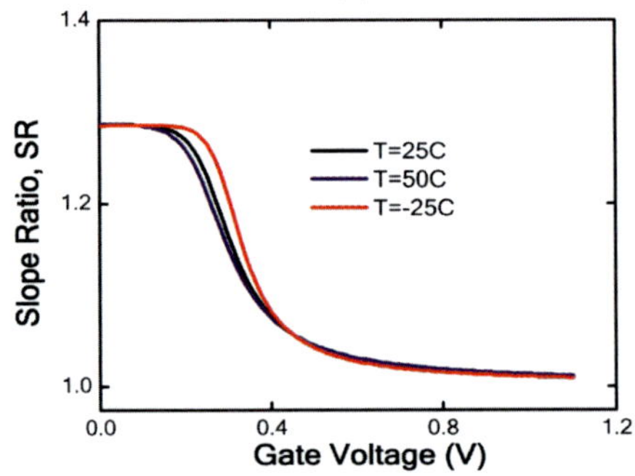


(b)

FIGURE 11.3 Slope ratio test: (a) Plot of slope ratio (SR) as defined in (11.4) (b) temperature-independent slope ratio test. Default model parameters are used along with **L**=**W**=1 μm.

the SHE model implementation is correct, then $I_{DS}$ obtained in these two steps should match. This is true for the BSIM-BULK model as shown in Fig. 11.4.

## 11.3 Symmetry test

MOSFET is essentially a symmetric device, i.e., the amount of current that flows in the device does not change if the polarity of voltage applied at the drain and source is reversed. A MOSFET compact model must not only be symmetric but also ensure that higher derivatives of the current around $V_{ds} = 0\,\text{V}$ exist. This is also true for terminal charges and their derivatives. The DC symmetry features of a compact model are tested through the Gummel Symmetry Test (GST), whose setup is shown in Fig. 11.5(a) [6]. Equal but opposite voltages are applied at the drain ($V_x$) and source (-$V_x$) terminals while keeping body and gate voltage fixed. $V_x$ is then swept from negative to positive and terminal currents, and deriva-

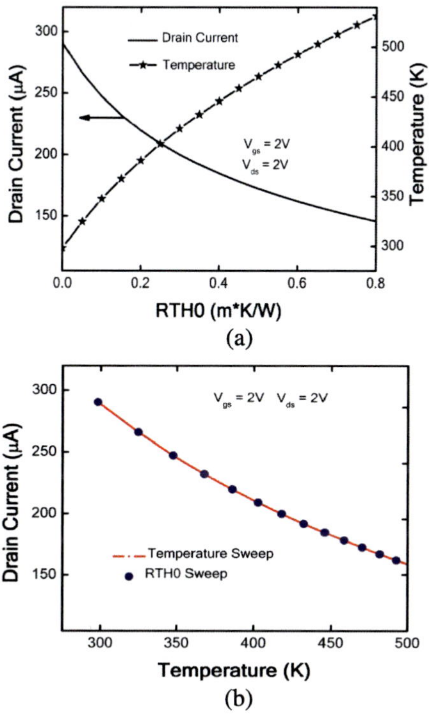


FIGURE 11.4 Self-heating test: (a) plot of drain current and temperature vs. **RTH0** at fixed bias condition. **SHMOD**=1 (b) Drain current vs. ambient temperature for **SHMOD**=0. This current matches the current obtained in Fig. (a), which verifies the correct implementation of the self-heating-effect compact model. Default model parameters are used with **L**=**W**=1 μm. $V_{DS} = V_{GS} = 2\,\mathrm{V}$.

tives are observed. Fig. 11.6(a) and (b) show GST results for weak inversion and strong inversion, respectively. Default model parameters are used for this test. Note that the current and higher order derivatives (shown here up to fourth order) are smooth and continuous. Therefore the BSIM-BULK compact model successfully passes GST.

### 11.3.1 Modified symmetry test

Like terminal current, charges, as well as capacitance of the MOSFET, are also symmetric. Modified version of the GST in Fig. 11.5(a) was developed by McAndrew et al. to test transcapacitance in the presence of substrate and gate currents [7]. The test setup is shown in Fig. 11.5(b). In-phase ac excitation is applied at the source and drain terminals, and terminal currents (represented by $^+$ superscript) are observed. Next, anti-phase ac excitation is applied, and terminal currents (represented by $^-$ superscript) are observed. Imaginary part of the AC terminal current for

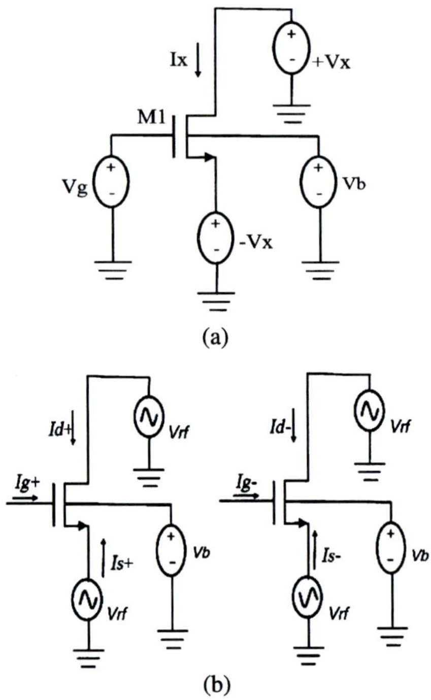


FIGURE 11.5 Gummel symmetry test setup: Equal but opposite voltages are applied at the drain ($V_x$) and source (-$V_x$) terminals while keeping body and gate voltage fixed. $V_x$ is then swept from negative to positive and terminal currents are observed. (b) Modified symmetry test setup for transcapacitance.

in-phase and anti-phase conditions are given as:

$$i_g^+ = 2\pi f(C_{gs} + C_{gd}) \tag{11.6}$$

$$i_g^- = 2\pi f(C_{gs} - C_{gd}) \tag{11.7}$$

The ratio

$$\delta_{cg} = \frac{i_g^-}{i_g^+} = \frac{C_{gs} + C_{gd}}{C_{gs} - C_{gd}} \tag{11.8}$$

should be an odd function of voltage $V_x$. Likewise, $\delta_{cgsd}$ should also be an odd function of $V_x$. Fig. 11.7 shows $\delta_{cg}$ and $\delta_{cgsd}$ of the BSIM-BULK model and their derivatives. It is evident from the figure that the model passes symmetry test for the transcapacitance, as the curves are smooth and continuous around $V_x$=0 V.

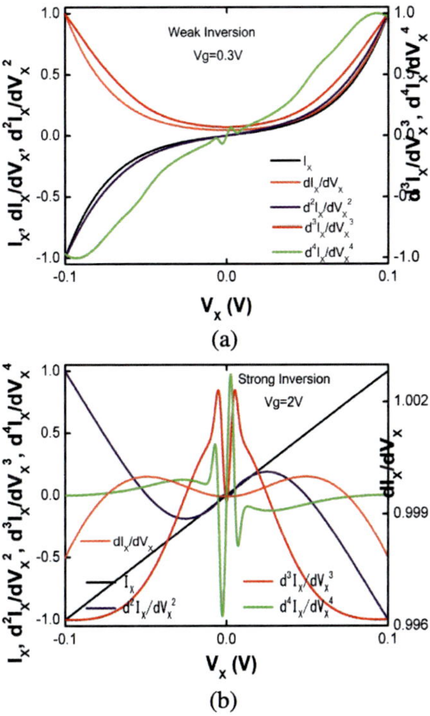


FIGURE 11.6 Gummel symmetry test: (a) Weak inversion at $V_{gs} = 0.3$ V (b) Strong inversion at $V_{gs} = 2$ V. Current and higher-order derivatives are smooth and continuous.

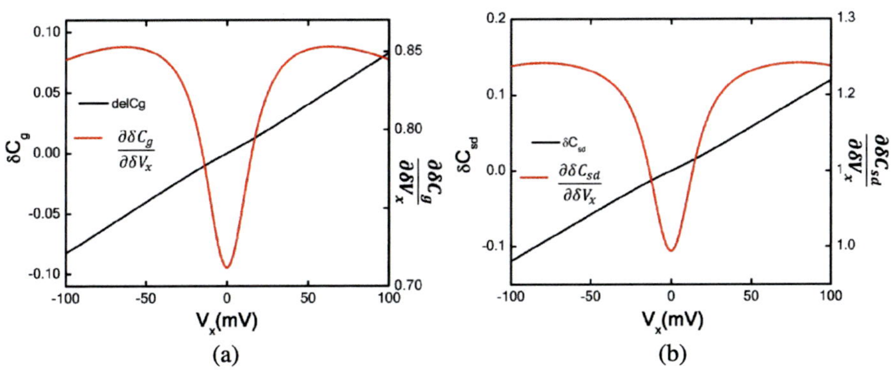


FIGURE 11.7 Modified symmetry test for transcapacitance: (a) $\delta C_g$ (b) $\delta C_{sd}$ vs. $V_x$ as obtained from the BSIM-BULK model. $\delta C_g$ and $\delta C_{sd}$ and their derivative are smooth, continuous, and symmetric around $V_x$=0.

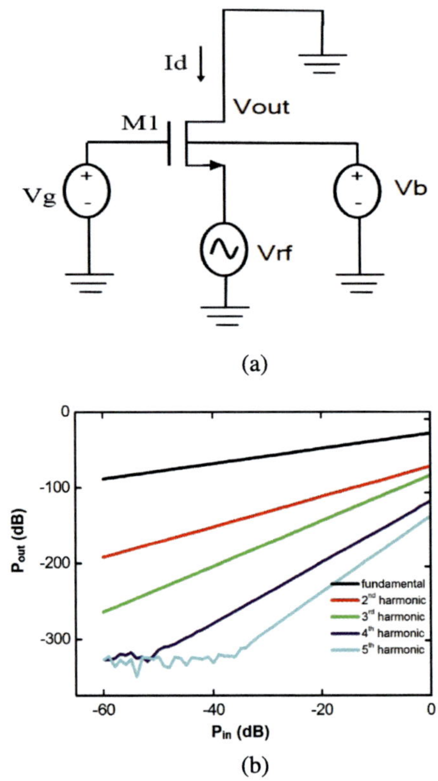


FIGURE 11.8 Harmonic balance test: (a) test setup (b) BSIM-BULK model results. The BSIM-BULK model produces correct slopes of the harmonics since slope of *n*th harmonic component is '*n*' times the slope of fundamental component.

## 11.4 Harmonic balance test

Distortion in analog RF and microwave circuits is analyzed using the harmonic balance simulations. Mathematically, *n*th harmonic should be proportional to *n*th power of the input signal [8,9]. Several compact models deviate from the theoretical prediction due to discontinuity in the higher-order derivatives at $V_{DS} = 0\,\text{V}$. Such incorrect and nonphysical harmonic balance simulation is detrimental for RF circuit design. To test the physical nature of the BSIM-BULK model for RF simulations, the harmonic balance test is performed using the setup shown in Fig. 11.8(a). The harmonic content of the drain current is plotted as a function of the input signal in Fig. 11.8(b). It is clearly evident that the model correctly reproduces the slope of different harmonic components as the slope of *n*th harmonic is '*n*' times the slope of the fundamental component.

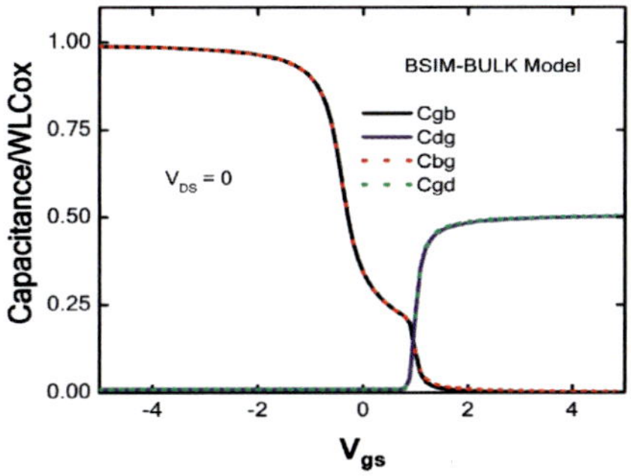


FIGURE 11.9 Core model results: plot of trans-capacitance Cgb, Cbg, Cgd and Cdg vs. gate voltage at $V_{ds} = 0$. Capacitance behavior is smooth across biases. Default parameters used along with L=W=1 µm, **CVMOD**=1, **RDSMOD**=2, **ADOS**=0, **NGATE**=0.

## 11.5 Capacitance reciprocity test

The capacitance between node i and j ($C_{ij}$) in a symmetric device, like a MOSFET, follows the following property:

$$C_{ij} = \delta_{ij}\frac{\partial Q_i}{\partial V_j}; \quad \delta_{ij} = -1 \text{ for } i \neq j \text{ otherwise } 1 \tag{11.9}$$

here $Q_i$ and $V_j$ are the terminal charge at node i and voltage at node j, respectively. At $V_{ds} = 0$, symmetry of the device requires capacitance to be reciprocal, i.e., $C_{ij} = C_{ji}$. Fig. 11.9(a) shows a plot of Cgb Cbg, Cgd, and Cdg vs. gate voltage for $V_{ds} = 0$ obtained using the BSIM-BULK model. As clearly seen, the model behaves physically and produces reciprocal capacitance at $V_{ds} = 0$.

## References

[1] A. Dutta, S. Sirohi, T. Ethirajan, H. Agarwal, Y.S. Chauhan, R.Q. Williams, BSIM6 – benchmarking the next-generation MOSFET model for RF applications, in: VLSI Design and 2014 13th International Conference on Embedded Systems, 27th International Conference on, Jan 2014.

[2] G. Gildenblat, Compact Modeling: Principles, Techniques and Applications, Springer, 2010.

[3] Y.S. Chauhan, S. Venugopalan, M.A. Chalkiadaki, M.A.U. Karim, H. Agarwal, S. Khandelwal, N. Paydavosi, J.P. Duarte, C.C. Enz, A.M. Niknejad, C. Hu, BSIM6: analog and RF compact model for bulk MOSFET, IEEE Transactions on Electron Devices 61 (2) (Feb 2014) 234–244.

[4] H. Agarwal, S. Venugopalan, M.-A. Chalkiadaki, N. Paydavosi, J.P. Duarte, S. Agnihotri, C. Yadav, P. Kushwaha, Y.S. Chauhan, C.C. Enz, A. Niknejad, C. Hu, Recent enhancements in BSIM6 bulk MOSFET model, in: Simulation of Semiconductor Processes and Devices (SISPAD), International Conference on, Sept 2013.

[5] Q. Chen, Z.-Y. Wu, R.Y. Su, J.-S. Goo, C. Thuruthiyil, M. Radwin, N. Subba, S. Suryagandh, T. Ly, V. Wason, J.X. An, A.B. Icel, Extraction of self-heating free I-V curves including the substrate current of PD SOI MOSFETs, in: 2007 IEEE International Conference on Microelectronic Test Structures, 2007, pp. 272–275, https://doi.org/10.1109/ICMTS.2007.374498.
[6] C. McAndrew, H. Gummel, K. Singhal, Benchmarks for compact MOSFET models, in: Proc. SEMATECH Compact Models Workshop, 1995.
[7] C. McAndrew, Validation of MOSFET model source-drain symmetry, IEEE Transactions on Electron Devices 53 (9) (2006) 2202–2206.
[8] N. Scheinberg, A. Pinkhasov, A computer simulation model for simulating distortion in FET resistors, IEEE Transactions on Computer-Aided Design of Integrated Circuits and Systems 19 (9) (2000) 981–989, https://doi.org/10.1109/43.863638.
[9] P. Bendix, P. Rakers, P. Wagh, L. Lemaitre, W. Grabinski, C.C. McAndrew, X. Gu, G. Gildenblat, RF distortion analysis with compact MOSFET models, in: IEEE Custom Integrated Circuits Conference, 2004, pp. 9–12.

# Index

Made in the USA
Las Vegas, NV
16 November 2024

11939658R00162